UNDERGROUND SENSING

UNDERGROUND SENSING

Monitoring and Hazard Detection for Environment and Infrastructure

Edited by

SIBEL PAMUKCU

LIANG CHENG

ACADEMIC PRESS
An imprint of Elsevier

Academic Press is an imprint of Elsevier
125 London Wall, London EC2Y 5AS, United Kingdom
525 B Street, Suite 1800, San Diego, CA 92101-4495, United States
50 Hampshire Street, 5th Floor, Cambridge, MA 02139, United States
The Boulevard, Langford Lane, Kidlington, Oxford OX5 1GB, United Kingdom

Notices

Knowledge and best practice in this field are constantly changing. As new research and experience
broaden our understanding, changes in research methods, professional practices, or medical
treatment may become necessary.

Practitioners and researchers must always rely on their own experience and knowledge in evaluating
and using any information, methods, compounds, or experiments described herein. In using such
information or methods they should be mindful of their own safety and the safety of others,
including parties for whom they have a professional responsibility.

To the fullest extent of the law, neither the Publisher nor the authors, contributors, or editors,
assume any liability for any injury and/or damage to persons or property as a matter of products
liability, negligence or otherwise, or from any use or operation of any methods, products,
instructions, or ideas contained in the material herein.

Library of Congress Cataloging-in-Publication Data
A catalog record for this book is available from the Library of Congress

British Library Cataloguing-in-Publication Data
A catalogue record for this book is available from the British Library

ISBN: 978-0-12-803139-1

For information on all Academic Press publications
visit our website at https://www.elsevier.com/books-and-journals

  Working together
to grow libraries in
developing countries

www.elsevier.com • www.bookaid.org

Publisher: Matthew Deans
Acquisition Editor: Ken McCombs
Editorial Project Manager: Jennifer Pierce
Production Project Manager: Julie-Ann Stansfield, Sruthi Satheesh
Designer: Greg Harris

Typeset by VTeX

CONTENTS

LIST OF CONTRIBUTORS

Traian E. Abrudan
University of Oxford, Oxford, United Kingdom

Aaron Bradshaw
University of Rhode Island, Kingston, RI, USA

Erika Bustos
Centro de Investigación y Desarrollo Tecnológico en Electroquímica, S.C., Querétaro, Mexico

Liang Cheng
Lehigh University, Bethlehem, PA, USA

Nicholas de Battista
University of Cambridge, Cambridge, United Kingdom

Vanessa Di Murro
University of Cambridge, Cambridge, United Kingdom

Mohammed Elshafie
University of Cambridge, Cambridge, United Kingdom

J.A. Garcia
Instituto Tecnológico de Atitalaquia, Atitalaquia, Mexico
Centro de Investigación y Desarrollo Tecnológico en Electroquímica S.C., Parque Tecnológico Querétaro s/n, Sanfandila, Pedro Escobedo, Querétaro, Mexico
Centro de Investigaciones en Óptica A.C., León, Guanajuato, Mexico

Russell A. Green
Virginia Tech, Blacksburg, VA, USA

C.Y. Gue
University of Cambridge, Cambridge, United Kingdom

Qi Han
Colorado School of Mines, Golden, CO, USA

Tissa H. Illangasekare
Colorado School of Mines, Golden, CO, USA

Magued Iskander
New York University, NY, USA

Anura P. Jayasumana
Colorado State University, Fort Collins, CO, USA

Cedric Kechavarzi
University of Cambridge, Cambridge, United Kingdom

Orfeas Kypris
University of Oxford, Oxford, United Kingdom

Rui Li
CSSC Marine Technology Co., Ltd., Shanghai, China

Jerome P. Lynch
University of Michigan, Ann Arbor, MI, USA

Andrew Markham
University of Oxford, Oxford, United Kingdom

Radoslaw L. Michalowski
University of Michigan, Ann Arbor, MI, USA

David Monzon-Hernandez Sr.
Centro de Investigaciones en Óptica A.C., León, Guanajuato, Mexico

Srinivasa S. Nadukuru
University of Michigan, Ann Arbor, MI, USA

Sean M. O'Connor
University of Michigan, Ann Arbor, MI, USA

Sibel Pamukcu
Lehigh University, Bethlehem, PA, USA

Feng Pan
Beihang University, Beijing, China

Loizos Pelecanos
University of Cambridge, Cambridge, United Kingdom

Mesut Pervizpour
Lehigh University, Bethlehem, PA, USA

Mohammad Pour-Ghaz
North Carolina State University, Raleigh, NC, USA

Abdul Salam
University of Nebraska-Lincoln, Lincoln, NE, USA

Kenichi Soga
University of California, Berkeley, CA, USA

Niki Trigoni
University of Oxford, Oxford, United Kingdom

Mehmet C. Vuran
University of Nebraska-Lincoln, Lincoln, NE, USA

W. Jason Weiss
Oregon State University, Corvallis, OR, USA

Christian Wietfeld
Communication Networks Institute (CNI), TU Dortmund University, Dortmund, Germany

Michael Williamson
University of Cambridge, Cambridge, United Kingdom

Tian Xia
University of Vermont, Burlington, VT, USA

Wen Xiao
Beihang University, Beijing, China

Xiaosu Yi
Beihang University, Beijing, China

Suk-Un Yoon
Samsung Electronics, Suwon, South Korea

PREFACE

This book is intended to compile theoretical and practical knowledge about underground sensing applicable to monitoring buried infrastructure's integrity, and detecting and tracking underground events and hazards. Specific topics covered include acoustic, electromagnetic and optical sensing and monitoring methods, ground penetration radar, EM-based wireless underground sensor networks, fiber optic sensing and sensor networks, wireless signal networks, magneto-inductive underground tracking, integration of UAVs with underground sensing, etc. The book is useful to researchers and R&D engineers in academia, industry, and government who are interested in advancing the underground sensing techniques and developing new applications to meet the needs of the modern society.

The motivation for this project solidified after encountering substantial number of inquiries from researchers and practicing engineers for an authoritative text on underground sensing following a keynote talk on the subject at the 2014 annual Geo-Congress of ASCE. It appeared that most of the existing works remained either technology-specific or application-specific, and were primarily found in academic papers distributed over multiple disciplines. The main goal of the editors of this book project was to provide a devoted reference that could aggregate essential information from multiple disciplines in useful and categorical distributions. We hope that this book will be useful to readers as a starting point for further guided research, as they determine the best sensing technique or suite of techniques in their underground projects, or jump-start their research and development on the subject matter using the appropriate tools and paradigms.

This book is designed to be used in the context of providing a comprehensive review of existing underground sensing technologies, and as a "tool box" for researchers and practitioners to help address underground sensing and monitoring problems and predict future requirements for sustainable sensing technologies. The first goal of the text is to bring to the target audience the technical and practical knowledge of existing underground sensing and monitoring methods based on the classification of their functionality. The second goal is to introduce emerging technologies and applications of sensing for environmental and geohazards in subsurface – focusing on sensing platforms that can enable fully distributed measurements. The third goal is to explore the implications of advanced sensing paradigms in underground, such as hybrid-sensing that can meet demands for preemptive and sustainable response to underground hazards.

This book manifests the multidisciplinary nature of underground sensing. The areas of expertise of the contributors for this book and their contributed contents span from civil and environmental engineering, electrical engineering including opto-electronics engineering, computer engineering to computer science, chemistry, and other underground-sensing related science disciplines. We want to express our gratitude to all the contributors for their comprehensive and highly valuable contributions to make this book a truly multidisciplinary work to achieve the above-mentioned goals.

We also want to thank André Wolff for his insight in getting this book project off the ground. We appreciate all members of the editorial and production team of Elsevier, including but not limited to, Ken McCombs, Jennifer Pierce, Mariana Kuhl, Sruthi Satheesh, Julie-Ann Stansfield and Narmatha Mohan. Without their efforts, this book could not have been completed.

Sibel Pamukcu, Liang Cheng

CHAPTER 1

Introduction and Overview of Underground Sensing for Sustainable Response

Sibel Pamukcu, Liang Cheng, Mesut Pervizpour
Lehigh University, Bethlehem, PA, USA

1.1 UNDERGROUND SENSING FOR ENVIRONMENTAL, ECONOMIC, AND SOCIAL SUSTAINABILITY

Nationwide there is a pervasive issue of aging infrastructure and concerns about transparency in government and industry decision-making around our key systems – water, electricity, natural gas, oil, and transportation infrastructure (ASCE, 2017). For example, 10 year (2007–2016) average of pipeline incidents was reported as 286 with 13 fatalities, 64 injuries, and over $474 million property damage per year in 2017 by USDOT.[1] Many of the over 450,000 brownfield sites in US remain as potential sources of pollution to human and environmental exposure. Many areas in urban centers, particularly in highly industrialized zones, contain persistent sources of contamination due to past disposal practices. The regulatory discussions for environmental pollution are driven by point concentration measurements, which require monitoring mass flux of the contaminants into the environment for risk and long-term sustainability assessments (EPA, 2012; NRC, 2011; ITRC, 2010). Approximately 20% of land in continental US is underlain by "karst terrain" susceptible to sinkhole events. These events tend to occur when water-drainage and storage patterns are altered, resulting in loss of life and substantial property damage (Kuniansky et al., 2016). These potential hazards and related incidents can undercut consumer confidence, create fear, subsequently jumpstart debate on infrastructure, environment, and the societal factors that influence government and industry decision-making. Consumers and citizens may become fearful of what they do not know – for example, how can they protect themselves and their

---

[1] http://www.phmsa.dot.gov/pipeline/library/data-stats/pipelineincidenttrends.

Underground Sensing.
DOI: http://dx.doi.org/10.1016/B978-0-12-803139-1.00001-1

families from gas pipeline leaks or sinkhole occurrences? Further, confusion around science and acceptance/understanding of scientific principles continues to muddy the waters around issues of environment and industry.

Using a web-based survey of expert opinions and extensive literature review, Becerik-Gerber et al. (2014) identified recently ten civil engineering challenges that can be addressed with further data sensing and analysis for sustainable development. Among the challenges presented were increased soil and coastal erosion, inadequate water quality, untapped and depleting groundwater, and poor and degrading infrastructure. Underground, where most of our lifeline infrastructure (i.e., all utilities that transport water, sewage, oil, gas, chemicals, electric power, communications, and mass media content), tunnels (i.e., all tunnels that transport and/or store water or sewage and to tunnels for hydropower, traffic, rail, freight) and storage and containment spaces (i.e., those used for dry storage that provide climate or security isolation) are placed, is one of the most essential spaces for sustainable development (Koo et al., 2009; NRC, 2013; Koch et al., 2013; McBratney et al., 2013; Hunt et al., 2014; Iten et al., 2015; Correia et al., 2016). Further, underground is where soils, sediments, and aquifers that provide the essential resources to mankind and nature alike reside. Most fresh water reserves are in groundwater form (8–10 million km^3) in direct contact with an aquifer overlaid by soil (Shah et al., 2007). The physical and chemical stability, storage and filtering functionality, and transformation of chemical compounds and sustaining biodiversity capability of these soils are integral parts of groundwater and food security (Godfray et al., 2010). Resilient, well-maintained, and well-performing underground infrastructure, and well-protected soil and groundwater are essential for sustainable development.

Indiscriminate geo-events such as sink holes, landslides, slope failures, pollutant plume migrations are a few of many geohazards that threaten the underground infrastructure and soil layers beneath with potentially catastrophic environmental and societal impact and economical losses. In order to plan and execute long-term risk management and respond preemptively to hazards, it is essential to acquire time and space-continuous information at strategic facilities and locations. The post event forensic evidence often suggests that monitoring geo-events to track symptomatic spatiotemporal variations underground that lead up to the event can help mitigate them (Dunncliff, 1993). There is a vast selection of underground monitoring/sensing techniques that rely on advanced technologies of geophysics, electromagnetics, and optics to track symptomatic events (i.e., distortion, displacement, cracks, changes in liquid and gas content, strain or temperature) potentially leading up to a hazard underground. Managing the performance and safety of geographically distributed infrastructure subject to low probability/high consequence events underground is arduous and costly. These distributed systems, including water supplies, levees, gas and liquid fuel supply networks may cover thousands of km^2 of area and be subject to numerous different failure mechanisms along their linear positioning and duration of lifetime. Although the existing approaches face some distinct challenges in realizing fully spatiotemporal continuous information underground for broad applications (Olhoeft, 2003; Vuran and Silva, 2009; Akyildiz and Vuran, 2010; Hofinghoff and Overmeyer, 2013), they offer the most sustainable solutions for underground health monitoring and damage detection up to date, as discussed in much detail in subsequent chapters of this book. Additionally, with continuous advancement of sensor technologies, particularly when used in a synergistic or hybrid manner, such assemblages can also serve as effective security and warning systems to man-made or natural hazards, providing a means for adaptive control of the built environment underground that fit the needs of a particular site.

It is well understood that underground offers multiple resources and vast spaces that can support new urban sustainability patterns (Doyle et al., 2016). The sustainability challenges depend on our ability for better planning and execution of energy-consuming processes, such as construction, maintenance, and adaptive control of the underground facilities. An essential enabler in this process is our ability to better characterize and monitor critical subsurface features. There are sophisticated surface and wellbore methods for geological characterization of the underground and to determine the physical, geochemical, and mechanical properties of the natural and man-made features. Nevertheless, these methods can fall short in their capabilities to cover large enough volumes and areas with sufficiently high spatial and temporal resolution underground. DOE recently described four pillars of their "New Subsurface Signals"[2] theme of research needs for underground development, including new sensing approaches and integration of multiscale, multitype data. The focus in these research and development needs statement is the advancement of underground sensing methods to enable complete and unbiased monitoring of geo-events by alleviating sparse spatial and temporal sampling of data.

[2] http://eesa.lbl.gov/subter/the-4-pillars/.

1.2 SUSTAINABILITY AND INDICATORS

According to an EPA Executive Order,[3] "sustainability" and "sustainable" mean *to create and maintain conditions, under which humans and nature can exist in productive harmony, that permit fulfilling the social, economic, and other requirements of present and future generations.* The concept of sustainability is based on the interdependence between human societies and the natural environment, supported by the three "pillars" of environmental, economic, and social influence. International business community has focused on the challenge of achieving sustainability (WBCSD, 2011) by adopting global environmental management system standards, such as ISO 14001.[4] These standards specify performance indicators for selection, monitoring, and verification of sustainability.

In the matter of "underground sensing for sustainable response" we not only encompass the act of information gathering using sustainable tools and methodologies, but also the act of germane use of that information to respond to the perceived hazard without compromising the three pillars of sustainable development – the environment, society, and economics. Most underground events are hidden from direct observation, often spatially distributed and transient with small symptomatic events leading up to the final event (i.e., progressive structural deterioration leading up to a pipeline failure). Hence, it is essential to gather strategic, pertinent, and adequately distributed (i.e., with optimum resolution to avoid blind spots) information to accurately assess the hazard for a measured and sustainable response. A specific definition of sustainability is given by Gibson (2006) as the "comprehensive, farsighted, critical and integrated approaches on important policies, plans, programs, and projects." Ideally then, for an underground sensing program to be sustainable it should be "comprehensive" and "integrated", possibly encompassing more than one technology and tool, "farsighted" and "critical", possibly being able to not only detect the time and location of an hazard but also detect and track the onset of events or conditions leading up to a likely hazard.

Multiple reports and models are available which provide a framework for implementing sustainability in policies, plans, programs, and projects

---

[3] Executive Order 13514; Federal Leadership in Environmental, Energy, and Economic Performance; signed on October 5, 2009.
[4] International Organization for Standardization. https://www.iso.org/standard/60857.html.

(NRC, 2011; EPA, 2012). In here, we focus on "sustainability" of underground sensing projects which lead to "sustainable response" to underground hazards. The underground hazards can be described as the natural and man-made events that impact negatively the health of the *underground environment* and the *underground infrastructure*. A recent report on sustainable development of underground (NRC, 2013) identify directions and the critical needs of research and development in underground engineering that support sustainable development. The report discussed how utilization of the underground, armed with geologic site characterization and geotechnical monitoring to predict geologic environment and ground response, among others, could maximize the resiliency and sustainability of urban development and address issues related to the impacts of climate change on urban environment. The report also discussed at length the need for smart underground structures and advanced systems and sensors that relate their status of health and safety.

The indicators of sustainability are those that support systems of adaptive management and social learning by integration of monitoring and reporting on environmental, economic and social conditions (Kates et al., 2001). For example, average concentration of blood lead (Pb) in humans is an indicator of both environmental exposures and possible impairment of human health; as is a change in industrial employment because of green chemistry innovations an indicator of both natural resource protection and economic development (Singh et al., 2009). Similarly, increased social and economical development because of better risk management, preemptive measures, and better informed public is an indicator of higher level of consumer confidence and environmental protection. Integrating information gathered from underground sensing of the environment and health monitoring of critical infrastructure into adaptive management and social learning can support sustainable development at multiple levels, including new paradigms in urban development (NRC, 2013; Doyle et al., 2016).

The attributes of *sustainability indicators* can be selected from among the following list of criteria (Fiksel, 2012):
- *Relevant* to the interests of the intended audiences
- *Meaningful* in terms of clarity, comprehensibility and transparency
- *Objective* in terms of measurement techniques and verifiability
- *Effective* for supporting benchmarking and monitoring over time
- *Comprehensive* in providing evaluation of progress with respect to sustainability goals

- *Consistent* across different sites or communities
- *Practical* in allowing cost-effective implementation and building on existing data
- *Actionable*, so that practical steps can be taken to address contributing factors
- *Transferable and scalable*, so that they are adaptable at different levels
- *Intergenerational*, reflecting distribution of costs and benefits among different generations
- *Durable*, so that they have long-term relevance

Most sustainability indicators are guided by a need and for practicality, therefore a small number of selected key criteria covering the important aspects of sustainability for the specific problem at hand can provide effective performance measurement. Sustainability is measured and evaluated using both qualitative and quantitative tools including risk, life-cycle, and benefit-cost assessments (Correia, 2015; Basu et al., 2014). A sustainable underground sensing program that addresses a multitude of the criteria listed above is also a *resilient* system that is (i) agile, quick, and both spatially and temporally distributed to produce fast and localized data of impending event as well as the actual event and/or damage; (ii) minimally invasive and durable, requiring minimal energy and maintenance and for sustained operation over long durations. Resilience of a system is its ability to respond and adapt to change in the environment (NRC, 2013). Building resilience into an underground sensor system implies removing or minimizing vulnerabilities to its essential parts that the system would be able to withstand extreme events and deliver service functionality quickly following such an event.

Hybrid sensing systems of multiple components with minimal cross-component interdependencies could provide the required resilience as well as the agility and durability attributed in venerable sustainability indicators. A hybrid sensing system can be integrated systems of conventional and advanced sensing and monitoring methods, including wellbore, probe or sampling platform based, surface and near surface imaging, acoustic, wireless and low-frequency radio, magnetic induction and optical techniques. Furthermore, new dimensions of sensing such as remote sensing (i.e., hyperspectral imaging), crowd sensing (i.e., social media) and unmanned airborne vehicle (UAV) can assist and complement the requirements of a sustainable sensing program for underground. Among the new paradigms of sensing, crowd sensing can perform as a community-involved warning system that proactively detects and locates hazard by monitoring the on-set of events or conditions leading up to a likely hazard. Crowd sensing, integrated with advanced sensor data is a form hybrid sensing system that can be helpful for the involved community (i.e., city, state, federal gov-ernment or industry) to make agile mitigation decisions for public safety and perform faster damage management. The time and space-continuous sensor networks are poised to be key components of the emerging In-ternet of Things (IoT) paradigm. With sensors deployed for underground event tracking, the event can be envisioned as a "thing" being monitored, tracked, modeled, and learned from using the Internet of Things (IoT) infrastructure.

Most of the sensing methodologies mentioned above are addressed indi-vidually in much detail in the following chapters in this book. The need to develop sustainable sensing programs that address time-variant sustainabil-ity of the underground environment and the underground infrastructure are also discussed in a number of the following chapters. For example, in Chapter 4 the authors articulate that the globe's groundwater as a source of freshwater supply is threatened from overdraft and contamination due to agricultural and industrial activity. They also make reference to the fact that many processes occurring in the subsurface, including leakages from geologically sequestrated carbon and hydraulic fracturing contribute to greenhouse gas (GHG) loading of the atmosphere that contributes to global climate change. Hence, sustainable monitoring of the subsurface is of critical importance in managing the subsurface to address existing and emerging problems in the environment. The authors of Chapter 4 also discuss a hybrid sensing approach of using wide area distributed sensor net-works to identify the "hot spots" in the subsurface where further sampling may be directed for better characterization and accurate measurement of target indices. They introduce a network optimization model approach by which an acceptable measurement resolution can be obtained within a given error bound while reducing the number of sensors to be used for a sustainable underground environmental monitoring. Similarly, in Chap-ters 6.1 and 7.3 the authors discuss the importance and benefits of using distributed and hybrid sensing systems to better understand the state and behavior of underground features and geotechnical structures and pinpoint localized problems such as cavity collapses, nonuniformly distributed soil–structure interaction loads, and joint movements in underground structures, including tunnels, shafts and pipelines.

1.3 OVERVIEW OF UNDERGROUND SENSING AND MONITORING

Managing the performance and safety of spatially distributed infrastructure underground, including water, gas and liquid fuel supply networks subject to low probability/high consequence hazards can be arduous and costly (ISSMGE, 2009; NRC, 2013; Doyle et al., 2016; Delmastro et al., 2016). Since we cannot readily detect an approaching hazard or observe tell-tale signs of degradation in hidden infrastructure, sensors have to be our "remote eyes" in the underground environments. Ideally, underground sensors should detect hazard by monitoring the onset of events leading up to a likely hazard or, in other words, act as spatially and temporally continuous warning systems.

A versatile underground sensing system should be (i) easy to embed and/or capable of sensing deeply, (ii) easy to operate, energy efficient and of low-complexity, (iii) accurate and reliable for long-term, continuous operation with self-organization, self-healing and self-referencing capabilities (as they apply), (iv) scaleable, (v) sensitive to preemptive warnings of sightless geo- and environmental hazards, and (vi) capable of reporting in real-time. The majority of the currently available underground sensor and monitoring techniques provide time-ordered point measurements at strategically placed devices (Dunnicliff, 1993). Point monitoring works well if a plausible event location is known, in which case one may settle for recording vast amounts of benign data until the hazard event occurs. While most are in the research and development stage, there are limited applications of time and space continuous monitoring techniques for underground infrastructure and space (Glisic, 2014; Liu et al., 2016). A number of these techniques and applications are discussed in great detail in subsequent chapters of this book. New concepts and paradigms based on passively powered and/or on-demand activated, embeddable sensor platforms are being developed that can deliver temporally and spatially continuous sensing capable of measuring an event signal where and when it might occur. These sensor platforms can enable time and space continuous data to map the evolution of a hazard as it triggers calibrated signals at a remote station, as depicted in Fig. 1.1.

1.3.1 Current Technologies for Underground Environmental and Geotechnical Monitoring

The availability of different sensing techniques has become increasingly important in health monitoring of earthen infrastructure (i.e., levees, slopes)

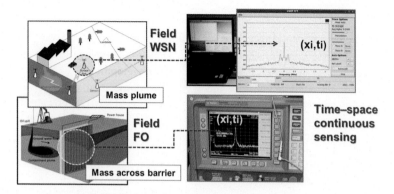

Figure 1.1 Conceptual application of time-space continuous global sensing using wireless sensor networks (WSN) and fiber-optic (FO) sensing

and underground civil infrastructure (i.e., pipelines, tunnels) and the environmental health and mitigation (Bell et al., 2001; Bhalla et al., 2005; Akyildiz and Vuran, 2010; Akyildiz et al., 2009; Buratti et al., 2009; NRC, 2013). Typical areas of underground sensor and monitoring applications along with citation list of a few recent related research and development are provided in Table 1.1.

Almost all sensing applications listed in Table 1.1 can be classified into two categories: (i) event detection and (ii) spatial process estimation. In event detection case, sensors are deployed to detect an event, such as fire, displacement, or vibration. In spatial process estimation case, spatial distribution of a given physical phenomenon, such as the concentration of chemicals in spill plume or the groundwater temperature variations in a wide area, is estimated based on the samples taken by sensors that are typically placed in random positions (Dardari et al., 2007). Most sensing applications deployed in the underground can function in both manners of operation. Typical examples of events and spatial processes that can be monitored by these methods include earth-slope inclination, landslides, strong ground motion, water and mineral content of soil for agriculture, oil leakage from an oil reservoir, and soil–structure interactions (Furlani et al., 2005; Glaser et al., 2005; Terzis et al., 2006; Chen, 2013; Abdoun et al., 2007; Whelan and Janoyan, 2007; Akyildiz and Vuran, 2010). Examples of environmental measurements by underground sensing include temperature of a bioreactor (Nasipuri et al., 2006), conductivity of a contaminant plume (Han et al., 2008; Loden et al., 2009), water quality (Shantaram et al., 2005; Donovan et al., 2008), and dissolved organic content in water and sedi-

Table 1.1 Commonly used sensing and monitoring applications

Category	Description of Applications	Some related recent studies, and development
Environmental Hazards	Monitoring presence and concentration of toxic substances in groundwater and sediments; monitoring transport of spill plume; monitoring landfill gas and pipeline leakages	Korostynska et al. (2013), Bavusi et al. (2013), Lowe et al. (2013), Van Meirvenne et al. (2014), Klavarioti et al. (2014), Versteeg et al. (2014), McNamara et al. (2013), Barnwall et al. (2014), Power et al. (2015), Bidmanova et al. (2016), Cozzolino (2016)
Geohazards	Monitoring slope stability, landslides, subsidence, liquefaction, and erosion; monitoring soil water distribution	Read et al. (2013), Alamdar et al. (2015), Ramesh (2014), Zhu et al. (2014), Shukla et al. (2014), Segalini et al. (2015), Hanifah et al. (2016), Suo et al. (2016), Dikmen et al. (2016), Hong et al. (2016); Dong et al. (2016); McVay et al. (2016)
Underground Shafts and Open Spaces	Monitoring vital environmental signs in tunnels, mineshafts and open underground spaces; monitoring tunnel shaft displacements and pore-water pressures	Klar et al. (2014), Dohare et al. (2015), Debliquy et al. (2015), Zhao et al. (2016), Ranjan et al. (2016), Mishra et al. (2016)
Underground Infrastructure	Monitoring the health of subsurface structures; water, sewerage, oil and gas pipelines, underground tanks and power distribution	Challener et al. (2013), Kang et al. (2013), BenSaleh et al. (2013), Tang and Wu (2014), Almazyad et al. (2014), Ferdinand (2014), Vidács and Vida (2015), Obeid et al. (2016), Rashid and Rehmani (2016), Gong et al. (2016), Li and Peng (2016), Sheltami et al. (2016)

Table 1.1 (*continued*)

Category	Description of Applications	Some related recent studies, and development
Agriculture	Monitoring underground soil conditions, such as water and mineral content for irrigation and fertilization	Mulla (2013), Nor et al. (2015), Rossel and Bouma (2016), Laskar and Mukherjee (2016), Ye et al. (2016)
Field/Home	Monitoring lawns, gardens, sports fields, and septic tanks	Capella et al. (2014), Yoon et al. (2012), Bose et al. (2016), Mamidi et al. (2017)
Security	Monitoring the aboveground presence and movement of people or objects; monitoring border security	Sun et al. (2011a), Prasanna and Rao (2012), Wu et al. (2014), Klar et al. (2014), Giannoukos et al. (2016), Fraga-Lamas et al. (2016)

ment columns (semipermeable membrane device (SPMD) and solid phase microextraction (SPME) samplers) (Tomaszewski and Luthy, 2008; Oen et al., 2011; Schubauer-Berigan et al., 2012).

There is a vast selection of advanced underground monitoring/sensing techniques that rely on electromagnetic wave (EM) propagation in subsurface. These geophysical methods have been used successfully to image subsurface geological features, detect soil and groundwater contamination, plume migration and changes in physical properties of soil, such as water content and salinity. Examples of these methods include remote electromagnetic induction (Sudduth et al., 2001; Reedy and Scanlon, 2003; Sun and Akyildiz, 2010a, 2010b), time domain reflectometry (TDR) (Robinson et al., 2003; Drnevich et al., 2007), ground penetrating radar (GPR) and active microwave remote sensing (Lunt et al., 2005; Soldovieri et al., 2009; Grote et al., 2010; Benedetto and Tedeschi, 2015; Xu et al., 2014; Bertolla et al., 2014; Zeng et al., 2015), borehole geophysics such as electromagnetic tomography, resistivity and NMR tomography and induced polarization (Kemna et al., 2004; Johnson et al., 2010; Schmidt-Hattenberger et al., 2011; Revil et al., 2012; Kiaalhosseini, 2016), and surface acoustic wave (SAW) wireless passive radio frequency (RF) method (Huang et al., 2010; Droit et al., 2012).

Table 1.2 Underground environmental and geotechnical sensing and monitoring technologies with functionality and applications

MONITORING TECHNOLOGIES http://www.clu-in.org/technologies/ (EPA, 2013)	FUNCTIONALITY[+]				COMMON APPLICATIONS Media/ Target	DELIVERY SYSTEM
	Diagnostics Assessment	Tracking Security	Sensing			
			Point Sensing	Global Sensing		
Laser Induced Fluorescence (LIF)	XX	X			Soil & Sediments/ NAPL distribution	Probe Platforms
X-Ray Fluorescence (XRF)	XX	X			Soil & Sediments/ NAPL distribution	Probe Platforms
Surface Acoustic Wave (SAWS)	XX	X			Soil, Sediments/ VOCs, SVOCs, PCBs	Probes w/GC, EM detectors
Membrane Interface Probes (MIP)	XX	X			Soil, Sediments/ VOCs	Probes w/GC detectors
Soil/ Gas/Water Samplers	XX	X			Soil, Sediment, Groundwater/ Organic & Inorganic content	Probe Platforms
Infrared (IR) spectroscopy	X	XX			Soil, Water, Air/ NAPL content	Probes w/ IR detectors
Open Path Techniques (UV-DOAS, OP-FTIR,LIDAR, TDLs)	X	XX			Air, Landfill Gasses/ Organic and inorganic contaminant content	Open path optical, EM transmission platforms
Electrical Resistivity Tomography, ERT	X	XX			Soil, Sediment, Groundwater/ Organic & inorganic content	Probes w/EM detectors
Ground Penetrating Radar, GPR	X	XX			Soil, Rock/ Organic content, anomalies	Surface EM detectors
Time Domain Reflectometry, TDR	X	XX			Soil, Sediment, Groundwater/ Organic & inorganic content	Probes w/EM detectors
Passive samplers	XX	X	X		Sediments, Surface & Groundwater/ VOCs, SVOCs, metals, anions	Deployed in water/soil column
Magnetometers/ Radio Frequency/ Acoustic Sensors	X	XX	X	X	Soil, Sediments, Rock/ Organic contaminant content, anomalies	Deployed in water/soil column; w/EM detectors
Fiber Optic Chemical Sensors	XX	X	XX	X	Soil, Sediment, Water/PAHs, VOCs, SVOCs, metals, anions	Deployed in water/soil column; Probes w/optical detectors
Wireless Sensor Networks	XX	XX	XX	X	Soil, Sediment, Water/Various contaminants, movement, anomalies	Deployed in well, water, sediment, soil column/EM detectors

[+] XX – primary functionality; X – secondary functionality

1.3.2 Environmental Underground Sensing and Monitoring

1.3.2.1 Overview

Table 1.2 summarizes a list of monitoring/sensing techniques deployed for underground environmental and geotechnical investigations. Short descriptions and some attributes of these technologies are provided in Table 1.3.

Table 1.3 Description of the underground environmental and geotechnical sensing and monitoring technologies given in Table 1.2

Technology	Basic Description and Typical Applications
Laser Induced Fluorescence (LIF)	• Fluorescence is the reemitted light energy at a lower level for a given wavelength. Aromatic molecules provide unique florescence spectra. Most ultraviolet filtering systems capture multiwavelength spectra. The conversion to a calibrated color-coded chart permits identification of the analyte and its proportion. • Real-time in situ field screening of NAPLs and PAHs provide distribution of subsurface contamination. It is deployed on probe rigs, and measurements conducted at every 2–3 cm stations. LIF is suitable for qualitative screening with detection limits from 10 to 1000 mg/kg.
X-Ray Fluorescence (XRF)	• A photoelectric effect is used for analyzing the fluorescent X-rays emitted from sample subjected to X-ray. Unique energy is emitted by each element at a given wavelength. The characteristic emission and intensity allow determination of the element and its proportion for quantitative analysis. Detection limits of 5–10 ppm are typical for metals. • Field portable systems available for measurement of metals and other elements with atomic weights from 19 (K) to 92 (U) (e.g., arsenic, selenium, iodine, bromine).
Surface Acoustic Wave (SAWs)	• Microelectromechanical systems (MEMS) relying on modulation of surface acoustic waves for sensing. They operate as passive (no power) remote (wireless) sensors and may be used as a one-port device that are directly affected by the measurand or a two-port device that are electrically loaded by a conventional sensor. • Environmental monitoring utilizes SAW as a sublayer coated with a selective detecting film (i.e., polymer layer) designed for a specific analyte.
Membrane Interface Probes (MIP)	• Deployed with probe platforms, it is used to detect VOCs from contaminants as gasses from heating adjacent soil and/or groundwater penetrates the membrane and is carried to ground surface.
Soil/Soil Gas/Ground-water Samplers	• Cone penetrometer for larger depths (CPT) and percussion hammer systems for shallower depth delivery of various soil sampling systems such as; piston-activated, latch-activated, and dual-tube systems. • Both systems can be equipped with continuous or discrete gas sampling tools for access to VOCs.

Table 1.3 (*continued*)

Technology	Basic Description and Typical Applications
Infrared (IR) Spectroscopy	• The measurements are based on the characteristic IR absorption spectra of the functional groups in the analyte molecules. The spectra and the amplitude of absorbance identify the compound and specify its proportions. • VOCs in water, total hydrocarbons in soils and water are measured. The system is delivered by direct probe methods. Typical detection limits are 10–100 ppm in soil.
Open Path Technologies (UV-DOAS, OP-FTIR, LIDAR, TDLs)	• Remote measurements using EM waves traveling in air. These methods detect gases and vapors created over typical hotspots (e.g., capped landfills, refineries, embedded tanks, natural gas pipeline). They provide low detection limits, and operate interference free.
Electrical Resistivity Tomography (ERT)	• Geophysical method used to construct topographical images of the subsurface from measured resistivity values. It can be conducted from the surface or by electrodes in boreholes. • ERT is used for determination of groundwater table, water leakage in concrete dams, soil moisture content, embedded barrels or tanks, plume and contaminant mapping, locating water bearing fracture zones in bedrock, cave or tunnel detection, bedrock mapping in karst terrain.
Ground Penetrating Radar (GPR)	• Geophysical tool which provides high resolution spatial imaging of shallow depth using high frequency pulsed EM waves of 10–1000 MHz. • The method allows mapping geologic topography, buried pipes, tanks, cables, and subsurface landfill boundaries.
Time Domain Reflectome-try (TDR)	• The reflection from an impulse is analyzed for duration, shape and magnitude, to determine target indices and properties of the system under the impulse. • TDR methods are used in electrical and optical systems to measure organic and inorganic compounds present in water and soil columns.
Passive samplers	• Allow free flow of contaminant molecules into a sampling interface. Equilibrium kinetics determines the sampling rate. The samplers are lowered within a screened interval of a borehole and collected periodically to determine time averaged concentration of target substance.

Table 1.3 (*continued*)

Technology	Basic Description and Typical Applications
Magnetome-ters/Radio Frequency/ Acoustic Sensors	• Magnetic methods are suitable for subsurface iron, nickel and alloys, and mostly used for locating source of leaking embedded ferrous drums and other. • NMR and RFID sensors are two typically operating in the radio frequency range. • The changes in acoustic (i.e., sound) waves traveling through a material are analyzed to obtain the changes in the desired physical or chemical phenomena (e.g., detection of VOCs).
Fiber-Optic Chemical and Physical Sensors	• FO chemical sensors can operate in direct mode (through variations in optical property of sample) or reagent-mediated mode based on absorption, fluorescence, and resonance concepts. Optical wave guides are immune to electrical and magnetic interference and can easily reach small spaces and operate in high temperature environments. • FO sensors are utilized to detect multiple indices including strain, temperature, pH and select chemical content.
Wireless Sensor Networks	• Wireless sensors are used in two modes, one where wireless component is only a transmitter to a detecting (sensing) interface, and other where wireless signal itself (power and frequency) is utilized as the means for detection. The wireless sensors can provide networked-distributed measurement capability of various physical and chemical indices in the underground.

The techniques are categorized by functionality as: (i) diagnostics/assessment, (ii) tracking/security, (iii) point sensing, and (iv) time-space continuous sensing. While most of the technologies listed function as diagnostics/assessment and tracking systems, only a few can be categorized as sensing systems that deliver either time-ordered point sensing at predesignated locations, or time-space continuous sensing.

Environmental monitoring systems and technologies are utilized to detect, identify, and quantify the existence of contaminants, track their source and progression, help in selection of remediation technologies, and verify their compliance with the limits set by the governing agencies. An important challenge in developing sensing, detection, and measurement plans is the identification of measurement locations. The potential subsurface contamination from point sources (single localized contamination sources), such as liquid containing impoundment ponds, landfills, storage tanks or injection wells, are widely mitigated by monitoring the electrical resistance and electrical potential tomography using embedded mesh of wires and fiber-optic systems or mobile sensors (Ritchie et al., 2000; Oh et al., 2007). The monitoring of non-point sources of pollutants (distributed over a large area) with potential impact on contaminating vital resources such as watersheds and drinking water sources is challenging. One such example has been monitoring sewers connected to a watershed and the impact of such a network (Kang et al., 2013). Remote sensing and use of GIS in non-point pollutant modeling has aided in similar investigations since late 1980s (Gilliland and Potter, 1987; Basnyat et al., 2000; Corwin, 1996; Corwin et al., 1998; Chowdary et al., 2004 and Shen et al., 2013).

The planning for identifying sensors and monitoring locations for a spill or leakage is often carried out by fusing the existing site data with other information such as contaminant concentration distribution, flow rate and direction, and rain intensity to extent and location of pollutants and develop an effective sensor array design. Groundwater modeling and simulations provide estimates of the extent, concentration, and representative arrival times of contaminants at points of interest, which can be used in design of sensor suits and their layouts. Nie et al. (2012) provide a comprehensive overview of non-point source pollution modeling and estimation and an extensive list of groundwater, fate and transport modeling is available in USGS[5] and EPA[6] documents. Ojha et al. (2015) discuss issues of subsurface heterogeneity, spatial and temporal scales, lack of data, and the impact on these features on modeling and simulation. Table 1.2 shows a wide range of technologies applicable to environmental sensing underground. Six main design concepts for soil sensors were identified by Adamchuk et al. (2004), which include electrical and electromagnetic, optical and radiometrics, mechanical, acoustic, pneumatic and electrochemical. The electrical resistivity tomography (ERT) and a wide range of electromagnetic wave (EM) methods have been utilized extensively in environmental studies of subsurface for diagnostic and/or tracking purposes of contamination. These methods range from non-contact remote measurements (e.g., remote magnetic induction spectroscopy, active microwave sensing, etc.) to surface and direct probe platforms (e.g., TDR, GPR, SAWs, etc.) and borehole geophysical applications.

[5] http://water.usgs.gov/software/lists/groundwater/.
[6] http://www2.epa.gov/exposure-assessment-models/groundwater.

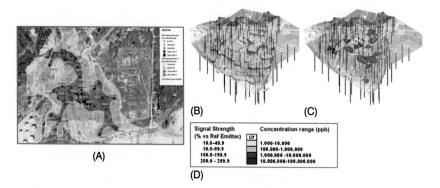

Figure 1.2 Laser induced fluorescence (LIF) based diagnostic/assessment monitoring in field (courtesy of Global Remediation Technologies, Inc.)

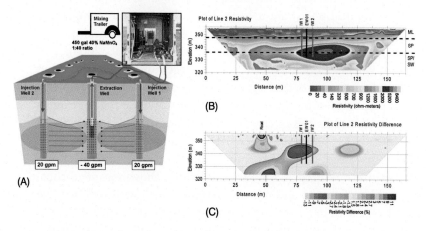

Figure 1.3 Electrical resistivity tomography of a contaminated site before and after in situ oxidation treatment with permanganate (Comfort et al., 2009)

Fig. 1.2 shows an example of laser induced fluorescence (LIF) method used to detect spatial distribution of NAPL contamination at a former petroleum refinery and storage site (GRTI, 2013). In this technique, a probe with a laser light source is driven into the subsurface at a constant rate and the fluorescence spectra of light emitting molecules of the encountered analyte (i.e., residual and non-aqueous phase hydrocarbons) is recorded and correlated to concentration. A 3D mapping of the NAPL concentration distribution is then possible as shown in Fig. 1.2. The electrical resistivity and electromagnetic wave techniques have been utilized in detection, quantification, and fate of NAPLs (non-aqueous phase liquids) including hydrocarbons (e.g., BTEX – benzene, toluene, ethylene, xylene) and chlorinated solvents (e.g., trichlorethene), and also heavy metals (Power et al., 2015). Fig. 1.3 shows an example of electrical resistivity imaging/tomography (ERI/ERT) used to monitor and facilitate an in situ chemical oxidation (ISCO) demonstration designed to treat RDX (hexhydro-1,3,5-trinitro-1,3,5-triazine) and trichlorethene (TCE) with permanganate at the Nebraska Ordinance Plant (Comfort et al., 2009). The figure shows the ERT of contaminated cross-section before and after treatment using three central wells to pump and recirculate the contaminated and treated analyte. These techniques often require geochemical sampling conducted in tandem with the in situ procedures that allow verification of the changing properties for both soil and groundwater.

Molecularly imprinted polymer films, grafted to an electrical or optical substrate (i.e., optical fibers) are used to detect the presence and quantity of a target chemical as the polymer film conductivity or optical properties change absorbing the surrounding chemical (Phillips et al., 2003). This layer or coat works as a transducer when in direct contact with the target chemical, and triggers an optical signal either due to optical loss or change in physical properties of the fiber. Some examples of these polymer interfaces are polyelectrolyte gels of cross-linked three dimensional networks of monomers that possess high swelling capability due to solvent sorption, pH or temperature (Siegel 1993; Dong and Pamukcu, 2012). Biosensors are also developed and utilized as a mean of detection. Biosensors are a combination of biotechnology and microelectronics (Singh et al., 2008). The analytes interact with bio-components through inhibiting metabolic activity or distorting existing conformation of molecules. The resulting change in the biochemical signal is detected by bio-recognition and converted to a signal by the transducer (electrochemical, optical, or thermal).

Recent developments in spectrometry have transitioned the laboratory methods to field scale applications (Borsdorf et al., 2011; Ren et al., 2014; Lian et al., 2015). Open path technologies such as differential optical absorption (DOAS), Fourier transform infrared (FTIR), tunable diode lasers (TDLs), Raman spectroscopy (RS), and LIDAR are remote methods effectively utilized in detecting contaminants in air or ground surface. Heavy metals (i.e., arsenic), pesticides, hydrocarbons and chlorinated solvents are a few of the highly toxic substances that can dwell in soil and water. There is wealth of studies conducted to develop sensors and sensor technologies to rapidly and accurately detect and quantify these substances in situ. A few recent ones are discussed below.

Du et al. (2014) studied a rapid and sensitive SERS (surface-enhanced Raman spectroscopy) detection and quantification of arsenic species using a mobile sensor. Sinha et al. (2014) discussed implanting an MEM based piezo-resistive microcantilever sensor in a monitoring cavity for onsite detection of arsenic. Kaur et al. (2015) presented the existing approaches and development schemes for arsenic biosensors based on recombinant whole cells or arsenic binding proteins. Although whole-cell based biosensors have been effectively utilized in measurements of arsenic in soils and groundwater (Merulla et al., 2013), they are limited by the incubation period and the low detection limits. Farahi et al. (2012) discuss the important issues and the critical need for sensor development for safety of food and water supplies. A review of terahertz spectroscopy methods in particular with application to identification, analysis, and quantification in agricultural products is provided by Qin et al. (2012). The application of a planar electromagnetic sensor array as an in situ measurement for rapid detection of nitrate and sulfate is studied by Nor et al. (2015). In their study, wavelet analysis was used to obtain the mean features and energy of the signals from sensor outputs to calibrate an ANN model that classify and quantify nitrate and sulfate contamination. A two-layered system consisting of a passive substrate base that vibrates and serves as an antenna and a polymer-based detection layer were used to develop a suit of SAW sensors. Newman (2012) illustrated the application of such sensor arrays in detection of organophosphate pesticide in solutions, where detection was accomplished due to resulting viscoelastic changes in the polymer coating with the mass loading. Hyperspectral methods utilize also various spectroscopy approaches in detection and quantification of pesticides in food and water supplies (Mulla, 2013; Bonifazi et al., 2014). Organic compounds such as aromatic hydrocarbons, chlorinated solvents, and PAHs have been investigated by a wide range of in situ detection and measurement methodologies, including SAW, ERT, GPR-4d, Near-IR spectroscopy, electromagnetic mapping and hyperspectral remote sensing (Power et al., 2015; Bender et al., 2013; Klavarioti et al., 2014; Mitsuhata et al., 2014; Noomen et al., 2015).

1.3.2.2 Wireless Underground Sensors and Networks

Wireless sensors and wireless sensor networks (WSNs) have found many applications for in situ environmental monitoring. In underground applications, wireless links between sensor nodes are established through electromagnetic signal propagation and wireless underground sensor networks are either infrastructure-based or ad hoc networks. There are two

categories of wireless underground sensor networks: one where wireless
links are used for data communications only, the other where wireless
links are used for both data communications and sensing functions because
electromagnetic signals may also be used to monitor the underground conditions. The common application configuration is that embedded sensors
transmit the collected data via wireless communications using radio signal as a carrier. Sensed data represents parochial distinctions taken from
a location where sensors are physically deployed. Even though wireless
underground sensor networks may offer advantages over traditional underground sensing systems in concealment, ease of deployment, timeliness
of data, reliability, and coverage density there are challenges in unearthing
and maintaining (e.g., recharging) the wireless sensors and propagating
wireless signals through dense underground substances. Researchers have
been addressing these challenges from the perspectives of power conservation, topology design, antenna design, and application scenarios. One
of the critical topics is the radio propagation model for underground
environments as it affects many design perspectives of power conservation, topology design, and antenna design (Akyildiz and Stuntebeck, 2006;
Vuran and Silva, 2009).

The WSNs have been proposed and used for (i) underground plume and
groundwater monitoring applications with automatic calibration of plume
or groundwater transport models to map and predict underground plume
or groundwater dynamics; (ii) monitoring temporal and spatial distribution
of soil water content and other soil properties over large areas for precision agriculture (iii) leakage detection (i.e., water, oil, gas) for underground
pipelines, and (iv) detection of greenhouse gases such as methane, carbon
dioxide from landfill facilities. Detailed coverage of WSNs for various applications underground are presented in subsequent chapters of this book
(e.g., Chapters 4, 5, 7.1 and 8). The following paragraphs describe briefly
a few research results representative of each of the four categories listed
above.

Precision Agriculture

Buratti et al. (2009) describe a case where a low-cost hardware and software WSN test-bed was developed for agricultural monitoring in Italy. The
platform was designed to provide a farmer with a periodic and punctual
monitoring of physical parameters (e.g., temperature, air pressure, humidity) for real-time control of cultivation. WSNs have been studied by others

to monitor underground soil conditions, such as water and mineral content distribution, to provide data for appropriate irrigation and fertilization (Sahota et al., 2011). Tooker et al. (2012) developed a novel underground sensor network test-bed equipped with soil moisture sensors and mobile data harvesting unit with improved underground and cellular communication capabilities. Real-time soil moisture data delivery was demonstrated over long communication distance. Dong et al. (2013) demonstrated a precision agriculture platform through the integration of a center pivot irrigation system with wireless underground sensor networks. The new system was expected to provide autonomous irrigation management capabilities by monitoring the soil conditions in real-time. The demonstration revealed that the wireless communication channel between soil and air was significantly affected by multiple spatiotemporal aspects, such as the location and burial depth of the sensors, soil texture and physical properties, soil moisture, and the vegetation canopy height.

Soil Water Distribution

The temporal and spatial distribution of soil water content can be monitored over large areas using WSNs (Ritsema et al., 2009; Dong and Vuran, 2010; Kerkez et al., 2012). Ritsema et al. (2009) reported acceptable performance from a wireless underground network system that continuously monitored soil water contents at different locations and soil depths across a golf course. Dong and Vuran (2010) used a WSN system to monitor soil moisture content and applied a spatiotemporal correlation based on rainfall, soil porosity, and vegetation root zone to analyze the discrepancies in soil moisture estimations. Kerkez et al. (2012) developed a wireless sensor network as part of a water balance instrument cluster that provided spatially representative measurements of soil moisture and matric potential. Yoon et al. (2011) showed that calibrated wireless signal strength variations can be used as indicators to sense changes in the underground environment by experiments in laboratory scale that simulate underground hazards (Yoon et al., 2012; Ghazanfari et al., 2012). An unequipped wireless signal network (e.g., wireless transceivers only, without a specific sensor to quantify a specific index) can return scaled change of an index between any two nodes over the duration of an event. In order to demonstrate the sensing potential of wireless signal networks, a water intrusion event was monitored using simultaneously a commercial soil moisture sensor suite and wireless signal network nodes (MICAz − 2.4 GHz). As shown in Fig. 1.4A, a transceiver at node S1 sent a packet of EM signal every 60 seconds while a transceiver

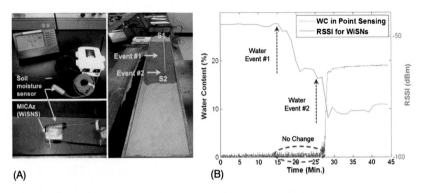

Figure 1.4 Comparative tracking of water intrusion event by WSiN and conventional soil moisture sensor in a floor scale test bed (Yoon, 2012). (A) Test-bed of WSiN. (B) Water intrusion traces

at node S2, located at 75 cm away from the sender, calculated the signal strength based on the received packet. The measured water contents (i.e., moisture sensor record) and the received signal strengths (RSSI) recorded by the wireless signal network at S2 are plotted in Fig. 1.4B. While the RSSI responded to the first and second water intrusions events instantaneously, the moisture sensor response was delayed until the second water intrusion event. The wireless signal network provided information on the "global change" between the two nodes well ahead of the point sensor, while the point sensor returned information only when the event reached its sensing location.

Plumes and Groundwater

A highly promising area for WSNs application underground is plume monitoring (Michaelides and Panayiotou, 2005; Han et al., 2008; Porta et al., 2009). Shepard et al. (2007) developed a floor scale chamber equipped with a wireless chemical sensor network to detect and track plumes of acetic acid to map and predict chemical plume dynamics. Han et al. (2008) developed a closed-loop contaminant-plume monitoring system that integrated WSN based monitoring with a numerical model of plume transport. Porta et al. (2009) tested a floor scale test bed of sand with wireless sensor nodes equipped with electrical conductivity probes. Loden et al. (2009) investigated application of WSNs in laboratory scale for subsurface groundwater plume monitoring with the objective of automatically calibrating the groundwater transport models. Similarly, Barnhart et al. (2010) captured transient plumes in a floor scale synthetic aquifer equipped with self-organizing WSN to calibrate a computational transport model. Laboratory and floor scale experiments, although successful for the most part, also demonstrated that sensor features (e.g., resolution, accuracy), and sensor calibration were equally critical to the success of plume monitoring using WSNs.

Landfill Gas

Landfill facilities contain a network of pipes connecting perforated wells placed into the waste body, while an applied negative pressure extracts the gases. Wireless sensor networks have been shown to be a viable option for landfill gas monitoring (Beirne et al., 2010; Collins et al., 2013). Autonomous monitoring platforms have been developed for perimeter landfill wells that can monitor gas migration through the waste body and into the surrounding soil around the landfill. Collins et al. (2013) developed a wireless sensor network for landfill gas monitoring that specifically senses methane, carbon dioxide, and the extraction pressure.

Pipeline Leakage

Wireless sensor networks have been shown to be a viable option for leakage detection in pipelines (Saftner et al., 2008). Yaacoub et al. (2011) introduced the application of an immersed mobile sensor in water pipe that communicated with relay nodes on the ground surface. Sun et al. (2011b) proposed leakage detection using magnetic induction (MI)-based WSN located both inside and around the underground pipelines. This technique allowed measurements from different

real-time.types of the sensors throughout the pipeline network to be reported in

1.3.3 Geotechnical Underground Sensing and Monitoring

In this book, Chapter 3 presents a comprehensive coverage of the conventional and advanced geotechnical monitoring and sensor tools and methods. In here we provide an overview of select geotechnical underground sensing applications, particularly those utilizing fiber-optic and wireless sensors.

Pipelines

Nikles et al. (2004) presented a leakage monitoring method for pipelines using distributed fiber optics sensing over a 55 km brine pipeline. The detection system was based on Brillouin time domain reflectometry

(BOTDR) measurement of temperature changes over the length of the pipeline. Sinha and Knight (2004) and Sinha and Fieguth (2006) developed an integrated system of pipeline health monitoring and guidelines for adaptive action. The system was designed to analyze scanned underground pipe images to determine defects (i.e., cracks, holes, joints, laterals, and collapsed surfaces) and identify the appropriate maintenance strategies. Koo and Ariaratnam (2006) combined the Ground Penetrating Radar (GPR) with Digital Scanning and Evaluation Technology (DSET) to monitor the condition of a large diameter PVC lined concrete pipeline. Wan and Mita (2010) compared two acoustic techniques for recognition of potential threats of heavy construction machines to underground pipelines. The methods were applied in actual site cases, with useful outcomes to prevent accidental breakage of underground lifeline infrastructures. Shin et al. (2010) proposed a new method for predicting the location of the impact forces caused by construction equipment in a large-diameter pipeline. The prediction error of the proposed method was much less than the conventional method. Inaudi and Glisic (2010) presented four application examples of fiber optic monitoring systems for pipelines, including distributed temperature monitoring, leakage detection, intrusion detection, and distributed strain and deformation monitoring. Hao et al. (2012) summarized various techniques for monitoring the condition of buried utility infrastructure, particularly water and sewage pipelines. Based on distributed fiber optic sensors, Glisic and Yao (2012) developed a method for detecting and localizing the pipeline damage and soil displacements after an earthquake. The method was validated through two large-scale tests. Kolbadinejad et al. (2013) used piezoelectric sensors to monitor the corrosion and crack of the pipelines made from steel and aluminum alloys. They developed an equivalent electrical model and performed an experiment to verify it. Whittle et al. (2013) introduced WaterWiSe, a platform that can provide real-time monitoring data from wireless sensor nodes and decision support for water distribution systems. Hydraulic, acoustic, and water quality parameters were measured and analyzed to predict water demand, hydraulic state, and abnormal events such as pipe bursts. Mirzaeia et al. (2013) showed that oil leakage of transportation pipelines could be detected and localized by Raman optical time domain reflectometer (ROTDR) and Brillouin optical time domain amplifier (BOTDA) sensors precisely. The transient response of BOTDA and ROTDR sensors were obtained through solution of the mass, energy, and heat transfer in soil and fiber cable. Hedan et al. (2014) monitored clay-rock strains as well as the opening and closing of desiccation cracks using digital image correlation (DIC). Huang et al. (2014) investigated use of long-period fiber gratings (LPFGs) for a corrosion-induced deterioration assessment of pipeline structures. A thin layer of nano-iron/silica particles dispersed polyurethane was coated on the outer surface of the LPFGs. Test results showed the proposed sensor could monitor both the initial and stable corrosion rate consistently and with multiple LPFGs in a single fiber, it thus was possible to provide a cost-effective corrosion-monitoring technique for pipeline corrosion monitoring. Arsénio et al. (2015) estimated the ground movement using radar satellite data and proposed a method to predict the pipeline failures due to ground movement.

Mines and Underground Spaces

Ge et al. (2007) monitored mine subsidence with differential interferometric synthetic aperture radar (DInSAR) technique which revealed mm-level resolution. Li and Liu (2007, 2009) designed a structure-aware self-adaptive WSN system (SASA) to rapidly detect structure variations in coal mines caused by underground collapses. The collapsed holes could be outlined and relocated sensor nodes in region could be reconfigured. The system reliability and validity of SASA was tested in a coal mine. Nomatungulula et al. (2014) developed a mobile and real-time gateway for processing data from wireless sensor nodes. The system was aimed to make the mine workers aware of dangerous condition such as gas explosions or mine collapses in time. Przyłucka et al. (2015) combined conventional and advanced DInSAR techniques to monitor the extent and the magnitude of mining subsidence, showing that the conventional DInSAR could detect the most active mining subsidence areas, while the advanced DInSAR could delimit residual mining subsidence in the surroundings. Li et al. (2016) summarized the tunnels that were equipped with structural health monitoring (SHM) systems in China and discussed two of them in detail. Debliquy et al. (2015) reviewed the existing optical fiber sensor methods, commercially available or under development in the field of air quality monitoring, in particular NO_2 that is representative of toxic automotive pollution, as well as flaming fire detection and combustible gas leak detection (in particular methane and hydrogen) in road tunnels and undercroft car parks.

Piles

Glisic et al. (2002) monitored the average strains in several segments over piles with long-gauge fiber optic sensors, showing the monitoring method allowed determination of the Young's modulus of the piles, the occurrence of cracks, the normal force distribution, as well as the curvature distribution, horizontal displacement, deformed shape, and damage localization. Kister et al. (2007) applied Bragg grating sensors to monitor both the strain and temperature of two reinforced concrete foundation piles. The strain distribution was observed along the entire length of the piles. Lu et al. (2012) monitored the strain of precast piles using a BOTDR-based technique. They analyzed distribution of the the axial force, side friction, end-bearing resistance, of the pile as well as the mechanism of pile–soil interactions based on the distributed strain data obtained by BOTDR sensing.

Tunnel

Soga et al. (2008) applied BOTDR technique to measure distributed strains of the old tunnel when a new tunnel was constructed nearby. Metje et al. (2008) developed a new optical fiber system, "Smart Rod", to monitor structural displacements and applied it to measure the displacements of a road tunnel lining. Li et al. (2008) monitored the surface strain of the second lining of a tunnel using the metal groove encapsulated FBG technique. They found the pressure of the wall rock was increased when rainy season came and the rainwater soaked into the slope. Li et al. (2009) developed a differential FBG strain sensor to monitor the stability of a tunnel, showing that the strains were related to back cavity in the backfilling period and became stable during the tunnel operation. Cheung et al. (2010) applied a trial strain-monitoring system using BOTDR technology to successfully monitor the joint movements of a tunnel section. Both studies showed BOTDR strain measurement of the tunnel walls can provide an increased safety level and help researchers determine the complex behavior of the underground tunnels. Bennett et al. (2010) presented two field trials of WSN systems in underground tunnel monitoring. The WSN system was designed to address various issues, such as reliability and long-term performance. Gue et al. (2015) monitored the strains of an existing cast iron tunnelling in both cross-sectional and longitudinal direction using the BOTDR technique. They also installed wireless linear potentiometric displacement transducers (LPDT) to measure the joint opening and closing so that the amount of strains in the linings and joints could be separated.

Hybrid Other Applications

Akyildiz and Stuntebeck (2006) were pioneers to introduce the concept and applications of wireless underground sensor network (WUSN) and demonstrated the challenges for WUSN design, such as power conservation, topology design, antenna design, and environmental extremes. They also presented the properties of the underground EM channel and the effect of various soil properties, particularly the water content, on this channel, showing low frequencies were able to propagate with lower losses through the underground but restricted the bandwidth available for data transmission. They suggested to develop the existing terrestrial WSN communication protocols to address these issues and proposed a cross-layer protocol solution. Stuntebeck et al. (2006) tested the packet error rate and the received signal strength between two underground sensors and between an underground sensor and an aboveground sensor, concluding that the terrestrial wireless sensor network solutions were possible but had limited applicability for wireless underground sensor networks. Sun and Akyildiz (2010a) developed a magnetic induction (MI) waveguide technique to create constant channel condition and reduce the path loss. When compared to the traditional EM wave system and the ordinary MI system, the transmission range of the MI waveguide system was significantly larger in underground environments. Sun and Akyildiz (2010b) also investigated the deployment of the MI waveguide to connect the wireless underground sensor networks. Habel and Krebber (2011) presented the use of several types of fiber-optic sensor technologies, such as FBG sensor arrays used to monitor anchor deformation, fiber Fabry–Perot interferometer sensors used as acoustic emission sensors to characterize the integrity and bearing capacity of large concrete piles, and concrete-embeddable pH sensors in steel-reinforced concrete structures to monitor the corrosion-free operation. Pei et al. (2012) developed FBG-based in-place inclinometers and installed them in a vertical borehole to monitor the movement of a slope. The concept was based on the classical beam theory and validated for long-term monitoring. Rucker et al. (2013a) applied satellite-based interferometry by synthetic aperture radar (InSAR) technique for land subsidence monitoring in three areas, which could identify possible impacts or threats to underground infrastructures, including a pipelines, tunnels, and coal mines. Owojaiye and Sun (2013) identified and discussed the vital design issues for deployment of WSN for underground infrastructure monitoring, especially pipeline monitoring, based on the survey of the WSN technologies. The design issues were classified into sensing modality, power efficiency, energy harvesting, network reliability, and localization. Rucker et al. (2013b) applied the electrical resistivity imaging technique for underground contaminant plume mapping beneath a tank farm. They planned to run the program for seven years after which to move to long-term monitoring according to site requirements. Gage et al. (2013) developed fiber-optically instrumented rock strain and temperatures trips (FROSTS) for monitoring the strain and temperature of rock mass surrounding underground openings. They tested the FROSTS in the laboratory and installed them 1250 m below the surface at the Sanford Underground Research Facility. Chawah et al. (2015) developed a pendulum borehole tiltmeter based on laser fiber interferometric displacement sensors and validated it in an underground facility. The tiltmeter had has several advantages, such as low cost, long functioning time, no energy required at the measurement point, and sensitivity to both static tilts and ground motions. Koch et al. (2015) summarized the current state of the visual inspection techniques used in precast concrete tunnels and underground concrete pipes. The achievements and limitations of existing methods were also presented. Zhou et al. (2015) presented a novel design for a wireless mobile platform to locate and gather data from different types of optical fiber sensors, thereby enabling the more effective integration of a number of such optical fiber sensors with an advanced mobile wireless sensor network (WSN). A fiber Bragg grating-based temperature sensor and an intrinsic pH optical fiber sensor were specially designed and integrated successfully into the optical fiber sensor module as an exemplar to investigate the performance of the integrated system based on the mobile WSN platform.

The existing systems with the potential of temporally and spatially continuous sensing capability for underground environments include wireless underground sensor networks (WUSN), and fiber-optic (FO) sensors. Their potentials for underground environmental hazard monitoring have been explored in pilot or floor scale operations and are yet to be deployed broadly in the field on a larger scale. The premise and operation of such sensors and their applications underground are covered extensively in the following chapters of this book.

REFERENCES

Abdoun, T., Abe, A., Bennett, V., Danisch, L., Sato, M., Tokimatsu, K., Ubilla, J., 2007. Wireless real time monitoring of soil and soil-structure systems. In: Proceedings of the Geo-Denver. Denver, CO, 18–21 February 2007.

Adamchuk, V.I., Hummel, J.W., Morgan, M.T., Upadhyaya, S.K., 2004. On-the-go soil sensors for precision agriculture. Comput. Electron. Agric. 44 (1), 71–91.

Akyildiz, I.F., Stuntebeck, E.P., 2006. Wireless underground sensor networks: research challenges. Ad Hoc Netw. 4 (6), 669–686.

Akyildiz, I.F., Sun, Z., Vuran, M.C., 2009. Signal propagation techniques for wireless underground communication networks. Phys. Commun. 2, 167–183.

Akyildiz, I.F., Vuran, M.C., 2010. Wireless underground sensor networks. In: Wireless Sensor Networks. John Wiley & Sons, Ltd., Chichester, UK. Chapter 17.

Alamdar, F., Kalantari, M., Rajabifard, A., 2015. An evaluation of integrating multisourced sensors for disaster management. Int. J. Digital Earth 8 (9), 727–749. http://dx.doi.org/10.1080/17538947.2014.927537.

Almazyad, A.S., Seddiq, Y.M., Alotaibi, A.M., Al-Nasheri, A.Y., BenSaleh, M.S., Obeid, A.M., Qasim, S.M., 2014. A proposed scalable design and simulation of wireless sensor network-based long-distance water pipeline leakage monitoring system. Sensors 14 (2), 3557–3577.

Arsénio, A.M., Dheenathayalan, P., Hanssen, R., Vreeburg, J., Rietveld, L., 2015. Pipe failure predictions in drinking water systems using satellite observations. Struct. Infrastruct. Eng. 11 (8), 1102–1111.

ASCE, 2017. Report card for America's infrastructure. www.asce.org/infrastructure.

Barnhart, K., Urteaga, I., Han, Q., Jayasumana, A., Illangasekare, T., 2010. On integrating groundwater transport models with wireless sensor networks. Ground Water 48 (5), 771–780.

Barnwal, R.P., Bharti, S., Misra, S., Obaidat, M.S., 2014. UCGNet: wireless sensor network-based active aquifer contamination monitoring and control system for underground coal gasification. Int. J. Commun. Syst. 30 (1), e2852. http://dx.doi.org/10.1002/dac.2852.

Basnyat, P., Teeter, L.D., Lockaby, B.G., Flynn, K.M., 2000. The use of remote sensing and GIS in watershed level analyses of non-point source pollution problems. For. Ecol. Manag. 128 (1–2), 65–73.

Basu, D., Misra, A., Puppala, A.A., 2014. Sustainability and geotechnical engineering: perspectives and review. Can. Geotech. J. 52 (1), 96–113.

Bavusi, M., Lapenna, V., Loperte, A., Gueguen, E., De Martino, G., Adurno, I., Catapano, I., Soldovieri, F., 2013. Groundwater monitoring and control by using electromagnetic sensing techniques. In: The Handbook of Environmental Chemistry, pp. 1–32.

Becerik-Gerber, B., Siddiqui, M.K., Brilakis, I., El-Anwar, O., El-Gohary, N., Mahfouz, T., Jog, G.M., Li, S., Kandil, A.A., 2014. Engineering grand challenges: opportunities for data sensing, information analysis, and knowledge discovery. J. Comput. Civ. Eng. 28 (4), 04014013. http://dx.doi.org/10.1061/(ASCE)CP.1943-5487.0000290.

Beirne, S., Kiernan, B., Fa, C., Foley, C., Corcoran, B., Smeaton, A.F., Diamond, D., 2010. Autonomous greenhouse gas measurement system for analysis of gas migration on landfill sites. In: IEEE Sensors Applications Symp, 2010. Limerick, Ireland, SAS 2010, pp. 143–148.

Bell, T.H., Barrow, B.J., Miller, J.T., 2001. Subsurface discrimination using electromagnetic induction sensors. IEEE Trans. Geosci. Remote 39, 1286–1293.

Benedetto, F., Tedeschi, A., 2015. Moisture content evaluation for road surfaces monitoring by GPR image and data processing on mobile platforms. In: The 3rd IEEE International Conference on Future Internet of Things and Cloud, pp. 602–607.

Bennett, P.J., Soga, K., Wassell, I., Fidler, P., Abe, K., Kobayashi, Y., Vanicek, M., 2010. Wireless sensor networks for underground railway applications: case studies in Prague and London. Smart Struct. Syst. 6 (5–6), 619–639.

BenSaleh, M.S., Qasim, S.M., Obeid, A.M., Garcia-Ortiz, A., 2013. A review on wireless sensor network for water pipeline monitoring applications. In: Proceedings of the International Conference on Collaboration Technologies and Systems (CTS'2013). San Diego, CA, USA, 20–24 May 2013, pp. 128–131.

Bertolla, L., Porsani, J.L., Soldovieri, F., Catapano, I., 2014. GPR-4d monitoring a controlled LNAPL spill in a masonry tank at USP, Brazil. J. Appl. Geophys. 103, 237–244.

Bhalla, S., Yang, Y.W., Zhao, J., Soh, C.K., 2005. Structural health monitoring of underground facilities-technological issues and challenges. Tunn. Undergr. Space Technol. 20, 487–500.

Bidmanova, S., Kotlanova, M., Rataj, T., Damborsky, J., Trtilek, M., Prokop, Z., 2016. Fluorescence-based biosensor for monitoring of environmental pollutants: from concept to field application. Biosens. Bioelectron. 84, 97–105. http://dx.doi.org/10.1016/j.bios.2015.12.010.

Bonifazi, G., Fabbri, A., Serranti, S., 2014. A hyperspectral imaging (HIS) approach for bio-digestate real time monitoring. In: Proc. SPIE Sensing for Agriculture and Food Quality and Safety VI, vol. 9108.

Borsdorf, H., Mayer, T., Zarejousheghani, M., Eiceman, G.A., 2011. Recent developments in ion mobility spectrometry. Appl. Spectrosc. Rev. 46 (6), 472–521. http://dx.doi.org/10.1080/05704928.2011.582658.

Bose, S., Mukherjee, N., Mistry, S., 2016. Environment monitoring in smart cities using virtual sensors. In: 2016 IEEE 4th International Conference on Future Internet of Things and Cloud. FiCloud. IEEE.

Buratti, C., Conti, A., Dardari, D., Verdone, R., 2009. An overview on wireless sensor networks technology and evolution. Sensors (Basel) 9 (9), 6869–6896.

Capella, J.V., Bonastre, A., Ors, R., Peris, M., 2014. A step forward in the in-line river monitoring of nitrate by means of a wireless sensor network. Sens. Actuators B, Chem. 195, 396–403.

Challener, W., Palit, S., Jones, R., Airey, L., Craddock, R., Knobloch, A., 2013. MOEMS pressure sensors for geothermal well monitoring. In: Proceedings, MOEMS and Miniaturized Systems XII. In: International Society for Optics and Photonics, vol. 8616.

Chawah, P., Chéry, J., Boudin, F., Cattoen, M., Seat, H.C., Plantier, G., Gaffet, S., 2015. A simple pendulum borehole tiltmeter based on a triaxial optical-fibre displacement sensor. Geophys. J. Int. 203 (2), 1026–1038.

Chen, X., 2013. Current developments in optical fiber technology. In: Harun, S.W., Arof, H. (Eds.), Optical Fiber Gratings for Chemical and Bio-Sensing. ISBN 978-953-51-1148-1. Chapter 8.

Cheung, L.L.K., Soga, K., Bennett, P.J., Kobayashi, Y., Amatya, B., Wright, P., 2010. Optical fibre strain measurement for tunnel lining monitoring. Proc. Inst. Civil Eng. Geotech. Eng. 163 (3), 119–130.

Chowdary, V.M., Kar, Y.S., Adiga, S., 2004. Modelling of non-point source pollution in a watershed using remote sensing and GIS. J. Indian Soc. Remote Sens. 32 (1), 59–73.

Collins, F., Orpen, D., McNamara, E., Fay, C., Diamond, D., 2013. Landfill gas monitoring network: development of a wireless sensor network platforms. In: Proc. of 2nd Int. Conf. on Sensor Networks. Sensornets 2013, Barcelona, Spain, pp. 222–228.

Comfort, S., Zlotnik, V., Halihan, T., 2009. Using electrical resistivity imaging to evaluate permanganate performance during an in situ treatment of a RDX-contaminated aquifer. ESTCP Project ER-0635.

Correia, A.G., 2015. Geotechnical engineering for sustainable transportation infrastructure. In: Proceedings of the XVI ECSMGE Geotechnical Engineering for Infrastructure and Development.

Correia, A.G., Winter, M.G., Puppala, A.J., 2016. A review of sustainable approaches in transport infrastructure. Transp. Geotech. 7, 21–28. http://dx.doi.org/10.1016/j.trgeo.2016.03.003.

Corwin, D.L., 1996. GIS Applications of Deterministic Solute Transport Models for Regional-Scale Assessment of Non-Point Source Pollutants in the Vadose Zone. Soil Science Society of America, pp. 69–100.

Corwin, D.L., Loague, K., Ellsworth, T.R., 1998. GIS-based modeling of non-point source pollutants in the vadose zone. J. Soil Water Conserv. 53 (1), 34–38.

Cozzolino, D., 2016. Near infrared spectroscopy as a tool to monitor contaminants in soil, sediments and water—state of the art, advantages and pitfalls. Trends Env. Anal. Chem. 9, 1–7.

Dardari, D., Conti, A., Buratti, C., Verdone, R., 2007. Mathematical evaluation of environmental monitoring estimation error through energy-efficient wireless sensor networks. IEEE Trans. Mob. Comput. 6, 790–803.

Debliquy, M., Lahem, D., Bueno-Martinez, A., Ravet, G., Renoirt, J.-M., Caucheteur, C., 2015. Review of the use of the optical fibers for safety applications in tunnels and car parks: pollution monitoring, fire and explosive gas detection. In: Sensing Technology: Current Status and Future Trends III. In: Smart Sensors, Measurement and Instrumentation, vol. 11. Springer International, pp. 1–24.

Delmastro, C., Lavagno, E., Schranz, L., 2016. Underground urbanism: master plans and sectorial plans. Tunn. Undergr. Space Technol. 55, 103–111.

Dikmen, U., Safak, E., Pinar, A., Edincliler, A., Erdik, M., 2016. Seismic monitoring system of marmaray submerged tube tunnel. J. Earthquake Tsunami 10, 1640011. http://dx.doi.org/10.1142/S179343111640011X.

Dohare, Y.S., Maity, T., Das, P.S., Paul, P.S., 2015. Wireless communication and environment monitoring in underground coal mines–review. IETE Tech. Rev. 32 (2), 140–150.

Dong, X., Vuran, M.C., 2010. Spatio-temporal soil moisture measurement with wireless underground sensor networks. In: Proc. of IFIP Annual Mediterranean Ad Hoc Networking Workshop 2010.

Dong, Y., Pamukcu, S., 2012. Preparation of polymer coated sands with reversible wettability triggered by temperature. In: GeoCongress 2012: GSP 225. ASCE, March 25–29, 2012, pp. 4446–4455.

Dong, X., Vuran, M.C., Irmak, S., 2013. Autonomous precision agriculture through integration of wireless underground sensor networks with center pivot irrigation systems. Ad Hoc Netw. 11 (7), 1975–1987.

Dong, X., Hu, X-Y., Shan, C-L., Li, R-H., 2016. Landslide monitoring in southwestern China via time-lapse electrical resistivity tomography. Appl. Geophys. 13 (1), 1–12. http://dx.doi.org/10.1007/s11770-016-0543-3.

Doyle, M.R., Thalmann, P., Parriaux, A., 2016. Underground potential for urban sustainability: mapping resources and their interactions with the deep city method. Sustainability 8 (9), 830. http://dx.doi.org/10.3390/su8090830.

Donovan, C., Dewan, A., Heoand, D., Beyenal, H., 2008. Batteryless, wireless sensor powered by a sediment microbial fuel cell. Environ. Sci. Technol. 42 (22), 8591–8596.

Drnevich, V.P., Zambrano, C.E., Huang, P.T., Jung, S., 2007. TDR technologies for soil identification and properties. In: Proc. of the 7th FMGM 2007: GSP 175 Int. Sym. on Field Measurements in Geomechanics. ASCE, Boston.

Droit, C., Friedt, J.-M., Goavec-Mérou, G., Ballandras, Martin G., Breschi, S.K., Bernard, J., Guyennet, H., 2012. Radiofrequency transceiver for probing SAW sensors and communicating through a wireless sensor network. In: SENSOR-COMM 2012: The Sixth International Conference on Sensor Technologies and Applications. ISBN 978-1-61208-207-3.

Du, J., Cui, J., Jing, C., 2014. Rapid in situ identification of arsenic species using a portable Fe_3O_4@Ag SERS sensor. Chem. Commun. 50, 347–349.

Dunnicliff, J., 1993. Geotechnical Instrumentation for Monitoring Field Performance, 2nd edn.. John Wiley & Sons, New York.

EPA, 2012. A framework for sustainability indicators at EPA. EPA/600/R/12/687/, October 2012, www.epa.gov.org.

Farahi, R.H., Passian, A., Tetard, L., Thundat, T., 2012. Critical issues in sensor science to aid food and water safety. Amer. Chem. Soc. Nano 6, 4548–4556. http://dx.doi.org/10.1021/nn204999j.

Ferdinand, P., 2014. The evolution of optical fiber sensors technologies during the 35 last years and their applications in structure health monitoring. In: Le Cam, Vincent, Mevel, Laurent, Schoefs, Franck (Eds.), EWSHM–7th European Workshop on Structural Health Monitoring. July 2014, Nantes, France.

Fiksel, J., 2012. A systems view of sustainability: the triple value model. Environ. Dev. 2, 138–141. http://dx.doi.org/10.1016/j.envdev.2012.03.015.

Fraga-Lamas, P., Fernández-Caramés, T.M., Suárez-Albela, M., Castedo, L., González-López, M., 2016. A review on internet of things for defense and public safety. Sensors 16 (10), 1644. http://dx.doi.org/10.3390/s16101644.

Furlani, K.M., Miller, P.K., Mooney, M.A., 2005. Evaluation of wireless sensor node for measuring slope inclination in geotechnical applications. In: Proc of the 22nd Int Symp on Automation and Robotics in Construction. Ferrara, Italy, pp. 11–14.

Gage, J.R., Fratta, D., Turner, A.L., Maclaughlin, M.M., Wang, H.F., 2013. Validation and implementation of a new method for monitoring in situ strain and temperature in rock masses using fiber-optically instrumented rock strain and temperature strips. Int. J. Rock Mech. Min. Sci. 61, 244–255.

Ge, L., Chang, H., Rizos, C., 2007. Mine subsidence monitoring using multi-source satellite sar images. Photogramm. Eng. Remote Sens. 73 (3), 259–266.

Ghazanfari, E., Pamukcu, S., Yoon, S., Suleiman, M.T., Cheng, L., 2012. Geotechnical sensing using electromagnetic attenuation between radio transceivers. Smart Mater. Struct. 21, 125017.

Giannoukos, S., Brkić, B., Taylor, S., Marshall, A., Verbeck, G.F., 2016. Chemical sniffing instrumentation for security applications. Chem. Rev. 116 (14), 8146–8172. http://dx.doi.org/10.1021/acs.chemrev.6b00065.

Gibson, R., 2006. Sustainability assessment: basic components of a practical approach. IAPA 24 (3), 170–182.

Gilliland, M.W., Potter, B., 1987. A geographic information system to predict non-point source pollution potential. J. Am. Water Resour. Assoc. 23 (2), 281–291. http://dx.doi.org/10.1111/j.1752-1688.1987.tb00807.x.

Glisic, B., Inaudi, D., Nan, C., 2002. Pile monitoring with fiber optic sensors during axial compression, pullout, and flexure tests, transportation research record. J. Trans. Res. Board 1808, 11–20.

Glisic, B., Yao, Y., 2012. Fiber optic method for health assessment of pipelines subjected to earthquake-induced ground movement. Struct. Health Monit. 11 (6), 696–711.

Glisic, B., 2014. Sensing solutions for assessing and monitoring pipeline systems. In: Wang, M.L., Lynch, J.P., Sohn, H. (Eds.), Sensor Technologies for Civil Infrastructures: Applications in Structural Health Monitoring, vol. 2. Woodhead Publishing, London, UK.

Glaser, S.D., Shoureshi, R., Pescovitz, D., 2005. Future sensing systems. Smart Struct. Syst. 1, 103–120.

Godfray, H.C.J., Beddington, J.R., Crute, I.R., Haddad, L., Lawrence, D., Muir, J.F., Pretty, J., Robinson, S., Thomas, S.M., Toulmin, C., 2010. Science 327 (5967), 812–818.

Gong, W., Suresh, M.A., Smith, L., Ostfeld, A., Stoleru, R., Rasekh, A., Banks, M.K., 2016. Mobile sensor networks for optimal leak and backflow detection and localization in municipal water networks. Environ. Model. Softw. 80 (C), 306–321.

Grote, K., Anger, C., Kelly, B., Hubbard, S., Rubin, Y., 2010. Characterization of soil water content variability and soil texture using GPR groundwave techniques. J. Environ. Eng. Geophys. 15 (3), 93–110.

GRTI, 2013. Global Remediation Technologies Inc., Remedial Investigation Technical Memorandum/Former Zephyr Naph-Sol Refinery, 1222 Holton Road, Muskegon, MI/ Prepared for: MDEQ – RRD, Grand Rapids, MI 49503, March 15, 2013.

Gue, C.Y., Wilcock, M., Alhaddad, M.M., Elshafie, M.Z.E.B., Soga, K., Mair, R.J., 2015. The monitoring of an existing cast iron tunnel with distributed fibre optic sensing (Dfos). J. Civil Struct. Health Monitor. 5 (5), 573–586.

Habel, W.R., Krebber, K., 2011. Fiber-optic sensor applications in civil and geotechnical engineering. Photon. Sens. 1 (3), 268–280.

Han, Q., Jayasumana, A.P., Illangaskare, T., Sakkai, T., 2008. A wireless sensor network based closed-loop system for subsurface contaminant plume monitoring. In: 22nd IEEE Int Symp on Parallel and Distributed Processes. Miami, FL, pp. 1–5.

Hao, T., Rogers, C.D.F., Metje, N., Chapman, D.N., Muggleton, J.M., Foo, K.Y., Saul, A.J., 2012. Condition assessment of the buried utility service infrastructure. Tunn. Undergr. Space Technol. 28, 331–344.

Hanifah, M.I.M., Omar, R.C., Khalid, N.H.N., Ismail, A., Mustapha, I.S., Baharuddin, I.N.Z., Roslan, R., Zalam, W.M.Z., 2016. Integrated geo hazard management system in cloud computing technology. In: IOP Conference Series: Materials Science and Engineering 160:1, p. 012081.

Hedan, S., Fauchille, A., Valle, V., Cabrera, J., Cosenza, P., 2014. One-year monitoring of desiccation cracks in tournemire argillite using digital image correlation. Int. J. Rock Mech. Min. Sci. 68, 22–35.

Hofinghoff, J.F., Overmeyer, L., 2013. Resistive loaded antenna for ground penetrating radar inside a bottom hole assembly. IEEE Trans. Antennas Propag. 61 (12), 6201–6205.

Hong, C.-Y., Zhang, Y.-F., Zhang, M.-X., Leung, L.M.G., Liu, L.-Q., 2016. Application of FBG sensors for geotechnical health monitoring, a review of sensor design, implementation methods and packaging techniques. Sens. Actuators A, Phys. 244, 184–197.

Huang, Y.-S., Chen, Y.-Y., Wu, T.-T., 2010. A passive wireless hydrogen surface acoustic wave sensor based on Pt-coated ZnO nanorods. Nanotechnology 21 (9), 5503.

Huang, Y., Liang, X., Azarmi, F., 2014. Innovative fiber optic sensors for pipeline corrosion monitoring. In: Pipelines 2014. ASCE.

Hunt, D.V.L., Nash, D., Rogers, C.D.F., 2014. Sustainable utility placement via multi-utility tunnels. Tunn. Undergr. Space Technol. 39, 15–26. http://dx.doi.org/10.1016/j.tust.2012.02.001.

Inaudi, D., Glisic, B., 2010. Long-range pipeline monitoring by distributed fiber optic sensing. J. Press. Vessel Technol. 132 (1), 11701.

ISSMGE, 2009. International society for soil mechanics and geotechnical engineering. In: Rao, V.V.S. (Ed.), Forensic Geotechnical Engineering, TC 40.

Iten, M., Spera, Z., Jeyapalan, J., Duckworth, G., Inaudi, D., Bao, X., Noether, N., Klar, A., Marshall, A., Glisic, B., Facchini, M., Jason, J., Elshafie, M., Kechavarzi, C., Miles, W., Rajah, S., Johnston, B., Allen, J., Lee, H., Leffler, S., Zadok, A., Hayward, P., Waterman, K., Artieres, O., 2015. Pipelines, pp. 1655–1666.

ITRC, 2010. Use and Measurement of Mass Flux and Mass Discharge. The Interstate Technology & Regulatory Council Integrated DNAPL Site Strategy Team, Washington, DC. 89 p.

Johnson, T.C., Versteeg, R.J., Ward, A., Day-Lewis, F.D., Revil, A., 2010. Improved hydrogeophysical characterization and monitoring through parallel modeling and inversion of time-domain resistivity and induced polarization data. Geophysics 75 (4), WA27–WA41.

Kang, O., Lee, S., Wasewar, K., Kim, M., Liu, H., Oh, T., Yoo, C., 2013. Determination of key sensor locations for non-point pollutant sources management in sewer network. Korean J. Chem. Eng. 30 (1), 20–26. http://dx.doi.org/10.1007/s11814-012-0108-y.

Kates, R.W., Clark, W.C., Corell, R., Hall, J.M., Jaeger, C.C., Lowe, I., McCarthy, J.J., Schellnhuber, H.J., Bolin, B., Dickson, N.M., Faucheux, S., Gallopin, G.C., Grübler, A., Huntley, B., Jäger, J., Jodha, N.S., Kasperson, R.E., Mabogunje, A., Matson, P., Mooney, H., Moore III, B., O'Riordan, T., Svedin, U., 2001. Sustainability science. Science 292 (5517), 641–642. http://dx.doi.org/10.1126/science.1059386.

Kaur, H., Kumar, R., Babu, J.N., Mittal, S., 2015. Advances in arsenic biosensor development – a comprehensive review. Biosens. Bioelectron. 64, 533–545.

Kemna, A., Binleyz, A., Slater, L., 2004. Crosshole IP imaging for engineering and environmental applications. Geophysics 69 (1), 97–107.

Kerkez, B., Glaser, S.D., Bales, R.C., Meadows, M.W., 2012. Design and performance of a wireless sensor network for catchment-scale snow and soil moisture measurements. J. Water Resour. Res. 48 (9), 1–18.

Kiaalhosseini, S., 2016. Third-Generation Site Characterization: Cryogenic Core Collection, Nuclear Magnetic Resonance, and Electrical Resistivity. Doctoral Dissertation. Department of Civil and Environmental Engineering, Colorado State University, Fort Collins, Colorado.

Kister, G., Winter, D., Gebremichael, Y.M., Leighton, J., Badcock, R.A., Tester, P.D., Fernando, G.F., 2007. Methodology and integrity monitoring of foundation concrete piles using Bragg grating optical fibre sensors. Eng. Struct. 29 (9), 2048–2055.

Klar, A., Dromy, I., Linker, R., 2014. Monitoring tunneling induced ground displacements using distributed fiber-optic sensing. Tunn. Undergr. Space Technol. 40, 141–150.

Klavarioti, M., Kostarelos, K., Pourjabbar, A., Ghandehari, M., 2014. In situ sensing of subsurface contamination—part I: near-infrared spectral characterization of alkanes, aromatics, and chlorinated hydrocarbons. Environ. Sci. Pollut. Res. 21 (9), 5849–5860.

Koch, A., McBratney, A., Adams, M., Field, D., Hill, R., Crawford, J., Minasny, B., Lal, R., Abbott, L., O'Donnell, A., Angers, D., Baldock, J., Barbier, E., Binkley, D., Parton, W.,

Wall, D.H., Bird, M., Bouma, J., Chenu, C., Flora, C.B., Goulding, K., Grunwald, S., Hempel, J., Jastrow, J., Lehmann, J., Lorenz, K., Morgan, C.L., Rice, C.W., Whitehead, D., Young, I., Zimmermann, M., 2013. Global Policy 4 (4), 434–441.

Koch, C., Georgieva, K., Kasireddy, V., Akinci, B., Fieguth, P., 2015. A review on computer vision based defect detection and condition assessment of concrete and asphalt civil infrastructure. Adv. Eng. Inform. 29 (2), 196–210.

Kolbadinejad, M., Zabihollah, A., Khayyat, A.A.A., Pour, M.O.M., 2013. An equivalent electrical circuit design for pipeline corrosion monitoring based on piezoelectric elements. J. Mech. Sci. Technol. 27 (3), 799–804.

Koo, D., Ariaratnam, S.T., 2006. Innovative method for assessment of underground sewer Pipe condition. Autom. Constr. 15 (4), 479–488.

Koo, D.-H., Ariaratnam, S.T., Kavazanjian Jr., E., 2009. Development of a sustainability assessment model for underground infrastructure projects. J. Can. Soc. Civil Eng. 36, 765–776.

Korostynska, O., Mason, A., Al-Shamma's, A.I., 2013. Flexible Electromagnetic Wave Sensors for Real-Time Assessment of Water Contaminants. Smart Sensors, Measurement and Instrumentation, vol. 7. Springer, pp. 99–115.

Kuniansky, E.L., Weary, D.J., Kaufmann, J.E., 2016. The current status of mapping karst areas and availability of public sinkhole-risk resources in karst terrains of the United States. Hydrogeol. J. 24 (3), 613–624. http://dx.doi.org/10.1007/s10040-015-1333-3.

Laskar, S., Mukherjee, S., 2016. Optical sensing methods for assessment of soil macronutrients and other properties for application in precision agriculture: a review. ADBU J. Eng. Technol. 4, 206–210.

Li, S., Peng, X., 2016. Safety monitoring of underground steel pipeline subjected to soil deformation using wireless inclinometers. J. Civil Struct. Health Monitor. 6 (4), 739–749.

Li, C., Zhao, Y.G., Liu, H., Wan, Z., Zhang, C., Rong, N., 2008. Monitoring second lining of tunnel with mounted fiber Bragg grating strain sensors. Autom. Constr. 17 (5), 641–644.

Li, C., Zhao, Y.G., Liu, H., Wan, Z., Xu, J.C., Xu, X.P., Chen, Y., 2009. Strain and back cavity of tunnel engineering surveyed by FBG strain sensors and geological radar. J. Intell. Mater. Syst. Struct. 20 (18), 2285–2289.

Li, H., Li, D.-S., Ren, L., Yi, T.-H., Jia, Z.-G., Li, K., 2016. Structural health monitoring of innovative civil engineering structures in Mainland China. Struct. Monitor. Mainten. 3 (1), 1–32.

Li, M., Liu, Y., 2007. Underground structure monitoring with wireless sensor networks. In: Proceedings of the 6th International Conference on Information Processing in Sensor Networks. IPSN '07, pp. 69–78.

Li, M., Liu, Y., 2009. Underground coal mine monitoring with wireless sensor networks. ACM Trans. Sens. Netw. 5 (2), 10. http://dx.doi.org/10.1145/1498915.1498916. 29 pages.

Lian, S., Ji, J., De-Jun, R., Hong-Bing, X., Zhen-Fu, L., Bo, G., 2015. Estimate of heavy metals in soil and streams using combined geochemistry and field spectroscopy in Wan-Sheng mining area, Chongqing, China. Int. J. Appl. Earth Observation Geoinform. 34, 1–9.

Liu, X., Jin, B., Bai, Q., Wang, Y., Wang, D., Wang, Y., 2016. Distributed fiber-optic sensors for vibration detection. Sensors 2016 (16), 1164. http://dx.doi.org/10.3390/s16081164.

Loden, P., Han, Q., Porta, L., Illangasekare, T., Jayasumana, A.P., 2009. A wireless sensor system for validation of real-time automatic calibration of groundwater transport models. J. Syst. Softw. 82, 1859–1868.

Lowe, T., Potts, M., Wood, D., Energy, D., 2013. A case history of comprehensive hydraulic fracturing monitoring in the canal woodford. In: SPE Annual Technical Conference and Exhibition. Society of Petroleum Engineers, pp. 1–15.

Lu, Y., Shi, B., Wei, G.Q., Chen, S.E., Zhang, D., 2012. Application of a distributed optical fiber sensing technique in monitoring the stress of precast piles. Smart Mater. Struct. 21 (11), 115011.

Lunt, I.A., Hubbard, S.S., Rubin, Y., 2005. Soil moisture content estimation using ground-penetrating radar reflection data. J. Hydrol. 307, 254–269.

Mamidi, S.R., Bukka, K., Haji-sheikh, M., Kocanda, M., Zinger, D., Taherinezahdi, M., 2017. Time domain reflectometer for measuring liquid waste levels in a septic system. In: Sensors for Everyday Life. Environmental and Food Engineering. Springer Int. Publ., pp. 157–177.

McBratney, A., Field, D.J., Koch, A., 2013. The dimensions of soil security. Geoderma 213, 203–213.

McNamara, E., Nardi Pinto, C., Collins, F., Fay, C., Fregonezi Paludetti, L., Zanoni Nubiato, K., Diamond, D., 2013. Autonomous remote gas sensor network platforms with applications in landfill, wastewater treatment and ambient air quality measurement. In: 3 Congresso Analitica Latin America. 24–26 Sept 2013, Sao Paulo, Brazil.

McVay, M.C., Khiem, T., Tran, S.J., Wasman Sullivan B., Siriwardane, D., 2016. Detection of Sinkholes or Anomalies Using Full Seismic Wave Fields: Phase II Final Report, FDOT Contract No.: BDV31-977-29, UF Contract No.: 117252, Florida Department of Transportation, 164 p.

Merulla, D., Buffi, N., Beggah, S., Truffer, F., Geiser, M., Renaud, P., van der Meer, J.R., 2013. Bioreporters and biosensors for arsenic detection. Biotechnological solutions for a world-wide pollution problem. Curr. Opin. Biotechnol. 24, 534–541. http://dx.doi.org/10.1016/j.copbio.2012.09.002.

Metje, N., Chapman, D.N., Rogers, C.D.F., Henderson, P., Beth, M., 2008. An optical fiber sensor system for remote displacement monitoring of structures – prototype tests in the laboratory. Struct. Health Monit. 7 (1), 51–63.

Michaelides, M.P., Panayiotou, C.G., 2005. Plume source position estimation using sensor networks. In: Proc. of IEEE International Symposium on Intelligent Control, Intelligent Control, pp. 731–736.

Mishra, P.K., Pratik, Kumar, M., 2016. Wireless sensor network: challenges in underground coal mines. In: Kumar Kamila, Narendra (Ed.), Handbook of Research on Wireless Sensor Network Trends, Technologies, and Applications, pp. 145–160.

Mirzaeia, A., Bahrampourb, A.R., Tarazc, M., Bahrampourd, A., Bahrampoure, M.J., Foroushanif, A.S.M., 2013. Transient response of buried oil pipelines fiber optic leak detector based on the distributed temperature measurement. Int. J. Heat Mass Transf. 65, 110–122.

Mitsuhata, Y., Ando, D., Imasato, T., Takagi, K., 2014. Characterization of organic-contaminated ground by a combination of electromagnetic mapping and direct-push in situ measurements. Near Surf. Geophys. 12, 613–621.

Mulla, D.J., 2013. Twenty five years of remote sensing in precision agriculture: key advances and remaining knowledge gaps. Biosyst. Eng. 114 (4), 358–371.

Nasipuri, A., Subramanian, K., Ogunro, V., Daniels, J., Hilger, H., 2006. Development of a wireless sensor network for monitoring a bioreactor landfill. GeoCongress 2006 (187), 1–6.

Newman, T., 2012. Analysis of the Detection of Organophospohate Pesticides in Aqueous Solutions Using Polymer-Coated SH-SAW Sensor Arrays. Master of Science Thesis, Marquette University, Milwaukee, Wisconsin.

Nie, J., Daniel, G., Benson, B.C., Zappi, M., 2012. Nonpoint source pollution. Water Environ. Res. 16, 1642–1657.

Nikles, M., Vogel, B.H., Briffod, F., Grosswig, S., Sauser, F., Luebbecke, S., Pfeiffer, T., 2004. Leakage detection using fiber optics distributed temperature monitoring. In: 11th SPIE Annual International Symposium on Smart Structures and Materials. San Diego, CA, pp. 18–25.

Nomatungulula, C.S., Ngandu, K.G., Rimer, S., Paul, B.S., Longe, O.M., Ouahada, K., 2014. Mobile sink wireless underground sensor communication monitor. In: Bissyandé, T., van Stam, G. (Eds.), e-Infrastructure and e-Services for Developing Countries. AFRICOMM 2013. In: Lecture Notes of the Institute for Computer Sciences, Social Informatics and Telecommunications Engineering, vol. 135. Springer, Cham.

Noomen, M., Hakkarainen, A., van der Meijde, M., van der Werff, H., 2015. Evaluating the feasibility of multitemporal hyperspectral remote sensing for monitoring bioremediation. Int. J. Appl. Earth Observation Geoinform. 34, 217–225.

Nor, A.S.M., Faramarzi, M., Yunus, M.A.M., Ibrahim, S., 2015. Nitrate and sulfate estimations in water sources using a planar electromagnetic sensor array and artificial neural network method. IEEE Sens. J. 15 (1), 497–504. http://dx.doi.org/10.1109/JSEN.2014.2347996.

NRC, 2011. Sustainability and the U.S. EPA. The National Academies Press, Washington DC, USA. ISBN 0-309-21252-9.

NRC, 2013. Underground Engineering for Sustainable Urban Development. ISBN 978-0-309-27824-9. https://doi.org/10.17226/14670.

Obeid, A.M., Karray, F., Jmal, M.W., Abid, M., Qasim, S.M., BenSaleh, M.S., 2016. Towards realisation of wireless sensor network-based water pipeline monitoring systems: a comprehensive review of techniques and platforms. IET Sci. Meas. Technol. 10 (5), 420–426.

Oen, A.M.P., Janssen, E.M.L., Cornelissen, G., Breedveld, G.D., Eek, E., Luthy, R.G., 2011. In situ measurement of PCB pore water concentration profiles in activated carbon-amended sediments using passive samplers. Environ. Sci. Technol. 45, 4053–4059.

Oh, M., Seo, M.W., Lee, S., Park, J., 2007. Applicability of grid-net detection system for landfill leachate and diesel fuel release in the subsurface. J. Contam. Hydrol. 96 (1–4), 69–82.

Ojha, R., Ramadas, M., Govindaraju, R., 2015. Current and future challenges in groundwater. I: Modeling and management of resources. In: Special Issue: Grand Challenges in Hydrology. J. Hydrol. Eng. 20, A4014007. http://dx.doi.org/10.1061/(ASCE)HE.1943-5584.0000928.

Olhoeft, G.R., 2003. Electromagnetic field and material properties in ground penetrating radar. In: Proceedings of the 2nd International Workshop on Advanced Ground Penetrating Radar, pp. 144–147.

Owojaiye, G., Sun, Y., 2013. Focal design issues affecting the deployment of wireless sensor networks for pipeline monitoring. Ad Hoc Netw. 11 (3), 1237–1253.

Pei, H., Yin, J., Zhu, H., Hong, C., Jin, W., Xu, D., 2012. Monitoring of lateral displacements of a slope using a series of special fibre Bragg grating-based in-place inclinometers. Meas. Sci. Technol. 23 (2), 25007.

Phillips, C., Jakusch, M., Steiner, H., Mizaikoff, B., Fedorov, A.G., 2003. Model-based optimal design of polymer-coated chemical sensors. Anal. Chem. 75, 1106–1115.

Porta, L., Illangasekare, T., Loden, P., Han, Q., Jayasumana, A.P., 2009. Continuous plume monitoring using wireless sensors: proof of concept in intermediate scale tank. J. Environ. Eng. 135, 831–838.

Power, C., Gerhard, J.I., Tsourlos, P., Soupios, P., Simyrdanis, K., Karaoulis, M., 2015. Improved time-lapse electrical resistivity tomography monitoring of dense non-aqueous phase liquids with surface-to-horizontal borehole arrays. J. Appl. Geophys. 112, 1–13.

Prasanna, S., Rao, S., 2012. An overview of wireless sensor networks applications and security. Int. J. Soft Comput. Eng. (IJSCE) 2 (2), 2231–2307.

Przyłucka, M., Herrera, G., Graniczny, M., Colombo, D., Béjar-Pizarro, M., 2015. Combination of conventional and advanced dinsar to monitor very fast mining subsidence with terrasar-X data: bytom city (Poland). Remote Sens. 7 (5), 5300–5328.

Qin, J., Ying, Y., Xie, L., 2012. The detection of agricultural products and food using terahertz spectroscopy: a review. Appl. Spectrosc. Rev. 48 (6), 439–457. http://dx.doi.org/10.1080/05704928.2012.745418.

Ramesh, M.V., 2014. Design, development, and deployment of a wireless sensor network for detection of landslides. Ad Hoc Netw. 13, 2–18.

Ranjan, A., Sahu, H.B., Misra, P., 2016. Wireless sensor networks: an emerging solution for underground mines. Int. J. Appl. Evolut. Comput. (IJAEC) 7 (4), 1–27.

Rashid, B., Rehmani, M.H., 2016. Applications of wireless sensor networks for urban areas: a survey. J. Netw. Comput. Appl. 60, 192–219.

Read, T., Bour, O., Bense, V., Le Borgne, T., Goderniaux, P., Klepikova, M.V., Boschero, V., 2013. Characterizing groundwater flow and heat transport in fractured rock using fiber-optic distributed temperature sensing. Geophys. Res. Lett. 40 (10), 2055–2059.

Reedy, R.C., Scanlon, R.B., 2003. Soil water content monitoring using electromagnetic induction. J. Geotech. Geoenv. Eng. 129, 1028–1039.

Ren, Z., Liu, G., Huang, Z., 2014. Design of a novel noninvasive spectrometer for pesticide residues monitor. In: Int. Symp. on Optoelectronic Tech. and Application 2014: Imaging Spectroscopy, and Telescopes and Large Optics. Beijing, China, p. 9298.

Revil, M., Karaoulis, M., Johnson, T., Kemna, A., 2012. Some low-frequency electrical methods for subsurface characterization. Hydrogeol. J. 15 (16), 011. http://dx.doi.org/10.1007/s10040.

Ritchie, L.J., Ferguson, C.P., Bessant, C., Saini, S., 2000. A ten channel fibre-optic device for distributed sensing of underground leakage. J. Environ. Monitor. 6 (2), 670–673. http://dx.doi.org/10.1039/B008210O.

Ritsema, C.J., Kuipers, H., Kleiboer, L., Elsen, E.V., Oostindie, K., Wesseling, J.G., Wolthuis, J.W., Havinga, P., 2009. A new wireless underground network system for continuous monitoring of soil water contents. J. Water Resour. Res. 45, W00D36.

Robinson, D.A., Jones, S.B., Wraith, J.M., Or, D., Friedman, S.P., 2003. A review of advances in dielectric and electrical conductivity measurement in soils using time domain reflectometry. Vadose Zone J. 2 (4), 444–475.

Rossel, R.A.V., Bouma, J., 2016. Soil sensing: a new paradigm for agriculture. Agric. Syst. 148, 71–74.

Rucker, M.L., Panda, B.B., Meyers, R.A., Lommler, J.C., 2013a. Using InSar to detect subsidence at brine wells, sinkhole sites, and mines. Carbonat. Evaporit. 28 (1–2), 141–147.

Rucker, D.F., Myers, D.A., Cubbage, B., Levitt, M.T., Noonan, G.E., Mcneill, M., Lober, R.W., 2013b. Surface geophysical exploration: developing noninvasive tools to monitor past leaks around Hanford's tank farms. Environ. Monit. Assess. 185 (1), 995–1010.

Saftner, D.A., Green, R.A., Hryciw, R.D., Lynch, J.P., Michalowski, R.L., 2008. Instrumentation for the NEESR sand aging field experiment. In: GeoCongress 2008. New Orleans, LA, pp. 525–532.

Sahota, H., Kumar, R., Kamal, A., 2011. A wireless sensor network for precision agriculture and its performance. J. Wirel. Commun. Mob. Comput. 11, 1628–1645.

Schmidt-Hattenberger, C., Bergmanna, P., Kießlinga, D., Krügera, K., Rückerb, C., Schüttc, H., 2011. Application of a vertical electrical resistivity array (VERA) for monitoring CO_2 migration at the Ketzin site: first performance evaluation. Energy Proc. 4, 3363–3370.

Schubauer-Berigan, J.P., Foote, E.A., Magar, V.S., 2012. Using SPMDs to assess natural recovery of PCB-contaminated sediments in Lake Hartwell, SC: I. A field test of new in-situ deployment methods. Soil. Sed. Contamin. 21, 82–100.

Segalini, A., Chiapponi, L., Pastarini, B., 2015. Application of modular underground monitoring system (MUMS) to landslides monitoring: evaluation and new insights. In: Engineering Geology for Society and Territory, vol. 2. Springer, pp. 121–124.

Shah, T., Burke, J., Villholth, K., Angelica, M., Custodio, E., Daibes, F., Hoogesteger, J., Giordano, M., Girman, J., van der Gun, J., Kendy, E., Kijne, J., Llamas, R., Masiyandama, M., Margat, J., Marin, L., Peck, J., Rozelle, S., Sharma, B., Vincent, L., Wang, J., 2007. Groundwater: a global assessment of scale and significance. In: Molden, D. (Ed.), Water for Food, Water for Life: A Comprehensive Assessment of Water Management in Agriculture Earthscan, London.

Shantaram, A., Beyenal, H., Raajan, R., Veluchamy, A., Lewandowski, Z., 2005. Wireless sensors powered by microbial fuel cells. Environ. Sci. Technol. 39, 5037–5042.

Sheltami, T.R., Bala, A., Shakshuki, E.M., 2016. Wireless sensor networks for leak detection in pipelines: a survey. J. Ambient Intell. Humaniz. Comput. 7 (3), 347–356.

Shen, Z.Y., Chen, L., Liao, Q., Liu, R.M., Huang, Q., 2013. A comprehensive study of the effect of GIS data on hydrology and non-point source pollution modeling. Agric. Water Manag. 118, 93–102.

Shepherd, R., Beirnea, S., Laub, K.T., Corcorana, B., Diamond, D., 2007. Monitoring chemical plumes in an environmental sensing chamber with a wireless chemical sensor network. Sens. Actuators B, Chem. 121 (1), 142–149.

Shin, Y.W., Kim, M.S., Lee, S.K., 2010. Identification of acoustic wave propagation in a duct line and its application to detection of impact source location based on signal processing. J. Mech. Sci. Technol. 24 (12), 2401–2411.

Shukla, S.K., Chaulya, S.K., Mandal, R., Kumar, B., Ranjan, P., Mishra, P.K., Sarmah, P.C., 2014. Real-time monitoring system for landslide prediction using wireless sensor networks. Int. J. Mod. Commun. Technol. Res. 2 (12), 14–19.

Siegel, R.A., 1993. Hydrophobic weak polyelectrolyte gels: studies of swelling equilibria and kinetics. Adv. Polym. Sci. 109, 233.

Singh, M., Verma, N., Garg, A.K., Redhu, N., 2008. Urea biosensors. Sens. Actuators B, Chem. 134, 345–351.

Singh, R.K., Murty, H.R., Gupta, S.K., Dikshit, A.K., 2009. An overview of sustainability assessment methodologies. Ecol. Indic. 15 (1), 281–299. http://dx.doi.org/10.1016/j.ecolind.2008.05.011.

Sinha, S.K., Knight, M.A., 2004. Intelligent system for condition monitoring of underground pipelines. Comput.-Aided Civ. Infrastruct. Eng. 19 (1), 42–53.

Sinha, S.K., Fieguth, P.W., 2006. Segmentation of buried concrete pipe images. Autom. Constr. 15 (1), 47–57.

Sinha, R., Oberoi, A., Tungala, H., Kapur, R., Kumar, A., 2014. An optimized piezoresistive microcantilever arsenic (III) sensor with CMOS-compatible active readout: towards in-situ subsurface characterization using microsystems technology. In: Proceedings of the 14th IEEE International Conference on Nanotechnology, pp. 196–200.

Soldovieri, F., Prisco, G., Persico, R., 2009. A strategy for the determination of the dielectric permittivity of a Lossy soil exploiting GPR surface measurements and a cooperative target. J. Appl. Geophys. 67, 288–295.

Soga, K., Mohamad, H., Bennett, P.J., 2008. Distributed fiber optics strain measurements for monitoring geotechnical structures. In: International Conference on Case Histories in Geotechnical Engineering 4. http://scholarsmine.mst.edu/icchge/6icchge/session14/4.

Stuntebeck, E.P., Pompili, D., Melodia, T., 2006. Wireless underground sensor networks using commodity terrestrial motes. In: 2nd IEEE Workshop on Wireless Mesh Networks, pp. 112–114.

Sudduth, K.A., Drummond, S.T., Kitchen, N.R., 2001. Accuracy issues in electromagnetic induction sensing of soil electrical conductivity for precision agriculture. J. Comput. Electron. Agricult. 31, 239–264.

Sun, Z., Akyildiz, I.F., 2010a. Magnetic induction communications for wireless underground sensor networks. IEEE Trans. Antennas Propag. 58 (7), 2426–2435.

Sun, Z., Akyildiz, I.F., 2010b. Deployment algorithms for wireless underground sensor networks using magnetic induction. In: Global Telecommunications Conference. GLOBECOM 2010. IEEE, pp. 1–5.

Sun, Z., Wang, P., Vuran, M.C., Al-Rodhaan, M.A., Al-Dhelaan, A.M., Akyildiz, I.F., 2011a. BorderSense: border patrol through advanced wireless sensor networks. Ad Hoc Netw. 9 (3), 468–477.

Sun, Z., Wang, P., Vuran, M.C., Al-Rodhaan, M.A., Al-Dhelaan, A.M., Akyildiz, I.F., 2011b. MISE-PIPE: magnetic induction-based wireless sensor networks for underground pipeline monitoring. J. Ad Hoc Netw. 2, 218–227.

Suo, W., Lu, Y., Shi, B., Zhu, H.-H., Wei, G., Jiang, H., 2016. Development and application of a fixed-point fiber-optic sensing cable for ground fissure monitoring. J. Civil Struct. Health Monitor. 6 (4), 715–724. http://dx.doi.org/10.1007/s13349-016-0192-5.

Tang, Z., Wu, W., 2014. A two-layer energy efficient framework using SAW sensor network for leakage detection monitoring water distribution system. In: Proceedings of the 20th International Conference on Automation and Computing. Bedfordshire, UK, 12–13 September 2014. Cranfield University, pp. 158–163.

Terzis, A., Anandarajah, A., Moore, K., Wang, K.J., 2006. Slip surface localization in wireless sensor networks for landslide prediction. In: 5th International Conference on Information Processing in Sensor Networks, pp. 109–116.

Tomaszewski, J.E., Luthy, R.G., 2008. Field deployment of polyethylene devices to measure PCB concentrations in pore water of contaminated sediment. Environ. Sci. Technol. 42, 6086–6091.

Tooker, J., Dong, X., Vuran, M.C., Irmak, S., 2012. Connecting soil to the cloud: a wireless underground sensor network testbed. SECON 2012, 79–81.

Van Meirvenne, M., Van De Vijver, E., Vandenhaute, L., Seuntjens, P., 2014. Investigating soil pollution with the aid of EMI and GPR measurements. In: Proc. of the 15th Int. Conference on Ground Penetrating Radar. GPR. IEEE, pp. 1006–1010.

Versteeg, R., Johnson, D., Henrie, A., Johnson, T., 2014. Cloud based electrical geophysical monitoring. In: 27th Annual Symposium on the Application of Geophysics to Engineering and Environmental Problems. SAGEEP, pp. 149–154.

Vidács, A., Vida, R., 2015. Wireless sensor network based technologies for critical infrastructure systems. In: Intelligent Monitoring, Control, and Security of Critical Infrastructure Systems. Springer, pp. 301–316.

Vuran, M.C., Silva, A.R., 2009. Communication through soil in wireless underground sensor networks – theory and practice. In: Ferrari, G. (Ed.), Sensor Networks: Where Theory Meets Practice. Springer.

Wan, C., Mita, A., 2010. Recognition of potential danger to buried pipelines based on sounds. Struct. Control Health Monit. 17 (3), 317–337.

Whelan, M.J., Janoyan, K.D., 2007. Wireless sensor network for monitoring of geostructural systems. In: Seventh International Symposium on Field Measurements in Geomechanics. Boston, MA. In: GSP, vol. 175, pp. 94–102.

Whittle, A.J., Allen, M., Preis, A., Iqbal, M., 2013. Sensor networks for monitoring and control of water distribution systems. In: 6th International Conference on Structural Health Monitoring of Intelligent Infrastructure. SHMII 2013, Hong Kong, December 9–11.

Wu, H., Wang, Z., Peng, F., Peng, Z., Li, X., Wu, Y., Rao, Y., 2014. Field test of a fully distributed fiber optic intrusion detection system for long-distance security monitoring of national borderline. Proc. SPIE 2014, 9157. http://dx.doi.org/10.1117/12.2058504.

WBCSD, 2011. Vision 2050. World business council for sustainable development. http://www.wbcsd.org/templates/.

Xu, X., Peng, S., Xia, Y., Ji, W., 2014. The development of a multi-channel GPR system for roadbed damage detection. Microelectron. J. 45 (11).

Yaacoub, E., Kadri, A., Abu-Dayya, A., 2011. An OFDMA communication protocol for wireless sensor networks used for leakage detection in underground water infrastructures. In: Proc. of 7th International Wireless Communications and Mobile Computing Conference, pp. 1894–1899.

Ye, Y., Hao, L., Liu, M., Wu, H., Zhang, X., Zhao, Z., 2016. Design of Farmland environment remote monitoring system based on zigbee wireless sensor network. In: Hung, J., Yen, N., Li, K.C. (Eds.), Frontier Computing. In: Lecture Notes in Electrical Engineering, vol. 375. Springer, Singapore.

Yoon, S.-U., Cheng, L., Ghazanfari, E., Pamukcu, S., Suleiman, M.T., 2011. A radio propagation model for wireless underground sensor networks. In: Proceedings of IEEE Globecom. Houston, TX, December 2011.

Yoon, S.-U., 2012. Wireless Signal Networks: A Proof of Concept for Subsurface Characterization and a System Design with Reconfigurable Radio. PhD Dissertation. Computer Science and Engineering, Lehigh University.

Yoon, S.-U., Ghazanfari, E., Cheng, L., Wang, Z., Pamukcu, S., Suleiman, M.T., 2012. Subsurface event detection and classification using wireless sensor networks. Sensors 12 (11), 14862–14886. http://dx.doi.org/10.3390/s121114862.

Zhou, B., Yang, S., Sun, T., Grattan, K.T.V., 2015. A novel wireless mobile platform to locate and gather data from optical fiber sensors integrated into a WSN. IEEE Sens. J. 15 (6). http://dx.doi.org/10.1109/JSEN.2015.2396040.

Zhao, Y., Zhang, N., Si, G., 2016. A fiber Bragg grating-based monitoring system for roof safety control in underground coal mining. Sensors 16 (10), 1759.

Zeng, Z., Li, J., Huang, L., Feng, X., Liu, F., 2015. Improving target detection accuracy based on multipolarization MIMO GPR. IEEE Trans. Geosci. Remote Sens. 53 (1), 15–24.

Zhu, H.H., Shi, B., Yan, J.F., Zhang, J., Zhang, C.C., Wang, B.J., 2014. Fiber Bragg grating-based performance monitoring of a slope model subjected to seepage. Smart Mater. Struct. 23 (9), 095027. http://dx.doi.org/10.1088/0964-1726/23/9/095027.

CHAPTER 2

Acoustic, Electromagnetic and Optical Sensing and Monitoring Methods

Wen Xiao*, Xiaosu Yi*, Feng Pan*, Rui Li[†], Tian Xia[‡]
*Beihang University, Beijing, China
[†]CSSC Marine Technology Co., Ltd., Shanghai, China
[‡]University of Vermont, Vermont, USA

SUBCHAPTER 2.1

Principles of Acoustic and Electromagnetic Sensing

Wen Xiao*, Xiaosu Yi*, Feng Pan*, Rui Li[†]
*Beihang University, Beijing, China
[†]CSSC Marine Technology Co., Ltd., Shanghai, China

2.1.1 INTRODUCTION

Underground measurement technology (UMT) is a collection of multiple methods that use multiple physical sensors including acoustical, gravitational, magnetic, electrical, electromagnetic, and optical sensors located either above ground or buried in the ground to measure the physical properties of the subsurface. By obtaining and analyzing these properties as well as any anomalies that may be present, a multitude of valuable information can be obtained, such as the presence and position of economically useful geological deposits (e.g., oil and gas resources, geothermal and groundwater reservoirs) (Demick et al., 2013), visualization of buried structures and geological features (Araya et al., 2012), and forecast warnings for potential geological hazards including seismicity, presence of solution cavities, or unstable rock or soil formations (Watts, 1997).

Underground Sensing.
DOI: http://dx.doi.org/10.1016/B978-0-12-803139-1.00002-3

	Field wave	Acoustic wave			Radio
		Infrasonic sound	Audible sound	Ultrasonic sound	
Sources	· Gravity field · Magnetic field · Temperature field	· Volcano · Geological movement · Nuclear explosion	· Artificial explosions or vibrations · Loudspeaker	· Structure crack · Rock crack · Pipeline leakage	· Emitted high-frequency radio wave
Applications	· Geology research · Geological hazard prediction	· Geology research · Geological hazard prediction · Nuclear explosion monitoring	· Mineral exploration · Tunnel geological hazards detection · Buried objects detection	· Mineral safety monitoring · Slope engineering · Pipeline leakage detection	· Imaging subsurface structures

Figure 2.1.1 Waves adopted in underground measurement and their applications

Various methods of UMTs have been reported in the literature. Most often in these methods, some form of "wave" (e.g., seismic wave, acoustic wave, electromagnetic wave) is adopted as the signal carrier for its propagation capability underground. Based on their frequency ranges, these "waves" can be classified into three types: field waves (<0.01 Hz), acoustic waves (0.01~500 MHz), and radio waves (i.e., electromagnetic waves). Furthermore, the acoustic wavescan be subdivided into infrasonic (0.01~20 Hz), audible (20 Hz~20 kHz) and ultrasonic sound (20 kHz~500 MHz) waves. Fig. 2.1.1 presents frequency range distribution of waves useful in geological and underground settings, and generation sources of such waves that travel in subsurface. For practical measurements, generation and transmission properties of "waves" in subsurface are closely related to the physical properties of the media they reside in. Hence, the transmission properties of waves through a medium can carry substantial information about that particular medium for targeted underground measurements, including geological movements, rock cracks, presence of a geological stratum, or a buried object.

2.1.1.1 Conventional Underground Measurement Methods

From Fig. 2.1.1 we can observe that there are several types of wave that correspond to useful measurement applications. Based on the classification of waves presented in this figure, a brief introduction of underground measurement methods are provided in the following section.

2.1.1.1.1 Physical Field Methods

In the domain of field wave, gravity and magnetic field detection techniques are two typical and widely adopted methods for profiling underground structures. Gravity field detection is suitable for monitoring changes in crustal movement over time to help predict earthquakes and water flow regimes (Demick et al., 2013; Araya et al., 2012). For example, Watts (1997) used a compilation of gravity anomaly data along the western margin of the British Isles in a gravity model to determine the manner in which the continental lithosphere responded to extension. Magnetic field detection technique is more suitable for shallow underground exploration. By using magnetic sensors to detect the magnetic fields generated by current-carrying cables, the spatial location distribution of these cables is determined (Jiang et al., 2014; Sun et al., 2014). In some applications of underground structure safety monitoring (e.g., coal mines or underground mass transit tunnels), temperature field detection is widely explored (Norman et al., 1992; Ji et al., 2010). For example, in coal mine safety monitoring, distributed temperature sensors can be used for unusual high temperature detection which provide prewarning of coal combustion (Poczesny et al., 2012).

2.1.1.1.2 Acoustic Methods

Acoustics is one of the popular techniques used in underground measurements, and many specific acoustic methods have been well researched and developed for practical applications. These methods can be classified into two categories based on the source of acoustic signal generation, namely those of passive type and active type.

In passive type methods, the acoustic signal is generated by the subsurface objects. For example, in the domain of infrasonic sound, an acoustic signal can be emitted by natural events including volcanic explosions, earthquakes, nuclear explosions, debris flows, and landslides (Zhu, 2014). By strategically placing acoustic sensors in the vicinity of the areas where such geological activities happen frequently, the emitted infrasonic signals can be captured and analyzed for ensuing geological activity and hazard prediction. Another passive acoustic signal generation example is the acoustic emission (AE), the phenomenon of energy release caused by a sudden change in material internal structure manifested as an acoustical wave. The sudden change could be the result of rock formation crack, underground pipeline leakage, fracture of inclusions in geological settings or structural transformations underground. The frequency domain of AE signal usually belongs to the audible and ultrasonic sound at frequencies ranging from 1 kHz to 100 MHz, while most of the energy emission happens in the range from 1 kHz to 1 MHz (Yang and Ma, 2006). Major applications of AE technique in underground measurements include underground structural health monitoring and failure localization (Sun and Li, 2010;

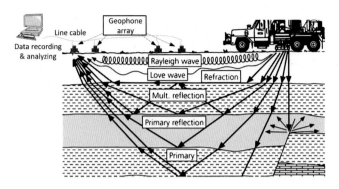

Figure 2.1.2 Schematic diagram of reflection seismology and various seismic waves (Sreechaka and Bordakov, 2008)

Kang and Yu, 2011) that use source location methods, which will be introduced in the following section.

In the active acoustic methods, the detected signal is generated by artificial explosions or vibrations and is sent into the earth's interior to estimate the properties of the earth's subsurface. A typical application is using reflection seismology to detect geological structure, as presented by the schematic diagram in Fig. 2.1.2 (Sreechaka and Bordakov, 2008). A seismic wave generated by a vibration stimulator sends waves of either wideband pulses or narrow band acoustic chirp signals into the earth's interior. Receivers' array (e.g., geophones or vibration sensors) suitably placed in a grid pattern on the ground surface detect the reflected and refracted seismic waves, which travel through multistrata. Analyzing the time of arrival and other signature properties of the received waves, some valuable information like the stratum structure and thickness, lithology, location of potential oil and gas formation layers can be extracted (Zhong et al., 2009; Yao et al., 2010).

2.1.1.1.3 Electrical and Electromagnetic Wave Methods

In many applications, the properties of the subsurface materials and their structure can be related to their electric or magnetic resistance. Hence, electrical or electromagnetic methods are often used to monitor the variations in the distribution of these resistivity values to extract temporal and spatial information about geological formations and subsurface structures. For example, electrical resistivity survey (ERS) is a technique used to image the subsurface structures by measuring the electrical resistivity distribution on the boundaries of different layers and formations (Danescu et al., 2013;

Figure 2.1.3 3D visualization of underground pipelines using GPR (Sun et al., 2012)

Wang et al., 2011). Similar to the acoustic method, substituting acoustic waves with electromagnetic waves provides another subsurface imaging approach for shallower depths in the subsurface. A representative technique is that of ground-penetrating radar (GPR), a well-accepted nondestructive detection technology. The GPR emits high-frequency radio waves pulses (in the range of 10 MHz~1 GHz) downward into the ground and detects the reflected signals from objects or material boundaries below. Subsequent analysis of the signals provides image scans of the subsurface in 2D or 3D format. The GPR method performs well in shallow underground detection, particularly in monitoring and 2D/3D visualization of underground pipelines as shown in Fig. 2.1.3 (Sun et al., 2012).

2.1.1.2 Conventional Devices Used for Underground Measurements

In the underground measurement methods introduced above, waves are used to transmit the signals back to the surface to be acquired and recorded by devices on the surface. The techniques and the devices used to acquire and subsequently analyze these wave signals are another important matter of discussion.

Gravity fields can be detected by a simple device known as gravimeter (as shown in Fig. 2.1.4) which uses a spring to counteract the force of gravity pulling on an object. By calibrating the change in the length of the spring to balance the gravitational pull, the local gravitational field of the earth can be measured. There are two types of gravimeters: absolute and relative. Absolute gravimeter measures the local gravity in absolute units. Relative gravimeter returns the ratio of the local gravity and a precalibrated gravity at a location where the gravity is known accurately (Childers, 2016).

Figure 2.1.4 Photo of an absolute gravimeter (Childers, 2016)

For electromagnetic wave based detection methods, like GPR, radar pulses are emitted from the transmitting antenna (T) of the device and the reflected signals are collected by the receiving antenna (R). By moving simultaneously the two antennas along the survey line, a time profile GPR image can be composed by sequential records (Sun et al., 2012). A discussion on the principle of imaging underground targets by using GPR will be introduced in the following section.

The basic principle of acoustic-wave based detection is sensing and recording the vibrations caused by transitive or reflective acoustic waves. A vibration sensor that detects motions on the earth surface is basically a spring–mass structure, in which the spring and mass are attached to a frame moving along with the earth's surface. Ground vibration can be measured by detecting the relative motion between the mass and the earth. For example, in a conventional electromagnetic sensor, a magnetic proof mass is attached to a spring and surrounded by fixed conducting coils. When the earth moves, proof mass moves correspondingly with the conducting coils, thus generating a current proportional to the ground motion. Unlike a conventional electromagnetic sensor, an optical vibration sensor can sense the vibration either by using the mass–spring structure to measure the relative motion of the mass, or by directly measuring the ground motion without an assisting structure. In optical vibration sensors, one or more characteristics (e.g., intensity, phase, wavelength, polarization state) of the transmitting light in an optical fiber are affected and modulated by the detected vibration. Optical vibration sensors can be most useful where susceptibility to electromagnetic fields, risk of current leakage or ignition in hazardous environments, and long-term requirement of power supply in distributed sensing may make both the electronic and magnetic sensors infeasible or unreliable.

Following the short introduction to methods and devices above, detailed discussions of some of the common acoustical, electric and electromagnetic methods widely used in underground sensing and measurements are presented in Sections 2.1.2 and 2.1.3, respectively. The optical sensing technology (OST), which potentially offers advantages over conventional measurements underground, is discussed in Section 2.1.4.

2.1.2 ACOUSTICAL MEASUREMENT METHODS—AMM

In AMM the desired information is obtained by analyzing the captured acoustic signals emitted or reflected from the target object. In underground measurements, such analysis may answer one or more of the following questions: (i) what generates the signals and for what kind of geological activity do the signals stand for; (ii) where is the acoustical source; (iii) what properties of the earth subsurface affect the transmitted acoustic waves, and (iv) where is the location of a buried object reflecting the acoustic signal. Four common acoustic measurement methods are discussed in the following sections, each providing an example of an answer to the questions above.

2.1.2.1 Direct Detection Method

In the direct detection method, acoustic sensors are located at the monitored area where they directly capture the acoustic signals emitted by the geological activity of interest (e.g., volcanic explosion, earthquake, landslide, rock crack). Because the acoustic signals emitted by a particular geological activity tend to have specific characteristics, by recognizing these characteristics, the geological activity can be recognized and its progress can be monitored. For example, the analysis of infrasonic pressure waves generated by earthquakes is essential to the understanding of earthquake dynamics (Wang and Shan, 2010).

However, in many cases, captured infrasonic signal may be contaminated with disturbance sources (e.g., long range seismicity, high impact weather activity such as hailstorm), which potentially raise the noise floor of the infrasound sensor and mask the critical signals essential for identification of the target event (Dickey and Mikhael, 2012). Hence, various signal processing methods to extract useful information from complicated noisy signals in order to recognize specific features of certain waveforms related a target event has gained emphasis in the direct detection method. Wang and Shan (2010) proposed dynamic fuzzy clustering to classify a set of infrasound time-varying multivariate signals, comparing their features to the captured precursor signal of an earthquake (Wang and Shan, 2010). To isolate the seismic response from infrasonic sensor, Dickey and Mikhael (2012) utilized a collocated seismometer to obtain a reference signal and then applied an adaptive filter to isolate the noise. This technique was applied to a real-world data set obtained from the October 2011 earthquake record in Turkey with good results (Dickey and Mikhael, 2012).

2.1.2.2 Acoustic Emission (AE) and Acoustic Source Location (ASL) Method

Acoustic emission is the phenomenon of transient elastic waves generated by the rapid release of energy from a localized source within a material. In underground measurements, the AE source is created by a defect (e.g., rock crack) or discontinuity (e.g., pressure discontinuity caused by pipeline leakage) that becomes active due to changes in operating loads, under overloading conditions or degradation of the material (e.g., stress corrosion cracking, corrosion fatigue, and hydrogen embrittlement) (Jiang and Zhao, 2008). The emitted acoustic signal is captured by an AE transducer array, and analyzed to detect the source location as well as recognize features of the defect or discontinuity (e.g., type, stage and criticality). This process is called the acoustical source location, ASL. The AE and ASL techniques have been widely used in some underground measurement applications, like subsurface pipeline leakage detection (Sun and Li, 2010), rock crack monitoring for coal mine safety monitoring (Zhang, 2012), and underwater concrete structures monitoring (Jiang and Zhao, 2008).

The principle of ASL method is explained through an example of pipeline leakage location detection practice, as depicted in Fig. 2.1.5 (Sun and Li, 2010). Two sensors are mounted on a pipe at a distance D. When a leakage (e.g., due to a crack or dislocation) occurs from the pipe, the acoustic waves are emitted from that location which then travel along the pipe in both directions at the same constant velocity. If one of the transducers (e.g., transducer #1) is closer to leakage location, the AE signal will be captured by transducer #1 first, and then by transducer #2 on the opposite side. It

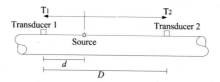

Figure 2.1.5 Schematic of pipeline leakage localization (Sun and Li, 2010)

can be then assumed safely that the leakage is located in the area between the two transducers which capture the earliest AE signals sequentially, and the leakage is closer to the transducer at which the AE signal arrives first. In reality, if the source location requires higher resolution, multiple evenly spaced transducers can be mounted along the pipe. Then any two of them detecting the signal earliest can give a better location resolution but still a rough estimate.

In order to obtain the exact position, one can use the time difference Δt between these two hits at the rough location area,

$$\Delta t = T_1 - T_2 \tag{2.1.1}$$

where T_1 and T_2 denote the time that the source signal arrives at the two sensors located on each side of the leak event. Then the source location can be determined by

$$d = \frac{1}{2}(D - \Delta t v) \tag{2.1.2}$$

where d is the unknown distance between AE source and the location of transducer #1; v is the acoustic wave velocity in pipe, which can be obtained from previous calibrations or from the known material properties of the pipe (e.g., elastic modulus, mass density). In order to obtain better location accuracy, algorithms based on basic cross-correlation (BCC), generalized cross-correlation (GCC), or other spectrum analysis methods have been be used to improve the estimation accuracy of the arrival time difference Δt, which is directly related to the estimation accuracy of the source location (Brennan and Gao, 2007).

Leakage location on a pipeline can be considered as a two–dimensional (2D) problem since AE signal travels linearly along the pipe. In contrast, the source location estimation turns out to be a three-dimensional (3D) problem in the case of rock failure detection, for instance, in coal mine safety monitoring. A schematic diagram explaining 3D source location method is shown in Fig. 2.1.6. Considering the source to be centered at unknown

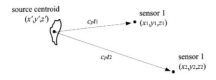

Figure 2.1.6 Schematic diagram of 3D source location (Li and Wan, 2010)

coordinates of (x', y', z'), an array of n sensors are employed, each centered at known coordinates of (x_1, y_1, z_1), (x_2, y_2, z_3), ..., (x_n, y_n, z_n). The medium where acoustical emission happens and travels is assumed to be homogeneous and isotropic for simplified computations.

In Fig. 2.1.6 (Li and Wan, 2010), the first wave arrival is at the ith sensor located at (x_i, y_i, z_i) when

$$\left(x' - x_i\right)^2 + \left(y' - y_i\right)^2 + \left(z' - z_i\right)^2 = \left(c_p t_i\right)^2 \tag{2.1.3}$$

where t_i is the time required for the first wave to reach the ith sensor, c_p is the longitudinal wave propagation velocity. For the n number of sensors in the array, n unique nonlinear equations will be obtained in this manner. Assuming t_0 to be the travel time to reach the sensor closest to the event source, and D_{t_i} to be the time difference between the closest and an ith sensor, one can determine $t_i = t_0 + D_{t_i}$. By solving the multivariate equation with four or more measured D_{t_i} values and sensors coordinates, the source location (x', y', z') can be determined (Li and Wan, 2010; Sun and Qian, 2010).

2.1.2.3 Reflection Seismology

Seismic waves are waves of low-frequency acoustic energy that travel through the earth's layers, which may be generated by extreme geological or man-made activity such as an earthquake, volcanic eruption, or an explosion. An exploration method in geophysics capturing and analyzing the seismic waves to estimate the material (e.g., stiffness) and structural (e.g., stratification) properties of the earth subsurface is called reflection seismology. In this method, a seismic source of energy generated by vibration stimulator sends seismic waves into the earth's interior. A small part of the down-going energy is reflected back to the surface on geological layer boundaries, but main fraction of the energy is transmitted by refraction and travels deeper where reflections and refractions occur at the subsequent layer boundaries, as shown in Fig. 2.1.7 (Sreechaka and Bordakov, 2008).

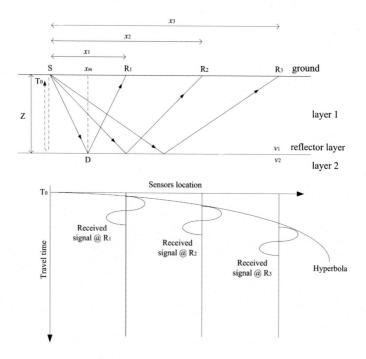

Figure 2.1.7 Schematic diagram of seismic reflection

Using vibration sensors array to detect the seismic wave energy reflected back to the ground surface, valuable information on properties of the earth subsurface can be extracted. This method has been widely applied to obtain geological profiles, and explore for oil and gas reservoirs in depths up to several kilometers (Hubscher and Gohlb, 2014).

Reflection seismology utilizes the basic time–distance curves, as depicted in Fig. 2.1.7 (Ulrych and Sacchi, 2005). In the top section of Fig. 2.1.7, a simple model of horizontal layer over a half-space is assumed, in which S is the seismic source; R_1, R_2 and R_3 are three receivers away from S at distances of x_1, x_2, x_3, respectively; and Z is the depth of the reflector layer.

The travel time $T(x)$ for primary reflection can be obtained as a function of the source–receiver distance x using the following expression:

$$T(x) = \sqrt{\left(\frac{2Z}{v_1}\right)^2 + \left(\frac{x}{v_1}\right)^2}$$

(2.1.4)

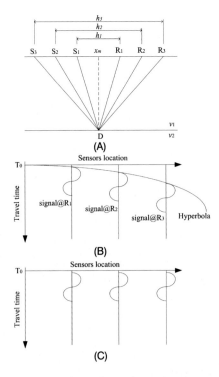

Figure 2.1.8 Schematic diagram of signal transformation process. (A) Sources and sensors in CMP. (B) Seismograms in CMP Gather. (C) Seismograms after NMO correction

where v_1 is the seismic wave transmission velocity in the layer 1. This equation can be transformed to the following:

$$\frac{-x^2}{T_0^2 v_1^2} + \frac{T(x)^2}{T_0^2} = 1 \qquad (2.1.5)$$

where $T_0 = 2Z/v_1$ is the two-way zero offset travel time. As shown in the lower part of Fig. 2.1.7, the curve of travel time $T(x)$, as a function of separated distance of source and receiver for a common source, x, is a hyperbola. The seismograms with common source and variable sensor location x are called common source gathers (CSG). However, often in practical applications an alternate experimental set up known as the common midpoint (CMP) gather is adopted for convenient computations. In the setup the midpoint location of the source–receiver is kept constant (as shown in Fig. 2.1.8A), and the travel time T is then written as:

$$T(h, x_m)^2 = T_0^2 + \left(\frac{h}{v_1}\right)^2 \qquad (2.1.6)$$

where h is the source–receiver offset and x_m is the midpoint position. The captured seismograms in CMP case are also hyperbolas, as shown in Fig. 2.1.8B. Figs. 2.1.8A–B show that, because of the different source–receiver offsets, the captured traces by different sensors on the same target D are separated by a series of travel time differences. These traces are corrected to a zero offset travel time using a data processing called normal movement (NMO) correction.

ΔT, the difference between travel time T and the travel time T_0 for zero-offset can be derived from Eq. (2.1.6) as follows:

$$\Delta T(h, x_m) = \sqrt{T_0^2 + \left(\frac{h}{v_1}\right)^2} - T_0. \qquad (2.1.7)$$

After applying NMO correction to CMP gather, the corrected traces are transformed to simulate a seismic experiment where each source-receiver pair shares the same midpoint and same offset ($h = 0$), as shown in Fig. 2.1.8C. Furthermore, stacking all the corrected traces together provides an image in time of the reflected layer's interior. The stacking process depresses the noise and helps construct the primary information (reflections) by superposition. If the wave transmission velocity is well known, a depth image can be obtained by conversion from seismic time record of stacked trace.

In order to better illustrate the process described above, an example from Hubscher and Gohlb (2014) is reproduced here to explain the basic data processing sequence in reflection seismology, as shown in Fig. 2.1.9 (Hubscher and Gohlb, 2014). Although this particular example refers to an application of marine reflection seismology, it can be considered analogous to an underground measurement case with horizontal layer boundaries.

Step 1: Choose the suitable seismic energy source and seismometer.

As described in the previous section, seismic wave used in the reflection seismology is generated by a seismic source, whose characteristics affect the performance of reflection seismology measurement. The main criteria of seismic source selection are its energy and frequency content. Yordkayhun and Suwan (2012) pointed that the energy content has to be high enough in order to acquire adequate information from the target depth, while the

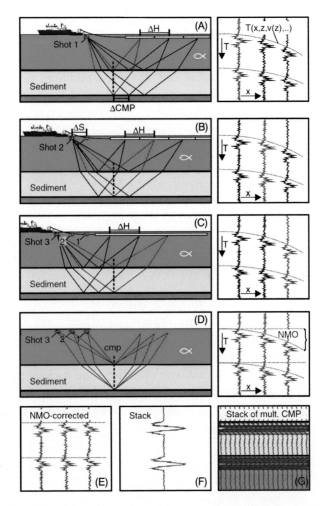

Figure 2.1.9 Schematic diagrams for reflection seismology (Hubscher and Gohlb, 2014)

frequency content has to be high and broad enough to provide adequate resolution for the subsurface imaging. Besides these major requirements, pulse coherency, source-generated noises, portability, safety and repeatability of the source must be taken into account (Yordkayhun and Suwan, 2012). The number and spatial separation of sources and seismometers are other important issues to be considered. In Fig. 2.1.9A, there are three pairs of source and seismometer, with separation distances ΔS for sources and ΔH for seismometers.

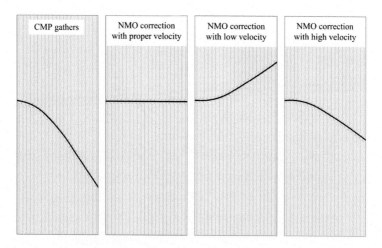

Figure 2.1.10 Influence of velocity on NMO correction

Step 2: Obtain reflected signals and preprocess the captured data.

When a seismic event happens, the seismic waves travel from the source point on surface, down the subsurface to the boundary and are reflected back to the surface. The reflected waves are captured by a sensor array when they reach the ground surface. These signals are then filtered and optimized, as shown on the right-hand side of Fig. 2.1.9A. As discussed above, the arrival time of three signals from the three source–seismometer pairs will exhibit a hyperbolic relationship with the sensors locations. If the seismic event spacing equals to the seismometers spacing (i.e., $\Delta S = \Delta H$), the same reflection point on the boundary is covered by the rays between the second and third source points and the seismometers, as indicated by the red lines in Figs. 2.1.9A–C. These are the CMP reflection waves. In data processing, the individual records sharing the same common midpoint in CMP gathers are sorted and shown in Fig. 2.1.9D.

Step 3: Wave transmission velocity analysis

Eqs. (2.1.6) and (2.1.7) reveal that the velocity of the reflection has great influence on NMO correction (Zhang et al., 2011). If the velocity is too low or too high, the NMO correction traces will be over-corrected or under-corrected, respectively, as shown in Fig. 2.1.10. Hence accurate velocity value is essential to obtain flat traces.

The $T^2 - h^2$ analysis is a practical method to determine the exact velocity value. By observing Eq. (2.1.6), plotting travel time expression for T^2

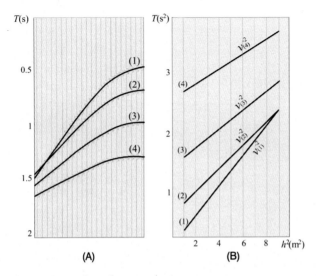

Figure 2.1.11 Example of $T^2 - h^2$ analysis. (A) Several CMP gathers. (B) Corresponding $T^2 - h^2$ curves

and h^2 results in a linear curve (Zhang et al., 2011). When different values for T^2 and h^2 are plotted, the slope can be used to determine v_1^2, as shown in Fig. 2.1.11. Besides the $T^2 - h^2$ analysis method, constant velocity panels (CVP) and constant velocity stacks (CVS) analysis of the velocity spectra are some of the other methods for accurate velocity determination.

Step 4: Stacking process and depth conversion

Using the accurate velocity and the NMO correction to CMP seismograms, traces can be flattened as shown in Fig. 2.1.9E. Then they are stacked to one single trace, as shown in Fig. 2.1.9F. Displaying all stacking traces along a profile gives an image of subsurface structure, as shown in Fig. 2.1.9G.

The application field of reflection seismology can be roughly categorized into three groups, each defined by its depth of investigation (Yilmaz, 2001). First, near-surface applications into depths of 1 km are typically used for environmental surveys and mineral exploration. Reflection seismology was applied for Bell Allard South Zn–Cu orebody exploration in northern Quebec (Milkereit and Eaton, 2003), whereby the size, shape, and composition of the orebody could be determined clearly by using the 3D seismic imaging results. The second category of reflection seismology use is in hydrocarbon exploration, where the detection depths can reach up to 10 km.

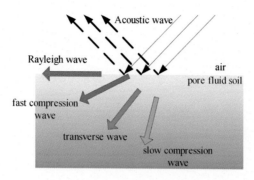

Figure 2.1.12 A wave generated by acoustic energy coupling into pore fluid soil

For example, legacy seismic data of Canada's High Arctic was collected for many years, which served for hydrocarbon exploration and discovery of 8 oil and 25 gas pools (Brent, 2009). The third category of use for the technique is in the earth's crustal studies. These include investigation into the structure and origin of the earth's crust at depths reaching 100 km. Zou and Zhou (2011) made a seismic investigation of the crustal structure beneath the Three Gorges Reservoir region in Central China, whereby an important crustal gradient was observed in the Moho region (Zou and Zhou, 2011).

2.1.2.4 Acoustic-to-Seismic (A/S) Coupling

When airborne sound is incident on the surface of soil or rock with pore fluid characteristic, the acoustic energy in the vibrating air above the soil surface couples with the air in the soil pores. This coupling generates four transmitting waves in soil: Rayleigh wave, transverse wave, fast and slow compression waves, as shown in Fig. 2.1.12 (Liu, 2008). The two compression waves described by Biot model (Biot, 1956) propagate simultaneously in both the fluid and the solid phase. The fast compression wave mainly propagates in the solid phase with less attenuation and dispersion. While the slow wave mainly propagates in fluids as a diffusion wave with high attenuation and dispersion. This phenomenon, known as the acoustic-to-seismic coupling, is well-documented in the literature (Liu, 2008; Biot, 1956).

The A/S coupling technique is mainly used for shallow underground detection (i.e., less than tens of meters). When there is a target object buried in shallow soils, the fast wave will directly pass through it with little or no reflection. In comparison, the slow wave will scatter from the target bound-

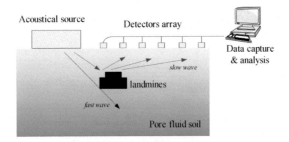

Figure 2.1.13 Schematic diagram of A/S coupling method for landmine detection

aries and will be reflected back to the surface. A vibration sensors' array on the ground surface can capture the reflected vibration signals, which can be further analyzed to determine the characteristics of the buried object (e.g., its depth, shape).

The most common application of A/S coupling is in landmine detection. A schematic diagram of a landmine detection system using A/S coupling is presented in Fig. 2.1.13. In this system, Liu (2008) used a loudspeaker as the acoustic source with sound pressure of 110 dB and a frequency range of 50 to 800 Hz. An electromagnetic seismometer array of five sensors was used to detect the reflected vibration signal on the ground surface. Figs. 2.1.14A and B show the obtained frequency spectra with and without buried landmine in the subsurface, respectively (Liu, 2008). The buried landmine reflects the acoustic wave in the frequency range of 400~700 Hz. The amplitude change of received signals in this frequency range clearly shows the presence of a buried landmine.

Taking advantage of the noncontact remote measurement technique, Sabatier and Ning (2001) conducted A/S coupling measurements for antitank landmine detection by using a laser Doppler vibrometer (LDV). A photograph of field measurement using LDV-based A/S mine detection system is presented in Fig. 2.1.15. When the experimental scans were transformed into an image form, the circular shape of the buried landmine (the red circle) could be observed (the right picture in Fig. 2.1.15). Subsequent analysis showed that the system achieved landmine detection with a probability of 95% in a field blind test (Sabatier and Ning, 2001).

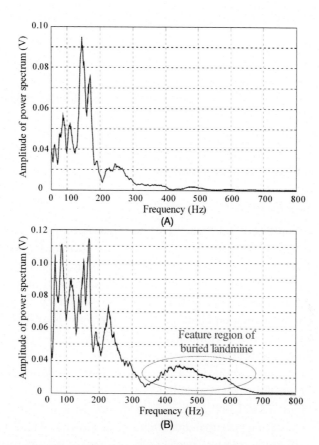

Figure 2.1.14 Frequency responses of obtained signal with and without buried landmine in soil (Liu, 2008). (A) Without buried landmine. (B) With buried landmine

Figure 2.1.15 Photograph of the LDV-based A/S mine detection system in field measurement (Sabatier and Ning, 2001)

2.1.3 ELECTRIC AND ELECTROMAGNETIC METHODS

The electric resistance distribution of the subsurface is measured and ana-lyzed for information in many underground exploration methods. Electric and electromagnetic methods are among the most versatile geophysical tools available for environmental, industrial, and archeological investiga-tions (Manstein and Manstein, 2015). In this section, a few commonly used electric and electromagnetic methods are discussed, with the emphasis on the ground-penetrating radar (GPR) as an example of a widely used electromagnetic method for underground exploration and detection.

2.1.3.1 Electrical Resistivity Surveys (ERS)

All materials, including soil and rock, have the intrinsic property of elec-trical resistivity that governs the relation between the current density that runs through the material and the gradient of the associated electrical po-tential. By introducing an electric current into the ground and measuring the resulting potential differences at the surface, the electrical resistivity of underground materials can be obtained and analyzed for several of their characteristics, including composition and depth (Aning and Sackey, 2014). This geophysical method is called electrical resistivity survey (ERS).

Having four electrodes is the minimum requirement in the measure-ment of electrical resistivity: two current electrodes (A and B) that are used to inject the current, I, and two potential electrodes (M and N) that are used to record the resulting potential difference, V, as shown in Fig. 2.1.16. The electrical resistivity, ρ, can be calculated as:

$$\rho = k\frac{V}{I} \tag{2.1.8}$$

where k is a geometric factor which depends on the arrangement of the electrodes (Morgan, 2001; Samouelian et al., 2005).

There are several types of electrode configuration array (e.g., Wenner, Pole–Pole, Dipole–Dipole, Wenner–Schlumberger), each corresponding to a different geometric factor (Samouelian et al., 2005). Samouelian et al. (2005) summarized and compared the characteristics of those array con-figurations, with their specific advantages and limitations. The choice of the array configuration often depends on the degree of heterogeneity to be mapped and also on the acceptable level of background noise.

Usually, for one-dimensional vertical (VES) or horizontal (HES) elec-trical sounding, having four electrodes is sufficient for the survey. In VES,

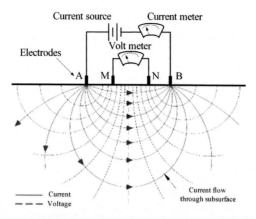

Figure 2.1.16 Concept illustration of ERS

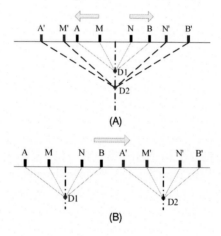

Figure 2.1.17 Electrodes arrangement for 1D ERS measurements. (A) Vertical electrical sounding. (B) Horizontal electrical sounding

the four electrodes are arranged in a line, at gradually increasing electrode spacing, as shown in Fig. 2.1.17A. As the spacing between the electrodes (A&M and N&B) increases, the detectable depth varies from D1 to D2. VES is used to detect the vertical resistivity variation and it is suitable for multihorizontal layer surveys. In HES, all four electrodes are shifted sideways in the same direction, and the detected point D1 moves to D2. The method is useful to detect the horizontal resistivity variation and suitable for multivertical layer surveys (Wang, 2013).

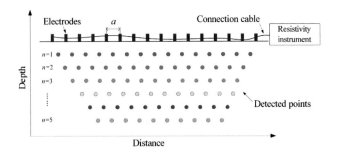

Figure 2.1.18 An arrangement of the electrodes for 2D measurement

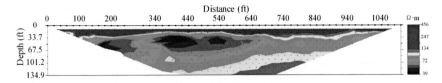

Figure 2.1.19 Collected resistivity data of Santee Basin displayed as a pseudo-section (Pierce et al., 2012)

Comparing the 1D survey from multihorizontal/vertical layer structures often reveals more complex formations with lateral and longitudinal changes over short distances. In such cases, two (2D) or three-dimensional (3D) survey techniques become more useful. A schematic diagram of 2D ERS setup is shown in Fig. 2.1.18 (Wang, 2013). The current and potential electrodes are maintained at a regular fixed distance from each other and are progressively moved along a line on the soil surface. Consider the spacing between two closed electrodes to be a, then pick four electrodes whose spacing is $L = na$ ($n = 1, 2, 3, \ldots$) as a group to make a VES system. Because the detecting depth of VES is related to electrodes spacing, resistivity at different depths can be measured by using different n-electrode groups, to detected points in different depths, as represented by different color circles in Fig. 2.1.18. It can be noticed that the larger the n-values, the larger the depths of investigation. A 2D electrical resistivity image (ERI) of Santee Basin in the United States captured by Pierce et al. (2012) produced an inverted triangle shape with depth, as described above and shown in Fig. 2.1.19 (Pierce et al., 2012).

There are two methods available to obtain a 3D electrical resistivity survey. The first method consists of building a 3D picture using a network of two-dimensional parallel sections. The other method consists of making the survey by a set of electrode arrangement in a square spread, as shown

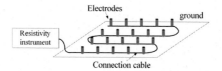

Figure 2.1.20 Square electrode arrangement in 3D ERS

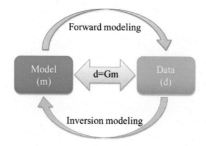

Figure 2.1.21 Concept illustration of forward and inversion modeling (Pierce et al., 2012)

in Fig. 2.1.20. Compared to 2D ERS, the 3D ERS method is more complicated and takes longer time with the required electrode arrangement, which makes it impractical for most field measurements. In the following section, a 2D ERS method is considered as an example to explain the data processing in more detail.

When the structure of the detected underground area is not complicated, the electrical resistivity data can be used directly for geophysical exploration. However, in some complex underground resistivity distribution cases, unique techniques known as inversion and forward modeling are used to process data to construct useful resistivity images.

Inversion and forward modeling is a common method for signal and system analysis in engineering and science, whose bias concept is illustrated in Fig. 2.1.21 (Pierce et al., 2012). In a measurement activity, a collected data set (d) related to a physical property model (M) of the measured object can be obtained, And this relationship can be expressed as a mathematical function G. Using a previously known model to predict a synthetic collected data set is called forward modeling. And using a collected data set to predict the model is called inversion modeling.

For ERS, it is an inversion modeling process to use the collected electrical resistivity data to generate the physical property model of underground area, and then to analyze the characteristics of underground materials and

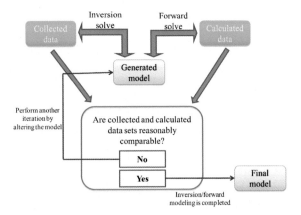

Figure 2.1.22 Flow chart of the process used in geophysical data inversion and forward modeling (Pierce et al., 2012)

structures. A simple inversion modeling process is shown in Fig. 2.1.22 (Pierce et al., 2012; Nostrand and Cook, 1966). Firstly, an initial generic model of the earth's resistivity is generated by using the collected data. And then by using the generated model and the forward solving process, a synthetic data set is calculated, which represents the expected data set from a geophysical survey to an earth's structure expressed as the generated model. The collected and calculated data sets are compared for equivalency. If the collected data and the calculated synthetic data sets do not agree, the generic model is modified by altering its parameters and generates a new synthetic data set for comparison. Each time the calculation and comparison is known as an iteration of the inversion. This iteration continues until the model accurately simulates the collected data, which indicates that the characteristics of underground area can be well represented by the final model. By using this inversion and forward modeling method, an ERS data processing of a survey area with bedrock in the Santee Basin of United States is shown in Fig. 2.1.23 (Pierce et al., 2012). The inversion obtained in this example was conducted using the forward modeling process to determine the resistivity distribution that best simulated the ERI data and property distribution.

2.1.3.2 Electromagnetic Induction (EMI) Method

Electromagnetic induction method is an effective approach to detect shallow buried conductive targets (e.g., metallic objects, pipes, and cables). A schematic diagram of its principle is shown in Fig. 2.1.24

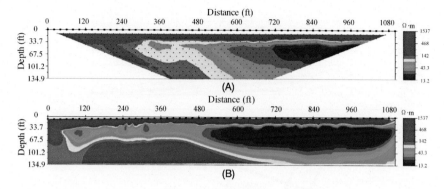

Figure 2.1.23 ERI observations on bedrock in Santee Basin (Pierce et al., 2012). (A) Collected resistivity data of survey area with bedrock. (B) The result of inversion process

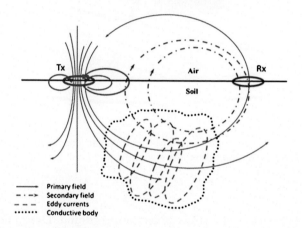

Figure 2.1.24 Simplified representation of the basic principle of electromagnetic induction (Manstein and Manstein, 2015)

(Manstein and Manstein, 2015). The detector of an EMI system consists of two adjacent electromagnetic induction sensors, a pair of transmitting and receiving coils (Tx & Rx) packaged together. In this measurement, an alternating magnetic field (primary field) is produced by passing a series of current pulses through the transmitting coil. An eddy–current field is generated in the target object by changing the frequency of the current pulses. Subsequently, a secondary magnetic field is created by the eddy–current field which transmits back to the ground surface and is detected by the receiving coil. For each exciting current, the amplified induction signals are processed in the receiving coil to analyze the electromagnetic induction

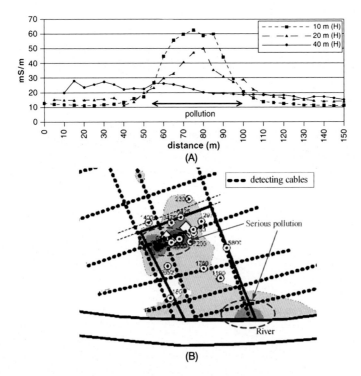

Figure 2.1.25 Pollution trace using EMI (Martens and Walraevens, 2009). (A) Electromagnetic measurements in a cable with three electromagnetic coil spacing. (B) Isolines of the apparent conductivity

response in the presence of the target object underground. Depending on the characteristics of this response signal, the position and the depth of the object is determined (Zhu, 1996).

A field example of using the EMI technique to trace soil and groundwater pollution caused by an abandoned salt storage was discussed by Martens and Walraevens (2009). In this study, the salt storage was located close to a river, and the detecting cables with multiple electromagnetic coils were placed over the site along several lines with equal spacing (see Fig. 2.1.25B). By increasing the electromagnetic coil-spacing to 10, 20, and 40 m, detection from different depths was possible, as shown in Fig. 2.1.25A. The conductivities increased in the range of $50 \sim 100$ m, indicating the presence of pollution due to dissolved salt. Meanwhile, the conductivity values decreased gradually with the increasing spacing between the electromagnetic coils which indicated downward permeation of pollution. Constructing

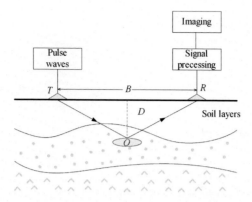

Figure 2.1.26 Principle of GPR measurements (Sun et al., 2012)

isolines of the conductivity on a map of two significant pollution locations could be observed clearly, as shown in Fig. 2.1.25B.

2.1.3.3 Ground-Penetrating Radar

Similar to reflection seismology, another method of subsurface exploration, known as ground-penetrating radar (GPR), works for shallower depths of detection by substituting acoustic waves with electromagnetic waves for carrier and reflected signals. This nondestructive method injects electromagnetic radiation into underground and detects the reflected signals from the subsurface features and objects to image them. Due to its easy operation, high automation, high resolution images, and reliable interpretation results, GPR has been used successfully in various fields for detection and mapping of landmines, municipal pipelines, and archaeological features (Sun et al., 2012).

Fig. 2.1.26 shows a schematic diagram of the GPR measurement principle (Sun et al., 2012). Radar pulses are emitted from transmitting antenna (T) and then are reflected by objects to a receiving antenna (R).

The travel time, t, of the incident and the reflected radar pulse can be calculated as:

$$t = \frac{TO + OR}{v} = \frac{2\sqrt{(B/2)^2 + D^2}}{v} \tag{2.1.9}$$

where B is the separation distance between T and R, D is the depth of object, and v is the speed of radar pulse. Superimposing T and R ($B = 0$),

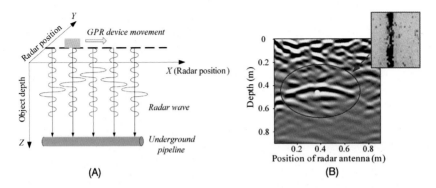

Figure 2.1.27 GPR scan along an underground pipe (Sun et al., 2012). (A) Schematic diagram of GPR scanning along an underground pipeline. (B) Image of underground pipeline

the computation for the depth of object can be simplified as:

$$D = tv/2. \tag{2.1.10}$$

Moving T and R simultaneously along survey lines, a time profile GPR image can be composed by sequential records. The detection, localization, and shape reconstruction of buried pipelines is one of the most widespread application fields of GPR diagnostics (Sun et al., 2012; Allred and Fausey, 2004). For example, Fig. 2.1.27A shows a schematic diagram of GPR scanning along an underground pipeline, while Fig. 2.1.27B shows an image of the scanned pipeline. The scan abscissa records the position of the radar antenna along the survey line, and the vertical axis is the depth of radar pulse penetration as calculated from its speed. The radar image clearly identifies the reflection of the metal pipe, indicated by the area in red circle in Fig. 2.1.27B. The GPR image reflecting a slice of the detected pipeline is constructed as shown in the right corner of the Figure (Sun et al., 2012).

In field applications, the dielectric permittivity of different subsurface layers will greatly affect the reflection coefficient and the speed of radar wave, hence the image accuracy. In order to obtain a high quality GPR profile, a prior field investigation should be carried out for accurate determination of the detection parameters. Additionally, methods such as high pass filtering mean and average gain adjustment of automatic compensation, and complex signal analysis can be used in data processing to improve the quality of a GPR profile (Allred and Fausey, 2004). Sun et al. (2012)

obtained a successful 3D display of buried pipelines, shown in Fig. 2.1.3, by using a radius estimation method and a technique of interactive transparent 3D visualization of the 2D GPR profiles.

Another aspect of GPR technology is that due to the high attenuation of the radar waves in soil and rock media, the GPR application is limited to probing mostly shallow depths. Hence, the combination of GPR with other underground measurement techniques can make up for its disadvantages and present more comprehensive underground detection and surveying solutions. For example, a combination of GPR and ERT was utilized to perform shallow characterization and describe deep macroscopic characteristics in road engineering in Tibet Bomi County Tongmai countryside, China (Guan et al., 2012). The construction project was located at the fault zone of Yigong–Palongzangbu region, which was also prone to landslides. First, an electrical resistivity survey was conducted to detect landslide zones. The resistivity profile displayed well the fault zone beneath the sliding body, as shown in Fig. 2.1.28A, but the slide slip surface (sliding plane) was not detected clearly due to the effect of groundwater enrichment in the fault zone. A GPR survey was then used to investigate the sliding mass closer to the surface. The landslide slip surface and the sliding body was observed clearly by a strong phase of reflection wave in the GPR image, as shown in Fig. 2.1.28B.

2.1.4 OPTICAL SENSING TECHNOLOGIES USED IN UNDERGROUND MEASUREMENT

The methods discussed in the previous sections focused on use of "waves" to detect various phenomenon underground. While using acoustic and electromagnetic waves offers many effective underground sensing and monitoring methods, the fast and accurate processing of the wave signals for accurate assessment of underground features and character is still an important area that requires research and advancement. Due to the added benefits they offer, optical sensing technologies (OSTs) in underground measurement applications have also become an important area of research and development. With new advancements in optical technologies, optical sensors have become increasingly versatile for underground measurements of all kinds including physical properties, vibration, strain/stress, temperature, and composition within the last decade.

Optical sensors satisfy a number of the requirements of underground detection, including:

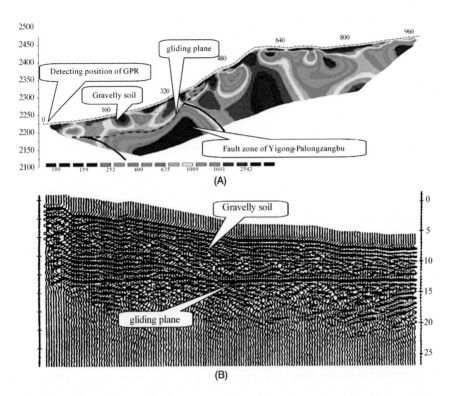

Figure 2.1.28 Observation on the fault zone of Yigong–Palongzangbu by using the combination of GPR and ERT (Guan et al., 2012). (A) Image of ERT method. (B) Image of GPR method

1. Convenient placement over longer distances

For most distributed fiber-optic sensors, the sensing element is the optical cable whose length could be hundreds of kilometers. In such configurations the sensing element can be located far away from the signal generator, which makes the fiber-optic sensor well suited for difficult-to-reach locations over long distances without the need for additional power supply. The fiber can be laid in or around the sensing area in any convenient pattern as the sensing takes place along the linear positioning of the fiber. This feature of the fiber greatly simplifies its working for some locations which can be arduous with discrete sensors for data acquisition.

2. Safety in hazardous environment

For most conventional sensors, detecting process and detected signals depend on the transmitting electrical signals which are easily affected by electromagnetic frequency (EMF) interferences. Optical fibers are immune to EMF environments and carry no risk of current leakage or ignition. These characteristics make the fiber optical sensor well suited for hazardous environments such as coal mines and gas wells (Poczesny et al., 2012).

3. High reliability and stability

Because the silica fiber is immune to humidity, corrosion, and high temperature, the signal guided in the fiber is prevented from disruption by such environmental factors. As a result, optical sensors have greater reliability and stability than most other sensors that may be affected by environmental factors. For example, field trials in coal mine safety monitoring have shown that the fiber–optic sensors can maintain accurate and reliable operation for two years without recalibration, which is significantly longer than for widely used catalyst type sensors requiring weekly calibration (Liu et al., 2013). The longevity and minimum calibration requirement renders optical sensors valuable for most underground applications.

4. Multiplexed sensing capability

Most optical sensors can detect several physical properties at the same time. For example, a Brillouin optical time domain reflectometry (BOTDR) system can simultaneously detect strain, vibration, or temperature on a fiber cable (Zhu et al., 2011). Fiber Bragg grating (FBG) elements also have the ability of sensing strain, stress, vibration, temperature on the same sensor assembly (Ferraroa and Natale, 2002). These sensing techniques will be introduced in detail in the following sections.

2.1.4.1 Vibration Measurement

Vibration is a commonly required parameter for detection in underground measurements. There are several OSTs, which can be used for vibration detection by either as single point sensing or distributed sensing over long distances (e.g., tens of kilometers).

2.1.4.1.1 Principles of Fiber Optic Vibration Sensing

The traditional fiber-optic vibration sensing (FOVS) technologies which are based on the modulation of light to detect vibration affecting the optical properties of the fiber guide include the intensity-modulation, phase-modulation and wavelength-modulation methods (Zeng et al., 2004).

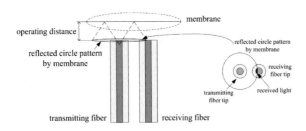

Figure 2.1.29 Principle diagram of intensity modulation type fiber-optic sensor (Li et al., 2010)

a) Intensity–modulation

The intensity modulation type fiber–optic sensing has the advantages of a simple structure and low cost, which is attractive for most applications in vibration detection. A common structure of intensity modulation type fiber-optical sensor is shown in Fig. 2.1.29, in which two fibers are aligned in parallel and a vibrating reflective membrane reflects the incident light from the transmitting fiber (TF) back to the receiving fiber (RF) (Li et al., 2010). Meanwhile, the membrane picks up the vibration signal and transfers it to its own movement, which then reflects the transmitted light to project a circular pattern on the receiving fiber tip. The overlap area determines the received light intensity by RF. The movement of the membrane modulates the received light intensity P_R by RF which can be expressed as:

$$P_R = \int 2\alpha r I(r)\, dr \qquad (2.1.11)$$

where α is a coefficient related to the physical parameters of the sensor (e.g., diameters of fiber and fiber core, distance between fiber tip and membrane), r is the integral radius of reflected circular pattern, and $I(r)$ is its radial intensity. Finally, the vibration signal is extracted by demodulating the received light intensity.

Different from the structure introduced above, the intensity modulation type sensing technologies used in underground vibration measurements are mainly based on direct coupling of light (Li et al., 2010; Kim and Feng, 2007) or the loss of light due to micro-bending/macro-bending of the fiber (Poczesny et al., 2012; Freal et al., 1987). Although they have different construction, their operation principles are similar, which are based on extracting the vibration signal by detecting the intensity change of the coupling light. A typical structure of a fiber-optic

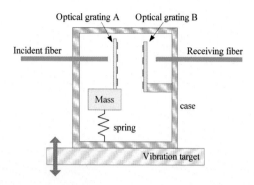

Figure 2.1.30 Vibration sensor with two transmitted grating panels (Kim and Feng, 2007)

vibration sensor based on direct coupling of light is shown in a schematic in Fig. 2.1.30 (Kim and Feng, 2007). Originally proposed for monitoring vibration of civil engineering structures in real time, the sensor is composed of two optical grating panels, A and B, which are attached to a mass and to the sensor case, respectively. A pair of optical fibers is fixed in the sensor case and aligned perpendicular to the optical grating panels. The light beam emerging from the incident fiber is received by the receiving fiber after passing through the two panels. The optical grating A is attached to a vibrating spring–mass and the optical grating B is attached to the sensor case. When the mass picks up the vibration motion, the transmitted light beam will either be received (when the transparent regions of grating panels coincide) or be cut off (when they do not coincide). Vibration can be detected by measuring the intensity variation of the transmitted light. This sensor was successfully applied to detect damage on a model concrete structural element.

Another vibration sensor, based on the micro-bending in the fiber, was developed and tested by Freal et al. for borehole deployment (Freal et al., 1987). The main features of this sensor are shown in Fig. 2.1.31. In this configuration the optical fiber is mounted between two deforming teeth to act as the spring component of a spring–mass vibration pick-up system. The moving deformer is considered the mass component. The vibration force moves the deformer, squeezing and bending further the mounted fiber in between. The induced micro-bend decreases the intensity of light propagating through the fiber. Due to the elasticity of the fiber the micro-bend recovers when the vibration force is released, and transmitted light intensity returns to original. Vibration is detected by detecting the transmitted light

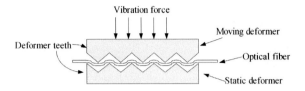

Figure 2.1.31 Schematic diagram of a micro-bend sensor (Freal et al., 1987)

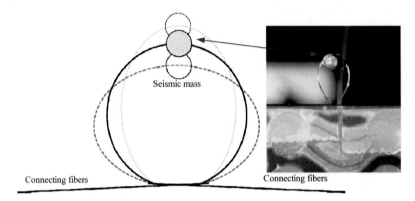

Figure 2.1.32 Scheme and photography of optical fiber macro-bend vibration sensor (Poczesny et al., 2012)

intensity variations. This sensor can detect accelerations as small as 5 μg at 1 Hz with a dynamic range in excess of 90 dB.

A macro-bend fiber-optic vibration sensor was introduced by Poczesny et al. (2012) as shown in Fig. 2.1.32. It consists of a single mode optical fiber bent into a loop of radius of a few millimeters with a small seismic mass (approximately 0.3 g) attached at the top of the loop. Vibration was measured by utilizing bend loss of the light intensity, which was proportional to the geometrical deformation of the loop caused by the vibrating mass attached.

b) Phase modulation

The phase modulation type fiber-optic sensing is based on the principle of light interferometer. A typical fiber-optic interferometer is shown in Fig. 2.1.33A. In this device, two light beams pass through two arms of the interferometer and interfere with each other at the second coupler (CP). The interference intensity can be expressed as:

$$I = I_1 + I_2 + 2\sqrt{I_1 I_2} \cos(\varphi_0 + \Delta\varphi(t)) \qquad (2.1.12)$$

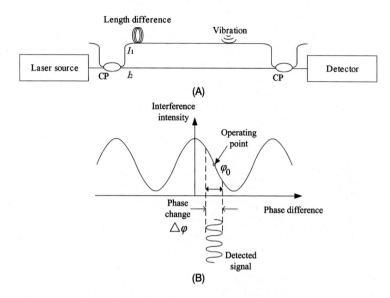

Figure 2.1.33 Principle diagram of phase modulation type FOVS. (A) Typical fiber-optic interferometer structure. (B) Principle of vibration signal extraction

where I_1, I_2 are the light intensities on two arms, φ_0 is the inherent phase difference induced by the length difference of the two fiber arms, and $\Delta\varphi$ is the phase change caused by vibration. The phase change occurs as the fiber's refractive index changes with the vibration strain induced on the fiber. From Eq. (2.1.12), the vibration detection sensitivity ($dI/d\varphi$) is related to a sinusoidal function of the inherent phase difference φ_0. In order to obtain the highest detecting sensitivity, the length difference of two arm fibers is optimized to set the operating point at $\varphi_0 = \pi/2$, as shown in Fig. 2.1.33B.

The phase-modulation type fiber-optic sensors are the most popular choice of device in underground optical sensing due to their reliable performance. The commonly known phase-modulation fiber-optic vibration sensing devices are Michelson interferometer, Fabry–Perot interferometer, Sagnac interferometer, and Mach–Zehnder interferometer. These are discussed below:

1. Michelson Interferometer

A popular single-point detection sensor used in underground measurements is a fiber-optic disk vibration sensor (FODVS) which is based on Michelson interferometer. It employs a push–pull mass–spring feature

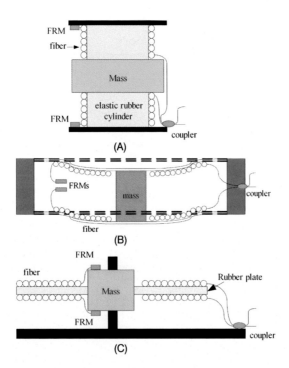

Figure 2.1.34 Three typical structures of FODVS. (A) Central support type (Zeng et al., 2004). (B) Edge support type (Wang et al., 2006). (C) Flexural disc type (Cranch and Nash, 2000)

that can be configured in three different ways: central support type, edgy support type, and flexural disk type, as sketched in Figs. 2.1.34A–C, respectively (Zeng et al., 2004; Wang et al., 2006; Cranch and Nash, 2000).

The basic sensing principle of an FODVS can be followed from the schematic in Fig. 2.1.34A (Zeng et al., 2004). In this configuration, the sensing fibers on two interference arms are wound over an elastic rubber cylinder, which is a part of a push–pull mass–spring system composed of elastic rubber cylinders and a mass fixed through a coaxial pole. The inertia force of vibrating mass induces deflections on the elastic rubber cylinder, which in turn applies tension on the wound sensing fibers and changes their refractive index. The strain induced on the fiber causes an optical phase shift in the transmitted light. The magnitude of the detected vibration is then obtained by demodulating the optical phase shift signal. Two Faraday rotator mirrors (FRMs) are attached to the ends of two Michelson arms to eliminate the polarization

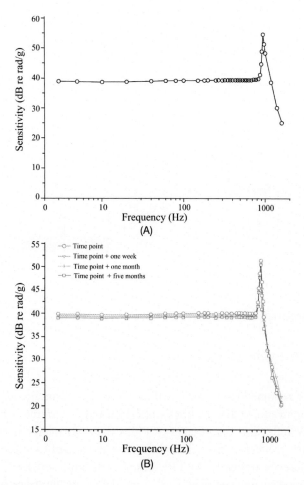

Figure 2.1.35 Characteristics test results of FOVDS element (Zeng et al., 2004). (A) FOVDS element sensitivity and frequency range. (B) Long term stability test

induced fading of interference intensity phenomenon (Cranch and Nash, 2000).

Zeng et al. (2004) designed an improved 3-component optical fiber accelerometer using the Michelson interferometer principles introduce above (Zeng et al., 2004). A series of experiments were setup to estimate the sensitivity, measurement frequency range, and longevity of their sensor. Some of the test results are shown in Fig. 2.1.35A–B. It is observed that, in a flat frequency response range of 3~800 Hz, sensor's sensitivity is 39 dB per rad/g (which is 89 rad/g). And in a five-month long stability test, the sensor's sensitivity drift is less than 1.8%.

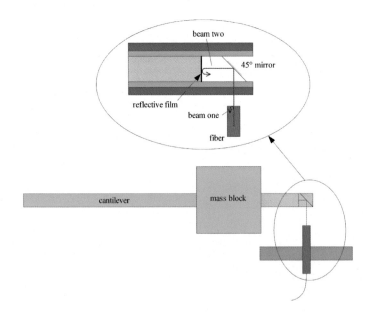

Figure 2.1.36 FP accelerators with 45 degree mirror (Han et al., 2014)

2. Fabry–Perot (F–P) Interferometer

Only a few types of fiber optical vibration sensor based on F–P interfer-ometer (FPI) are reported to be used in underground measurements (Han et al., 2014; Litter et al., 2010). In fiber-optic FPI, two interference beams are generated by reflection on surfaces of F–P cavity, normally consisting of a fiber tip and a reflective film. The difference in the light paths deter-mines the interference intensity. An FPI sensor with improved F–P cavity structure developed by Han et al. (2014) is used here to illustrate device operation, as shown in Fig. 2.1.36. A cantilever beam carrying a vibrating mass block was used to test the operation of the sensor. In this FPI device, beam 1 is the light reflected by the fiber tip and beam 2 is the light pro-jected onto a mirror oriented 45 degrees to its line and then reflected by the reflective film. When the cantilever–mass block vibrates, the reflective film moves along with the mass changing the path of the reflected beam 2, hence changing interference intensity correlated to vibration frequency. By inserting the 45° mirror, the light propagating direction is changed, leading to improved sensitivity. This device could sense vibration resonant frequency up to 400 Hz with a sensitivity of 0.042 rad/g.

Another fiber-optical FP interferometer using fiber Bragg grating (FBG) mirrors was proposed by Litter et al. (2010), and is shown in Fig. 2.1.37.

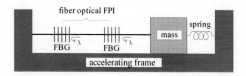

Figure 2.1.37 Sensing element of fiber optical FPI with FBG mirrors (Litter et al., 2010)

Fiber Bragg grating introduces periodic variation in the index of refraction along short sections in the fiber core (Ferraroa and Natale, 2002). It can reflect propagating light at certain wavelength λ along the optical fiber. The operational principle of this sensor is still based on the spring–mass structure, in which the mass is connected to the accelerating frame via both a spring and an FPI. Two fiber Bragg gratings (FBGs) written into the core of a single mode optical fiber were used as reflective mirrors to compose an FP cavity. Integration of FBG mirrors in fiber renders the device small and flexible for versatile uses (Litter et al., 2010).

3. Sagnac interferometer

Fiber-optic vibration sensors based on Sagnac interferometer are mainly used to detect rotational vibration, including ground particles' motion associated with the rotational effects caused by the earthquakes or by interaction of seismic waves with micro-morphic features of rocks (Jaroszewicz and Krajewski, 2007). The fiber-optic rotational vibrometer uses a fiber-optic gyroscope (FOG) for direct measurement of ground rotation component in near-field seismic events (Jaroszewicz and Krajewski, 2007). Fig. 2.1.38 shows a typical FOG configuration. Light from source passes through a coupler and is split into two beams propagating in opposite directions (clockwise light and counterclockwise light) in a coil of several kilometers of wound fiber. Due to the Sagnac effect, the beam traveling against the rotation experiences a slightly shorter path delay than the other beam, and then leads to a differential phase shift measured through interferometry. The shift of the interference pattern is detected by a photo-diode (PD) and helps extract the angular velocity by a signal process. Using fiber-optic rotational vibration sensors, Jaroszewicz and Krajewski (2007) measured the magnitude of seismic rotational events which were delayed in time with respect to the classical seismic waves occurring in earthquakes. Test results showed the sensitivity of the Sagnac interferometer based sensor on the order of 4.27×10^{-8} rad/s.

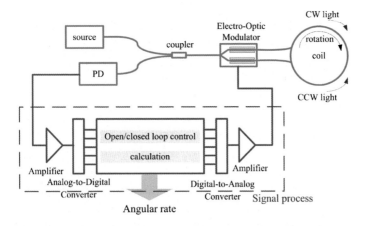

Figure 2.1.38 Fiber optical gyroscope configuration used in FORV

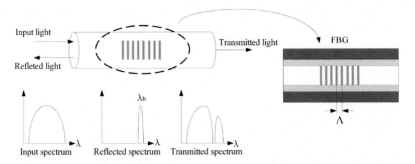

Figure 2.1.39 Depiction of the reflected and transmitted spectrum of FBG for a broadband light source (Ferraroa and Natale, 2002)

c) Wavelength modulation

Fiber bragging grating (FBG) vibration sensor is a typical wavelength modulation type fiber-optic vibration sensor. The operation of FBG is illustrated in Fig. 2.1.39, where periodic variation in the index of refraction along short sections in the fiber core causes phase shifts of the transmitted and reflected light useful for the detection of a physical phenomenon (Ferraroa and Natale, 2002). When light from a broadband source is coupled into the optical fiber with an FBG written in it, a narrow band of light with a characteristic center wavelength λ_b is reflected at the FBG location, while the rest of the spectrum is transmitted. The center wavelength of reflected light is

$$\lambda_b = 2n\Lambda \tag{2.1.13}$$

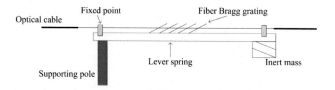

Figure 2.1.40 Design of sensor head (Zhang and Li, 2005)

where Λ is the grating pitch and n is the refractive index of fiber. The reflected wavelength shifts show linear response to the change of fiber grating's properties (e.g., grating length frequency, pitch). Therefore any external physical force applied to the gratings such as strain, pressure, temperature, or vibration that causes a change can be recovered from the measured wavelength shifts. The FBGs have been studied intensively as an optical sensor for various sensing applications for strain, pressure, and temperature.

The principle of vibration sensing by FBGs is as follows: as an external vibration applies strain to the fiber, the characteristic wavelength λ_b varies as a consequence of both the grating pitch change (due to the simple elastic elongation) and the refractive index change, which correlates to the magnitude of the vibration. The Bragg wavelength change is measured through the use of an optical spectrum analyzer or an optical interferometer.

Most examples of FBG based vibration sensing in underground comes from measurements conducted on structural components such as flexural beams (Ferraroa and Natale, 2002; Zhang and Li, 2005). Fig. 2.1.40 shows a sensor head in which an FBG is attached to one end of a lever spring beam (Zhang and Li, 2005), in which the vibration profiles were extracted from the wavelength shift measurements using FBG sensors attached to the flexural beam. The vibration introduces an acceleration change on the inert mass, causing the lever spring to stretch or compress and inducing strain variation on the FBG attached on the beam. The strains on the FBG could then be transformed into a wavelength shift through the optical measurement design and back to the vibration magnitude by a precalibration routine. These FBG sensors were used in an experiment in the shooting range of Army Fort Dix in New Jersey, whereby the presence of personnel or vehicle movement could be detected through vibration at the perimeter of a designated area.

There are four common interrogation methods used for reading out wavelength shifts: (a) direct spectroscopy using an optical setup with

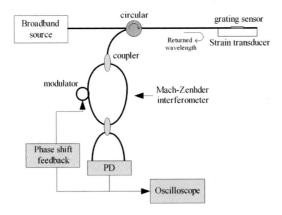

Figure 2.1.41 MZI wavelength discriminator for strain measurement (Chandra et al., 2016)

dispersive optical elements (e.g., grating, prism) and photo-detectors of fiber spectrum analyzers (e.g., linear array charge coupled device (CCD)); (b) passive optical filtering components (e.g., wavelength division multiplexing (WDM) couplers); (c) tracking tunable filters (e.g., FP cavity filter, acoustic-tuned optics filter, filter based on a tunable FBG); and (d) interferometers (Ferraroa and Natale, 2002). Chandra et al. (2016) designed a fiber Mach–Zehnder interferometer (MZI) to extract wavelength shift in strain transducer, as shown in Fig. 2.1.41. In this device, light from a broadband source is coupled into the fiber which transmits the light to the grating sensor. The wavelength component reflected from the grating sensor back along the fiber towards the source is tapped off using a coupler and fed to a Mach–Zehnder interferometer. This returned light becomes a source light into the interferometer with a wavelength λ. The interferometer output, or interference intensity $I(\lambda)$, can be expressed as:

$$I(\lambda) = A\{1 + \cos[\delta(\lambda) + \varphi]\} \qquad (2.1.14)$$

where A is proportional to the input intensity and system losses; $\delta(\lambda) = 2\pi nd/\lambda$, in which n is the refractive index of fiber and d is the length imbalance between two arms in MZI, φ is a bias phase offset of MZI (i.e., a slowly varying random parameter affected by environmental factors). When the wavelength of returned light shifts, the interference intensity changes and can be detected by a photodiode (PD). Alternatively, the interferometer can be used as a discriminator to detect the wavelength shifts at the source

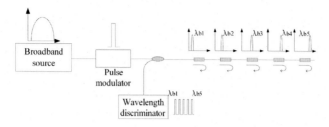

Figure 2.1.42 A TDR multiplexing configuration to interrogate multiple FBG sensors

formed by the strained grating. Using the set up described above an experimental resolution as low as 6 nanostrain/$\sqrt{\text{Hz}}$ has been reported (Chandra et al., 2016).

Another important issue in using FBG sensors in underground measurements is the requirement to multiplex a large number of sensors along a single optical fiber when distributed sensing is desired. Ferraroa and Natale (2002) introduced a time domain reflectometry (TDR) approach used for strain sensing network for geophysics application, as illustrated by the optical setup shown in Fig. 2.1.42. A pulsed light from a broadband light source is coupled into the optical fiber where a number of FBGs are written along the fiber with different Bragg wavelengths. The reflected signals composed of different reflected wavelengths, ranging from λ_{b1} to λ_{b5}, and their shifts are analyzed by a receiving device (e.g., wavelength discriminator) to extract the distributed strain. In this way, depending on the source bandwidth and the strain, it is possible to monitor tens of FBG sensors along the same fiber. For example, a broadband source ranging from 50 to 100 nm has the capability to monitor 10~20 sensors. Furthermore, with an additional combination of wavelength division multiplexing (WDM) with TDR, more than 100 sensors can be monitored along the same fiber (Ferraroa and Natale, 2002). The high number of sensors along the same optical fiber makes the FBG approach attractive for in-field and underground applications.

2.1.4.1.2 Distributed Sensing of Vibration

Except for the TDR array of FBGs, the system designs and sensors introduced above are mainly used for single point vibration detection. However, the capability of distributed sensing is one of the advantages for optical sensing technologies. Two typical fiber-optic distributed sensing designs, dual

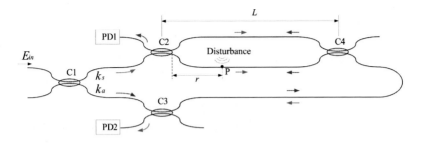

Figure 2.1.43 Structure of DMZI (Li and Xiao, 2015)

Mach–Zenhder interferometer (DMZI) and φ-OTDR will be introduced here.

a) Dual Mach–Zenhder interferometer

The general structure and operation of DMZI is illustrated in Fig. 2.1.43 (Li and Xiao, 2015). In this device, two light beams propagate in opposite directions (the figure shows clockwise light by a blue arrow and counterclockwise light by a red arrow) and interfere at couplers C2 and C3. The interference outputs are detected by photodiodes PD1 and PD2, respectively.

When there is a vibration disturbance at a distance of r away from coupler C2, the unique phase difference induced by the vibration travels along the distinguishing paths to PD1 and PD2. The path lengths for these light beams are r and $2L - r$ (where L is sensing cable length), respectively, which will cause an arrival time lag between the two channel signals. The interference signals, I_1 and I_2, on two detectors and the arrival time lag T can be expressed as:

$$I_1 = \frac{1}{8}I_0 + \frac{1}{8}I_0 \cos\left[(1 - K)\Delta\varphi\left(t - \frac{n(L - r)}{c}\right)\right], \qquad (2.1.15)$$

$$I_2 = \frac{1}{8}I_0 + \frac{1}{8}I_0 \cos\left[(1 - K)\Delta\varphi\left(t - \frac{n(L + r)}{c} + \pi\right)\right], \qquad (2.1.16)$$

$$T = 2n\frac{L - r}{c} \qquad (2.1.17)$$

where I_0 is the intensity of input light, λ is the wavelength, c is the velocity of light in vacuum, n is the refractive index of fiber core, and K is the

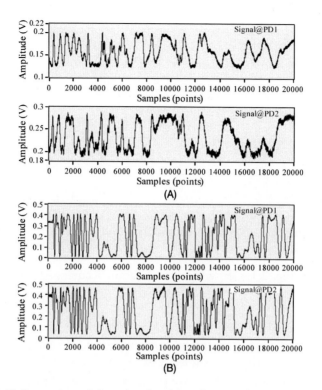

Figure 2.1.44 Comparison of detection signal waveforms with and without PIPS elimination (An et al., 2012). (A) Without PIPS elimination. (B) With PIPS elimination

interference contrast. Usually, by applying cross-correlation between two channel output interference signals, the time lag T can be measured exactly.

However, due to the phenomenon of polarization-induced fading in the sensing cable, correlation between the detection signals often degenerates, hence large signal locating errors arise in practical applications. This effect is known as polarization-induced phase shift (PIPS) (An et al., 2012). In order to solve this problem, PIPS elimination methods, such as controlling polarization state of input light source, are proposed (An et al., 2012; Chen et al., 2013). An et al. (2012) applied DMZI approach in an underground oil pipeline safety monitoring and prewarning system which covered a monitoring distance of 43 km. An original detection signal waveform without the PIPS elimination methods is shown in Fig. 2.1.44A. As observed, signals at PD1 and PD2 had a quite low correlation coefficient of 0.28, which led to large locating errors. Employing the PIPS elimination method improved the correlation coefficient of two signals to

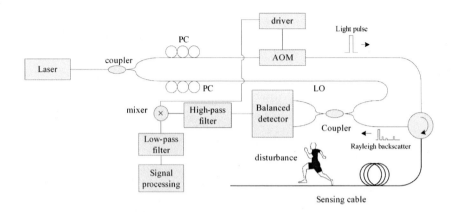

Figure 2.1.45 Structure of φ-OTDR system (Lu and Zhu, 2010)

0.95, as shown in Fig. 2.1.44B, whereby a location accuracy of 10 m was achieved.

b) φ-OTDR

The design of a φ-OTDR system is shown in Fig. 2.1.45 (Lu and Zhu, 2010). In this device, a continuous wave (CW) laser light with long coherent distance is split by a coupler into two parts, one on a signal path and the other on a local oscillator (LO) path. Then light component on the signal path is modulated by an acousto–optic modulator (AOM) to be an optical light pulse propagating into the sensing cable. Rayleigh backscattered signal is generated in the sensing cable which propagates back to interfere with the light component on the LO path. Passing through a coupler, the interfered signal is then detected by a balanced detector. When a disturbance exists, the Rayleigh backscatter at the disturbance position changes, which also reflects to the interference signal. Then by using a heterodyne detection, in which the driving frequency to AOM and the interference signal passing a high pass filter are sent to a mixer, the signal variation can be extracted. The deviating signal position on the interference signal trace corresponds to the disturbance position. Averaging several interference signal traces can reduce noise (e.g., phase noise of the laser, random polarization, thermal and shot noise) for better resolution of detection.

The φ-OTDR technique has obtained great success in long pipeline monitoring (Shi et al., 2014) and intrusion detection along large area perimeters (Juarez et al., 2005; Juarez and Taylor, 2007). Shi et al. (2014) applied a φ-OTDR system with a modified wavelet thresholding method to

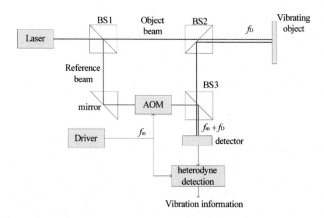

Figure 2.1.46 Model of the laser Doppler interferometer

a pipeline as an early warning system. Field experiments showed it to be effective reaching 40 m of position resolution over 23 km distance of sensing (Shi et al., 2014). Juarez and Taylor (2007) conducted a field demonstration of the distributed φ-OTDR sensor over a long cable with length of 19 km buried in desert terrain at the US Marine Corps Air Station in Yuma, Arizona. High sensitivity and consistent detection of intruders on foot and of vehicles traveling down a road near the cable line was achieved in real time (Juarez and Taylor, 2007).

2.1.4.1.3 Remote Sensing With Laser Doppler Technology

Laser Doppler vibrometer (LDV) has some advantages over traditional methods of vibration detection, including high sensitivity, large dynamic range, large frequency response range, and long operational distance. An LDV can be mounted on a plane, helicopter, or a fire balloon, which greatly increases its operating distance as a remote sensing device. Applying laser Doppler technology in underground measurements to achieve remote vibration sensing may offer additional advantages over traditional methods.

LDV is based on the principle of laser Doppler interferometry (Li et al., 2012, 2011). In an LDV heterodyning interferometer a coherent laser beam is divided into an object and a reference beam by a beam splitter BS1, as shown in Fig. 2.1.46. The object beam strikes a point on the target object and the light reflected from that point travels back to beam splitter BS2 and interferes with the reference beam, as the reference beam passes through a Bragg cell to obtain a frequency shift modulation f_m at BS3. If the target object is moving or vibrating, this mixing process produces an intensity fluctuation in the light. A detector converts the interference signal to a voltage fluctuation, which can be expressed as:

$$V = K\cos\left[2\pi\left(f_{\mathrm{m}} + f_{\mathrm{D}}\right)t\right], \tag{2.1.18}$$

$$f_{\mathrm{D}} = \frac{2v}{\lambda} \tag{2.1.19}$$

where K is the photoelectric conversion efficiency of detector, λ is the laser wavelength, f_{D} is Doppler frequency proportional to the velocity v of the object. A signal process of heterodyne detection is conducted to demodulate the Doppler frequency from which the vibration amplitude and frequency of the target object can be obtained.

Very few reports of application of LDV systems to underground measurements are available. Berni (1992, 1991) adopted LDV systems to remotely sense the motions of the earth's surface by locating inertial reflectors on the surface of the earth to enhance the signal. Experimental results showed successful measurements at a distance of 162 m within the frequency response range of 0~100 Hz and accuracy of 2.3 μm/s. Lei et al. (2006) added a new signal processing system to LDV for seismic exploration, which improved the resolution of vibration acquisition to 20 nm~2 mm within the linear frequency range of 1~1000 Hz.

2.1.4.2 Strain/Stress Measurement

2.1.4.2.1 FBG for Strain Sensing

FBG sensors are some of the most popular devices among fiber-optic strain/stress sensors, whose principle of operation was introduced earlier in the discussion of fiber-optic vibration sensors (FOVS). For strain measurements, instead of attaching the FBG assembly onto the target surface as was the case for FOVS, FBGs are often attached onto a well-characterized substrate, which is then mounted onto the target surface for strain detection. For example, Sato et al. (2001) attached FBG sensors on a bronze plate with adhesive and installed this device vertically into the ground. The deformation of the ground was detected as the deformation of the plate, and subsequently detected by the FBG strain sensors mounted on the plate. Wnuk and Mendez (2005) placed an FBG sensor arrays in a thin teflon tubes without any adhesion to the tubes' walls. The tubes were then embedded in glass–fiber-reinforced polymer rods to construct rockbolt sensors. These rockbolt sensors were installed in a test tunnel in Switzerland. The

FBG strain sensors performed well throughout a year-long strain monitoring.

In harsh environment applications, the thermal expansion and shrinkage of the substrate itself may induce extra strains on the FBG arrays if they are attached on the substrate with an adhesive material. The properties of the adhesive material should also be taken into consideration to minimize measurement errors. Therefore the mounting and packaging of FBG elements is an important area of development for accurate measurement of physical parameters using these systems. Wnuk and Mendez (2005) proposed and tested different FBG packaging designs, including use of a super alloy shim substrate which showed good strain transfer characteristics, oxidation resistance against changing temperature and moisture, as well as matched thermal expansion with the detected structural material. In their design a ceramic cement compound was preferred as the adhesive material than the more frequently used polymeric bonding compound.

Temperature and strain variations induce changes simultaneously in FBG spectra of the transmitted or reflected light as expressed in Eq. (2.1.20):

$$\frac{\Delta\lambda_b}{\lambda_b} = (\alpha + \xi)\Delta T + (1 - p_e)\varepsilon \qquad (2.1.20)$$

where λ_b is the center wavelength of reflected light by FBG and $\Delta\lambda_b$ is the center wavelength shift caused by temperature and strain variation, α is the thermal-expansion, ξ is the thermo-optical coefficients of FBG, ε is the longitudinal strain experienced by the optical fiber at FBG location, and p_e is the effective photo-elastic constant of the fiber core material.

Several techniques have been developed to eliminate the influence of temperature variation from strain/stress sensing, including the use of gratings with two main wavelengths, canceling of the grating thermal response, and use of reference gratings (free from the influence of deformation) measuring only temperature (Wnuk and Mendez, 2005; Mokhtar et al., 2012). Fig. 2.1.47A shows a sensor element package design with a reference grating. Reference grating is enclosed in a thin metal envelope to protect and isolate the grating from any mechanical stresses (Mokhtar et al., 2012). Figs. 2.1.47B–C show the strain and temperature calibration results of this particular strain sensor, where FBG1 is the test grating and FBG2 is the reference grating. It can be observed that the reference grating (FBG2) is free from strain influence but responsive only to temperature. The temperature calibration of FBG2 can be used to eliminate the influence of temperature on FBG1 sensing for strain.

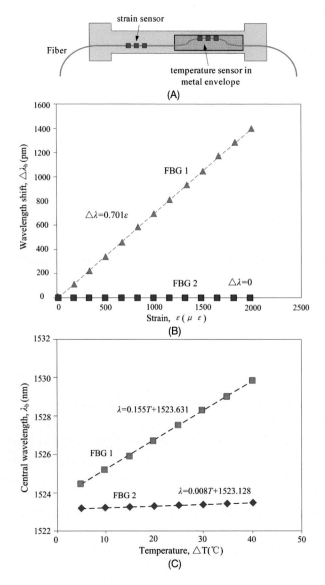

Figure 2.1.47 Strain sensor package design of using reference grating and its temperature/strain calibration (Mokhtar et al., 2012). (A) Schematic diagram of the FBG strain sensor packaging design. (B) Strain calibration. (C) Temperature calibration

2.1.4.2.2 BOTDR for Strain/Stress Sensing

Distributed optical fiber sensing based on BOTDR (Brillouin optical time domain reflectometry) has been an important development to achieve event

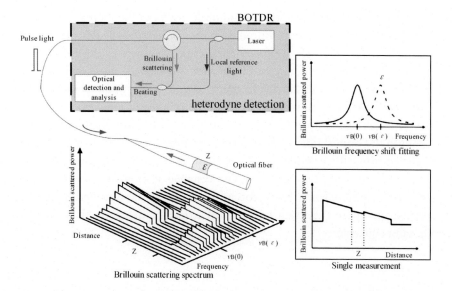

Figure 2.1.48 Schematic diagram of BOTDR (Zhang et al., 2004)

measurements at any point along a single optical fiber without the need to strategically place the sensing elements, such as the FBGs. When a strong optical power transmits along the optical fiber, a small fraction of light generates three kinds of scattered light as a result of the material defects and/or inclusion of noncrystalline materials in micro-dimension: Rayleigh scattering, Raman scattering, and Brillouin scattering. Among the three, Brillouin scattering is closely related to the straining of the fiber, therefore used as the source of the sensing signal in BOTDR systems of distributed strain sensing.

Eq. (2.1.21) expresses a relationship between the frequency shift variation of the Brillouin scattering light and the axial strain of the fiber:

$$f_B(\varepsilon) = f_B(0) + \frac{df_B(\varepsilon)}{d\varepsilon} \varepsilon \qquad (2.1.21)$$

where $f_B(\varepsilon)$ and $f_B(0)$ denote the Brillouin scattering light frequency when the optical fiber is under strain occurs and zero strain, respectively. ε is the fiber axial strain, and $df_B(\varepsilon)/d\varepsilon$ is its scale factor to Brillouin scattering frequency.

A BOTDR analyzer is used measure the frequency drift of Brillouin scattering due to a strain occurring at a section along an optical fiber, as shown in Fig. 2.1.48 (Zhang et al., 2004). In this device, a pulse of light is launched into one end of the optical fiber and the power of the spontaneous Brillouin backscattered light is measured by means of heterodyne detection in the time domain. A single measurement result is shown in the lower right corner. Then the frequency of the incident light is changed slightly and the same measurements are made repeatedly at many frequencies to obtain the Brillouin spectrum (left figure). The frequency that gives the peak power is calculated by fitting the spectrum to a Lorentzian curve at every point in the optical fiber (figure in the top right corner). The strain is then obtained from that frequency shift of the peak from the characteristic peak frequency by using Eq. (2.1.21). The distance Z from the position where the pulse light is launched to the position where the scattered light is generated, or where the strain had occurred, can be determined using the following equation:

$$Z = \frac{cT}{2n} \qquad (2.1.22)$$

where c is the light velocity in vacuum, n is the refractive index of the optical fiber, and T is the time interval between launching the pulsed light and receiving the scattered light at the end of the optical fiber. Soto et al. (2016) reported on BOTDR measurement of strains in El Teniente (Codelco) mine in Chile, where the measurement range was 49.4 m with a resolution is 0.1 m.

2.1.4.3 Temperature Measurement
2.1.4.3.1 FBG for Temperature Sensing

The FBGs detect thermal changes as their refractive index changes by the thermo-optic effect, and their grating pitch changes by thermal expansion (Xiang et al., 2011). All these changes influence the spectrum of the transmitted or the reflected light, and the FBG Bragg wavelength change correlates directly with the temperature change.

A typical design of FBG array used in distributed sensing of temperature is shown in Fig. 2.1.49. Light from a broadband light source is coupled into the optical fiber with the FBG array, in which each FBG has a characteristic nominal center wavelength. A series of narrow-band light with different center wavelengths will be reflected by the matching FBG. The reflected light is demodulated by an FBG interrogation, discussed in an earlier section (Ferraroa and Natale, 2002). Norman et al. (1992) designed a distributed FBG temperature sensor system for underground mass transit system in London where they achieved reliable thermal monitoring. Xiang

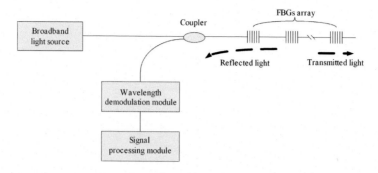

Figure 2.1.49 Traditional design of FBG distributed temperature system

et al. (2011) proposed a distributed FBG temperature sensor system with more than 20 temperature measurement points on a single-mode fiber. This system had the potential to expand to accommodate up to 1200 temperature measurement points.

Another major advantage of using FBGs in temperature sensing is their stability under harsh environmental conditions, such as elevated temperatures, which cover a range of temperatures above 100°C. With advancements in FBG manufacturing, such as optimizing doping levels (Dong et al., 1993), choosing suitable coatings, recording and characterization of different grating types, high temperature-stable FBGs up to 1000°C with improved spectral properties have been achieved (Bartelt et al., 2007).

2.1.4.3.2 Raman Scattering Based Fiber-Optic Temperature Sensing

As introduced above, Rayleigh scattering, Brillouin scattering, and Raman scattering occur when the laser pulse interacts with the fiber optic molecules. Raman scattering occurs due to the energy exchange of thermal vibration of the material molecules and the photon interactions. Since the number of vibrating molecules determines the temperature characteristics of the material, Raman scattering can be used to detect temperature (Yang and Zhu, 2010).

A typical design of Raman scattering based fiber-optic temperature sensing system is shown in Fig. 2.1.50 (Liu and Lei, 2012). When a laser pulse of frequency f_0 is sent into the fiber, due to Raman back-scattering the Stokes and the anti-Stokes lights appear. The intensity ratio of the Stokes light $I_s(T)$ and the anti-Stokes light $I_a(T)$ is correlated with temper-

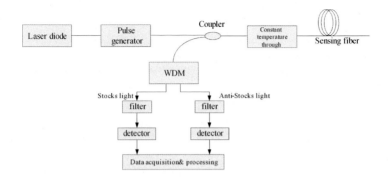

Figure 2.1.50 Fiber-optic Raman temperature sensing system (Liu and Lei, 2012)

ature. The relation is expressed as:

$$\frac{I_a(T)}{I_s(T)} = \left(\frac{f_a}{f_s}\right)^4 \exp\left[-\left(\frac{h\Delta f}{kT}\right)\right] \tag{2.1.23}$$

where f_s and f_a are the frequencies of the Stokes light and the anti-Stokes light, respectively, which can be expressed as $f_s = f_0 - \Delta f$, $f_a = f_0 + \Delta f$, and Δf is the Raman frequency shift, k is Boltzmann constant, and h is Planck constant. The intensity ratio is determined at a reference temperature T_0 as $I_a(T_0)/I_s(T_0)$ and is used as a calibration reference. Measuring the intensity ratio ($I_a(T)/I_s(T)$) at the detected temperature T then provides an expression for T as follows:

$$T = \frac{h\Delta f T_0}{h\Delta f - kT_0 \ln\left[\frac{I_a(T)I_s(T_0)}{I_a(T_0)I_s(T)}\right]}. \tag{2.1.24}$$

The fiber-optic Raman temperature sensing systems have been used widely in underground measurements. Liu and Lei (2012) applied the fiber-optic Raman temperature sensing system for Dongtan underground coal mine safety monitoring in China (Liu and Lei, 2012). The use of the distributed fiber optic temperature sensing system laid over 2500 meter length of various sections of the tunnel and mining area is illustrated with the record of temperature obtained in Fig. 2.1.51. A general increase in temperature could be observed when the measurement position moved deeper along the sensing fiber cable laid from monitor room to coal mine goaf.

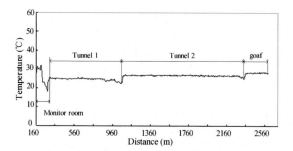

Figure 2.1.51 Temperature distribution measurement in Dongtan underground coal mine using a Raman scattering fiber-optic sensor system (Liu and Lei, 2012)

2.1.4.4 Gas Detection

Detection of flammable gases has always been a serious requirement in underground operations, including methane concentration measurement in tunnel or coal mine safety monitoring. Considering methane monitoring as an example, the traditional methane sensors (e.g., catalytic combustion sensors, semiconductor sensors, and piezoelectric sensors) have their own inherited safety weaknesses, such as the potential of generating sparks during detection and signal transmitting processes (Zhou and Chen, 2014). The no additional explosion risk requirement is always a technical challenge for most of the traditional gas sensors.

Optical sensors are well known for their safety in sensing flammable gases. The laser absorption spectroscopy (LAS) based technologies are the most popular ones used by the industry and have the ability of detecting many different gases, including CO, CO_2, SO_2, NO_2, NH_3, CH_4, C_2H_2, C_3H_8, etc.

The design of an LAS based gas detection system is shown in Fig. 2.1.52 (Yang and Liu, 2013). In this device, a narrow-band light source sends monochromatic light, at frequency f, to pass through a chamber filled with the gas being detected. Because of the interaction between the photons and molecules of the gas medium, part of the light will be absorbed and part scattered. The transmitted light intensity $I(f)$ is given by Beer–Lambert law:

$$I(f) = I_0(f) \exp(-\alpha(f)LC) \tag{2.1.25}$$

where $I_0(f)$ is the incident intensify from light source, $\alpha(f)$ is the absorption coefficient of the gas being detected to light at frequency f, L is the gas chamber length, and C is the gas concentration. Hence if L and $\alpha(f)$

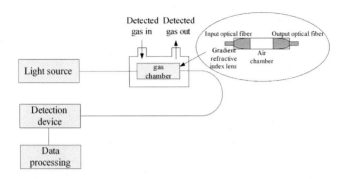

Figure 2.1.52 Spectrum absorption type optic gas sensor (Yang and Liu, 2013)

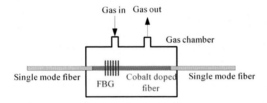

Figure 2.1.53 Novel fiber optic methane sensor based on thermal conductivity detection (Zhou et al., 2014)

are known, then by measuring $I_0(f)$ and $I(f)$, the concentration of gas, C, can be obtained. However, because of some disturbance factors (e.g., fluctuation of light intensity, instability of the light path, and other variations in measurement environment), it is often difficult to determine exactly the gas concentration. In practice this problem can be solved by: (1) using the frequency modulation technique (scanning the frequency of the light across the absorbing medium) to cancel out the noise generated by fluctuation of light intensity at the low frequency ranges (Han and Tian, 2009); (2) increasing the effective path length in the gas chamber, such as by placing specific elements inside the chamber to reflect light beam back and forth multiple times (Nelson et al., 2004).

In some applications, gases in narrow and small spaces need to be measured, such as in curved tunnels of coal mines or near pipeline routes. A novel fiber optic gas sensor was proposed for natural gas leakage along an underground pipeline in Zhou et al. (2014). In this device the sensing element was composed of a gas chamber, a cobalt doped fiber and a temperature sensing FBG, as illustrated in Fig. 2.1.53. The cobalt doped fiber can effectively transform strong optical power into heat by nonradiation processes (Davis and Digonnet, 2000). The heat is then diffused into the gas filled chamber and the temperature change is detected by the FBG sensor. Because the temperature around and in the cobalt doped fiber is sensitive to the conductivity coefficient of the surrounding gas, and in turn its concentration, the measurement of the temperature can directly be related to gas concentration and its variation (Zhou and Chen, 2014; Xia and Zhou, 2008). This sensor was applied to coal mine gas monitoring where methane concentrations could be measured up to 6.4% with concentration variation sensitivity of 1% (Xia and Zhou, 2008).

2.1.4.5 Examples of Practical Applications of Optical Sensor Technologies in Underground Measurements

In this section, a few practical applications of optical sensor technologies (OSTs) in underground measurement are introduced. In some cases, OSTs were multiplexed to sense several parameters simultaneously.

2.1.4.5.1 Earthquake Observation

Earthquake observation on the seafloor at Uchiura Bay in Japan (Shindo and Yposhikawa, 2001; Watanabe and Takahashi, 2004) was conducted using fiber-optic vibration sensor arrays based on the Michelson interferometer (FODVSs were discussed in Section 2.1.4.1.1). In this application, the wide-area earthquake observation network laid on the seafloor was basically composed of optical fiber, including the sensor head and the transmission line. A 3-axis fiber-optic sensor head with a pressure tank was installed on the seafloor at 37 m of depth. On November 3 1999, the system successfully detected and recorded an earthquake of M4.1, with epicenter West Side of Tokyo, according to Japan Meteorology Agency report. The records of this earthquake obtained by three different types of sensor, a moving coil type velocity meter, a piezo type accelerometer, and the fiber-optic accelerometer, are compared in Figs. 2.1.54A–C, respectively. Among the three, the noise level of the optic sensor was found to be less than 2 μg, lower than those of other reference sensors (Shindo and Yposhikawa, 2001). Subsequently, in 2003, a mega thrust earthquake (Tokachi-oki earthquake, MJMA 8.0) was recorded by the fiber-optic system again, which was the first ocean floor recording of an earthquake with magnitude eight in the world (Watanabe and Takahashi, 2004).

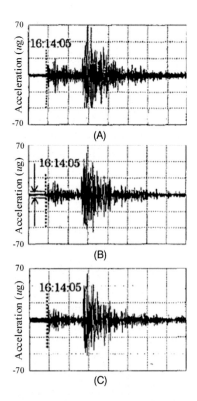

Figure 2.1.54 Comparison of observation results using three types of sensor on an earthquake (Shindo and Yposhikawa, 2001). (A) Moving coil type velocity meter. (B) Fiber optic accelerometer. (C) Piezo type accelerometer

2.1.4.5.2 Mineral Exploration

Owing to mineral exploration activity in increasingly deeper wells, durable sensors that can work in high-temperature and hostile environments are in high demand. Fiber-optic sensors meet most of requirements of these demanding applications with their high temperature capacity, multiplexed and distributed sensing and small space placement capabilities. The OSTs used in mineral exploration include three types: FOVS for seismic detection (Zhang and Li, 2009), distributed fiber-optic sensor array for downhole temperature detection (Zhang et al., 2009; Schroeder, 2002), and fiber-optical gyroscope (FOG) for directional measurement-while-drilling surveying (Ledroz et al., 2005).

Examples of FOVS used in seismic detection have been introduced previously. In this section, a distributed fiber-optic temperature sensor system

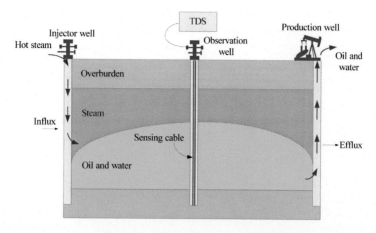

Figure 2.1.55 Scenarios for movement of injected steam (Zhang et al., 2009)

used in a steam–injected heavy-oil recovery well (Zhang et al., 2009), and an FOG based oil drilling navigation system (Ledroz et al., 2005) will be introduced. Fig. 2.1.55 shows a Raman scattering based fiber optic distributed temperature sensing system used in a steam-injected heavy-oil recovery well for monitoring the movement of injected steam. In a steam-injected heavy-oil recovery operation, hot steam is injected into a well at very high temperatures (exceeding 250°C). The oil is gradually heated to increase its flow-ability at a decreased viscosity. The hot flowing oil spreads laterally and is pumped out from the production well. Because of the high-temperature environment and long periods (weeks to months) of steam injection required to mobilize the oil, a Raman scattering based distributed fiber-optic temperature sensor is deemed most suitable to install in a steam flood observation well to monitor the whole process.

Fig. 2.1.56 shows the fiber optic temperature measurement distribution with depth and time along the gradual steam flow path (Zhang et al., 2009). It can be observed that there are three siltstone layers in the steam flow path. At the beginning of the monitoring period, the hot steam reaches the lower two siltstone layers at the bottom of the observation well. As the temperature at the bottom siltstone layers increases, the hot steam gradually moves upwards and breaks through the third siltstone layer after 15 months.

Horizontal measurement-while-drilling (MWD) is another practical technique in oil exploration and production due to its higher productivity and longevity compared to the traditional vertical drilling. Present MWD technology utilizes magnetic sensors and accelerometers to provide position

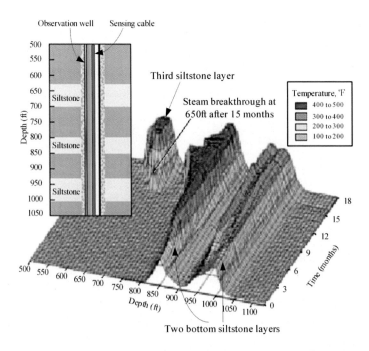

Figure 2.1.56 A realistic fiber-optic temperature monitoring in steam flood observation well (Zhang et al., 2009)

and direction to underground drilling operation. The magnetic surveying sensors are easily affected by a variety of external factors (e.g., downhole ore deposits and geomagnetic influences), which can lead to navigation errors. An excellent substitute to magnetic sensor is FOG, selected because of its relatively small size, high reliability, and relatively low susceptibility to vibration conditions and temperature dynamics. FOG is a rotational rate sensor which can be adapted into an inertial navigation setup to provide the direction and movement trace of drilling for oilfield downhole surveying. An FOG-inertial navigation unit (FOG-IMU) mounted in an inclinometer for oil logging is shown in Fig. 2.1.57. Ledroz et al. (2005) explored the feasibility of utilizing FOG-IMU in downhole environment for MWD processes, including performance test in alignment and real-time navigation, performance evaluation under the vibrations and high temperature conditions of drilling operations. Their analysis indicated that the integration of the FOG-IMU required minor changes in the presently used drilling tools, and resulted in lower costs and improved accuracy (Ledroz et al., 2005).

Inclinometer in drilling tools for oil logging

FOG-IMU

Figure 2.1.57 FOG-IMU mounted in inclinometer in drilling tools

2.1.4.5.3 Underground Pipeline Monitoring

Applications of OSTs in underground pipeline monitoring include two aspects, pipeline leakage detection and pipeline intrusion detection. Major accidents can result from internal leaks in pipelines that transport high-pressure contents. Leaks in pipelines can result in enormous financial loss to the industry and adversely affect public health. Hence, leak detection and localization is a major concern for researchers studying pipeline systems. As introduced in Section 2.1.2.2, acoustic emission (AE) is a practical method for leak detection and localization. Due to their higher sensitivity and wider frequency response range, FOVS arrays have been used to capture the acoustic waves generated when a leakage occurs to locate the leakage (Liu and Li, 2015; Jin and Zhang, 2014).

Fiber optic distributed vibration sensing system is another good choice for AE measurement. A distributed micro-vibration sensor based on DMZI was developed by using three single-mode fibers (two for sensing and one for signal transportation) laid side by side, to detect leakage noise along a pipeline. The principle of this measurement was introduced earlier in Section 2.1.4.1.2. Fig. 2.1.58 gives an example of leakage detection using a DMZI system in a pipeline of 50 km length (Shan and Wang, 2007). The traces of the detected interference signals from two channels of the DMZI and their correlation efficiency are shown in Fig. 2.1.58A. Using the correlation peak the leakage position was determined to be at 19.14 km mark, as shown in Fig. 2.1.58B. Comparing to the realistic leakage location of 19 km, the leakage positioning error was 140 m. The data plot in Fig. 2.1.58B also showed that the intensity of leakage increased with time.

An example of using DMZI system for pipeline intrusion detection has been introduced in Section 2.1.4.1.2, in which the underground oil pipeline safety monitoring and prewarning system covered a monitoring distance of 43 km, with intrusion positioning accuracy of 10 meters (An et al., 2012).

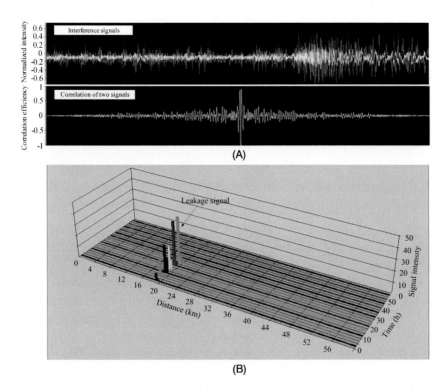

Figure 2.1.58 Leakage detecting and locating in pipeline (Shan and Wang, 2007). (A) Detected signals in two channels of DMZI. (B) 3D illustration of pipeline leakage

2.1.4.5.4 Geological Disaster Warning

Slope instability can lead to landslides and falling rocks which often invade transportation facilities and pose threat to lives and property. Many advanced techniques have been proposed and used for geological disaster detection and forecast warning, including OSTs.

An example of a distributed FBGs array strain sensor used to monitor slope stability in Yongtaiwen Highway (China) is discussed here (Guo, 2014). A schematic diagram of the slope and the installed sensing system is shown in Fig. 2.1.59. FBG sensors were used to monitor micro–changes in displacements and induced strains in soil at various sections of a slope. Securely packaged FBG sensors were mounted in rockbolts which were buried into the slope surface. As a collapse was triggered, soil and rock layers would tend to slide downward, leading to increased lateral and shear stresses on the rockbolts. The FBG arrays detected the progressive stress buildup on the rockbolts, hence acted as a

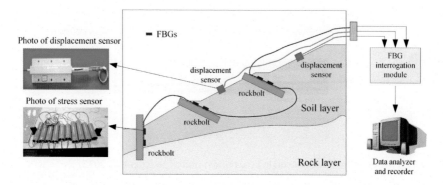

Figure 2.1.59 FBG sensors used in slope structure state monitoring (Guo, 2014)

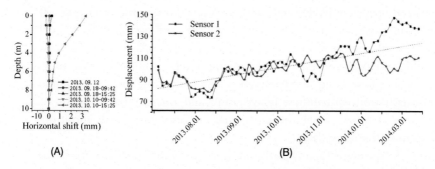

Figure 2.1.60 Slope stability monitoring result in Yongtaiwen Highway by using FBG based sensors (Guo, 2014). (A) FBG based rockbolt monitoring result. (B) FBG based displacement sensor monitoring result

warning system for impending slope instability. Moreover, displacement sensors, also based on FBGs, corroborated the movement of the soil and rock.

Fig. 2.1.60A shows deflection measurements by an FBG strain sensor array mounted on a rockbolt. The deflection monitoring was conducted over time along the length of the bolt, from surface to 10 m of depth. As observed, a clear horizontal shift occurred at 15:25, on Oct. 10th, which was found to be the triggering cause of a landslide that occurred at a later time. Fig. 2.1.60B shows the time trace of two FBG displacement sensors monitoring over six months. It clearly indicates increasing soil movement occurring over time as the response of one sensor increasingly deviates from the other passed the third month of monitoring.

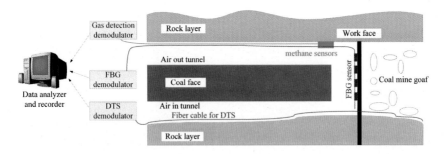

Figure 2.1.61 Fiber-optic sensors used in coal mine monitoring (Liu et al., 2013)

2.1.4.5.5 Coal Mine Safety Monitoring

It is crucial to have reliable safety monitoring systems which can give off early warning signs to prevent serious accidents and casualties from coal mining hazards such as methane explosion, fire, roof collapse, or rock burst (Liu et al., 2013). A good safety monitoring system should consist of multiple sensors monitoring key parameters simultaneously, such as methane gas concentration, distributed temperature and micro-seismic events or micro-stress changes, to help operators make informed decisions about the conditions in the mine daily.

The effectiveness of many coal mining safety monitoring systems based on traditional sensors (e.g., catalyst combustion sensor, inductive coil sensor, piezoelectric sensor) can be limited primarily by three factors (Liu et al., 2013): (1) poor accuracy and reliability; (2) inadequate distributed operation leading to blind monitoring areas; (3) potential safety hazard due to electric sparks generated by electric signal transmission. As discussed earlier, optical sensors can overcome some of these disadvantages.

Liu et al. (2013) introduced a series of fiber-optic sensors deployed in coal mine safety monitoring in Dongtan Coal Mine of Yankuang Group, China (Liu et al., 2013). The layout of the sensor system in the coal mine is shown in Fig. 2.1.61.

The fiber-optic methane sensors were based on laser absorption spectroscopy (see Section 2.1.4.4) and they were deployed for tunnel gas monitoring. Meanwhile, an electrical gas analyzer was also mounted next to a fiber optic methane sensor to compare performance. The concentration measurements from both sensors are shown in Fig. 2.1.62. It was demonstrated that the fiber-optic methane sensor had similar measurement performance as the electrical sensor. During a two-year field monitoring, fiber-optic gas sensor remained stable and accurate, which was a signifi-

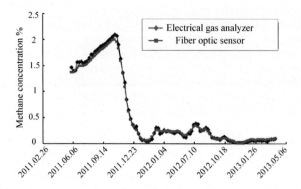

Figure 2.1.62 Long term field test data comparison of two gas sensor (Liu et al., 2013)

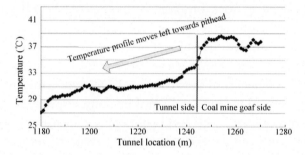

Figure 2.1.63 Field test data of temperature distribution side coal mine goaf (Liu et al., 2013)

cant improvement compared to the weekly calibration requirement of the conventional electrical sensor.

Besides methane gas monitoring, fiber-optic distributed temperature sensors were used to continuously monitor temperature in the goaf region of the mine for potential coal combustion hazard. Sensing cables were laid along the tunnel face. As the coal production continued, part of the cable network had to be left embedded inside the goaf hence the sensor would continue monitoring the temperature on both sides of the work face. Fig. 2.1.63 shows a temperature distribution profile obtained by the sensor network in the tunnel and the goaf region at a time. It can be observed that the temperature on the goaf side is higher than the tunnel side, and the temperature profile decreases towards the pithead as shown. This temperature distribution confirmed that the oxidization zone moves forward and in agreement with the goaf combustion theory (Liu et al., 2013).

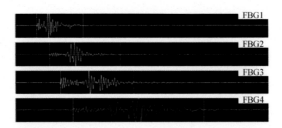

Figure 2.1.64 Micro-seismic signal detected by FBG sensors array (Liu et al., 2013)

In deep underground mining, gas, coal, or rock burst is another major safety hazard. A rock burst event creates acoustic emissions and micro-seismic activity caused by rock breakage or movement. An FBG based fiber–optic vibration sensor array laid on the coal goaf surface can help detect the location and magnitude of such a seismic event. In a field test, a simulated coal mine micro–seismic event source properties were detected by four FBG sensing elements, as shown in Fig. 2.1.64, where the horizontal and vertical axes stand for time and signal intensity, respectively. Using optimized least squares method for event positioning and analyzing its energy, burst events with energy levels above 100 J could be located with spatial resolution error of ±11 m (Liu et al., 2013).

2.1.5 CONCLUSIONS

Due to capabilities of transmission in underground, an introduction on wave based underground measurement methods, including field, acoustic, and electromagnetic waves was presented. Among them, acoustic measurement methods are the most widely used in practical applications. Therefore some typical acoustic measurement methods were discussed with more detail, including their principles of operation and application examples. Electrical and electromagnetic methods are also widely adopted in geological exploration and subsurface feature detection. Several routine approaches of these methods and their applications were discussed. Because of the multiple benefits and advantages offered over the traditional methods, optical sensing techniques were discussed with sufficient detail with their applications examples.

REFERENCES

Allred, B., Fausey, N., 2004. GPR detection of drainage pipes in farmlands. In: Slob, E. (Ed.), Proceeding of the 10th International Conference Ground Penetrating Radar. Delft, The Netherlands, 21–24 June, pp. 307–310.

An, Y., Feng, H., Zhou, Y., 2012. A control method to eliminate polarization-induced phase distortion in dual Mach–Zehnder fiber interferometer. In: Liu, G., Wang, Z. (Eds.), Proceeding of UKACC International Conference on Control. Cardiff, UK, 3–5 September, pp. 988–991.

Aning, A., Sackey, N., 2014. Electrical resistivity as a geophysical mapping tool; a case study of the new art department, Knust-Ghana. Int. J. Sci. Res. Publ. 4 (1), 1–7.

Araya, A., Kanazawa, T., Shinohara, M., 2012. Gravity gradiometer implemented in AUV for detection of seafloor massive sulfides. In: Proceedings of Oceans. Hampton Roads, US, 14–19 Oct., pp. 1–4.

Bartelt, H., Schuster, K., Unger, S., 2007. Single-pulse fiber Bragg gratings and specific coatings for use at elevated temperatures. Appl. Opt. 46 (17), 3417–3424.

Berni, A., Remote seismic sensing, US Patent: 5109362, 1992-4-28.

Berni, A., Remote seismic sensing, US Patent: 5070483, 1991-12-3.

Biot, M., 1956. Theory of propagation of elastic waves in a fluid-saturated porous solid. II: Higher frequency range. J. Acoust. Soc. Am. 28, 179–191.

Brennan, M., Gao, Y., 2007. On the relationship between time and frequency domain methods in time delay estimation for leak detection in water distribution pipes. J. Sound Vib. 304, 213–223.

Brent, T., 2009. Reflection seismic data from legacy hydrocarbon exploration of cenozoic and older basins of the Canadian High Arctic. Recorder 34 (9), 5–7.

Chandra, V., Tiwari, U., Das, B., 2016. Elimination of light intensity noise using dual-channel scheme for fiber MZI-based FBG sensor interrogation. IEEE Sens. J. 16 (8), 2431–2436.

Chen, Q., Liu, T., Liu, K., 2013. An elimination method of polarization-induced phase shift and fading in dual Mach–Zehnder interferometry disturbance sensing system. J. Lightwave Technol. 31 (19), 3135–3141.

Childers, V., 2016. Gravimetric measurement. In: Lastovicka, J. (Ed.), Geophysics and Geochemistry, vol. 3. UNESCO-EOLSS Joint Committee, EOLSS host, viewed on 20 June.

Cranch, G., Nash, P., 2000. High-responsivity fiber-optic flexural disk accelerometers. J. Lightwave Technol. 18 (9), 1233–1243.

Danescu, L., Morega, A., Morega, M., 2013. New concept of measurement apparatus for the in situ electrical resistivity of concrete structures. In: Proceeding of 8th International Symposium on Advanced Topics in Electrical Engineering. Bucharest, Romania, 23–25 May, pp. 1–6.

Davis, M., Digonnet, M., 2000. Measurements of thermal effects in fibers doped with cobalt and vanadium. J. Lightwave Technol. 18 (2), 161–165.

Demick, E., Luukkonen, K., Nonis, S., 2013. Design to improve the productivity and execution of gravity surveys. In: Donohue, S. (Ed.), Proceedings of the IEEE Systems and Information Engineering Design Symposium. Charlottesville, USA, April 26, pp. 179–183.

Dickey, J., Mikhael, W., 2012. An adaptive technique for isolating the seismic response of an infrasound sensor. In: Proceeding of 55th International Midwest Symposium on Circuits and Systems (MWSCAS). Boise, US, 5–8 Aug., pp. 1548–3746.

Dong, L., Archambault, J., Reekie, L., 1993. Bragg gratings in Ce3-doped fibers written by a single excimer pulse. Opt. Lett. 18 (11), 861–863.

Ferraroa, P., Natale, G., 2002. On the possible use of optical fiber Bragg gratings as strain sensors for geodynamical monitoring. Opt. Lasers Eng. 37, 115–130.

Freal, J., Zarobila, C., Davis, C., 1987. A microbend horizontal accelerometer for borehole deployment. J. Lightwave Technol. Lt-5 (7), 993–996.

Guan, H., Ye, X., Wang, S., 2012. Environmental geological applications of ground penetrating radar. In: Proceeding of 14th International Conference on Ground Penetrating Radar. Shanghai, China, 4–8 June, pp. 779–784.

Guo, Y., 2014. Research on Monitoring Technology and Application of High-Steep Slope and Rock Fall Based on Fiber Bragg Grating. Wuhan University of Technology, Wuhan, China.

Han, X., Tian, F., 2009. Research on methane gas sensor with spectrum absorption and signal processing method. In: Long, B., Li, W. (Eds.), Proceeding of IEEE Circuits and Systems International Conference on Testing and Diagnosis. Chengdu, China, 28–29 April, pp. 1–4.

Han, J., Zhang, W., Wang, Z., 2014. Fiber optical accelerometer based on 45 degrees Fabry–Perot cavity. In: Liu, T., Jiang, S., Neumann, N. (Eds.), Proc. SPIE, vol. 9274, Beijing, China, 9 October, p. 927418.

Hubscher, C., Gohlb, K., 2014. Reflection/Refraction Seismology. Springer Science & Business Media, Dordrecht.

Jaroszewicz, L., Krajewski, Z., 2007. Fiber-optic rotational seismometer as device for detection the seismic rotational events. In: Morawski, R. (Ed.), Proceeding of Instrumentation and Measurement Technology Conference. Warsaw, Poland, 1–3 May, pp. 1–5.

Ji, M., Jin, F., Zhao, X., 2010. Mine geological hazard multi-dimensional spatial data warehouse construction research. In: Liu, Y., Chen, A. (Eds.), Proceeding of 18th International Conference on Geoinformatics. Beijing, China, 18–20 June, pp. 1–5.

Jiang, A., Zhao, Y., 2008. Experimental study of acoustic emission characteristics of underwater concrete structures. In: Proceeding of Symposium on Piezoelectricity, Acoustic Waves, and Device Applications. Nanjing, China, 5–8 Dec., pp. 252–257.

Jiang, S., Han, C., Huang, T., 2014. Research on an underground electricity cable path detection system. In: Bilof, R. (Ed.), Proceedings of 7th International Conference on Intelligent Computation Technology and Automation. Changsha, China, 25–26 October, pp. 500–504.

Jin, H., Zhang, L., 2014. Integrated leakage detection and localization model for gas pipelines based on the acoustic wave method. J. Loss Prev. Process Ind. 27, 74–88.

Juarez, J., Taylor, H., 2007. Field test of a distributed fiber-optic intrusion sensor system for long perimeters. Appl. Opt. 46 (11), 1968–1971.

Juarez, J., Maier, E., Choi, K., 2005. Distributed fiber-optic intrusion sensor system. J. Lightwave Technol. 23 (6), 2081–2087.

Kang, Z., Yu, Y., 2011. Study on stress and strain and characteristics of acoustic emission in the process of rock failure. In: Wang, S. (Ed.), Proceeding of Second International Conference on Mechanic Automation and Control Engineering (MACE). Hohhot, China, 15–17 July, pp. 7737–7740.

Kim, D., Feng, M., 2007. Real-time structural health monitoring using a novel fiber-optic accelerometer system. IEEE Sens. J. 7 (4), 536–543.

Ledroz, A., Pecht, E., Cramer, D., 2005. FOG-based navigation in downhole environment during horizontal drilling utilizing a complete inertial measurement unit: directional measurement-while-drilling surveying. IEEE Trans. Instrum. Meas. 54 (5), 1997–2006.

Lei, S., Fang, S., Xu, L., 2006. Signal processing system for seismic exploration based on laser Doppler effect. In: Proceeding of 8th International Conference on Signal Processing. Beijing, China, 16–20 Nov., pp. 1–4.

Li, Y., Wan, Z., 2010. Design and analysis for acoustic emission source location algorithm of three-dimensional braided composites. In: Proceeding of International Conference on Measuring Technology and Mechatronics Automation (ICMTMA). Changsha, China, 13–14 March, pp. 918–920.

Li, R., Xiao, W. A simplified polarization control method to eliminate polarization in-duced distortion in dual Mach–Zehnder interferometry disturbance sensing system. In: Proceeding of International Conference on Optical Communications and Networks. Nanjing, China, 3–5 July, 2015, pp. 1–3.

Li, R., Madamopoulos, N., Xiao, W., 2010. Influence of membrane surface shape change on the performance characteristics of a fiber optic microphone. Appl. Opt. 49 (35), 6660–6667.

Li, R., Zhu, Z., Xiao, W., 2011. Vibration characteristics of various surfaces using an LDV for long-range voice acquisition coherence length lasers. IEEE Sens. J. 11 (6), 1415–1422.

Li, R., Zhu, Z., Xiao, W., 2012. Performance comparison of an all-fiber-based laser Doppler vibrometer for remote acoustical signal detection using short and long. Appl. Opt. 51 (21), 5011–5018.

Litter, I., Gray, M., Lam, T., 2010. Optical-fiber accelerometer array: nano-g infrasonic operation in a passive 100 km loop. IEEE Sens. J. 10 (6), 1117–1124.

Liu, Z., 2008. Studies on Landmine Detection Experimental System Based on Acoustic to Seismic Coupling. Tianjin University, Tianjin, China.

Liu, Y., Lei, T., 2012. Application of distributed optical fiber temperature sensing system based on Raman scattering in coal mine safety monitoring. In: Liu, C. (Ed.), Proceeding of Symposium on Photonics and Optoelectronics (SOPO). Shanghai, China, 21–23 May, pp. 1–4.

Liu, C., Li, Y., 2015. A new leak location method based on leakage acoustic waves for oil and gas pipelines. J. Loss Prev. Process Ind. 35, 236–246.

Liu, T., Wei, Y., Song, G., 2013. Advances of optical fiber sensors for coal mine safety monitoring applications. In: Das, S., Kumar, J., Das, K. (Eds.), Proceeding of International Conference on Microwave and Photonics (ICMAP). Dhanbad, India, 13–15 Dec., pp. 1–5.

Lu, Y., Zhu, T., 2010. Distributed vibration sensor based on coherent detection of phase-OTDR. J. Lightwave Technol. 28 (22), 3243–3249.

Manstein, Y., Manstein, A., 2015. Non-invasive measurements for shallow depth soil ex-ploration: development and application of an electromagnetic induction instrument. In: Catelani, M. (Ed.), Proceeding of IEEE International Conference on Instrumentation and Measurement Technology Conference. Pisa, Italy, 11–14 May, pp. 1395–1399.

Martens, K., Walraevens, K., 2009. Tracing soil and groundwater pollution with electro-magnetic profiling and geo-electrical investigations. In: Ritz, K., Dawson, L., Miller, D. (Eds.), Criminal and Environmental Soil Forensics. Springer, Berlin, pp. 181–194.

Milkereit, B., Eaton, D., 2003. 3D seismic imaging for mineral exploration. In: Proceeding of 12th International Workshop of Commission on Controlled-Source Seismology: Deep Seismic Methods. Mountain Lake, US, 8–11 October, pp. 1–4.

Mokhtar, M., Owens, K., Kwasny, J., Taylor, S., 2012. Fiber-optic strain sensor system with temperature compensation for arch bridge condition monitoring. IEEE Sens. J. 12 (5), 1470–1476.

Morgan, F., 2001. Self-Potential and Resistivity for the Detection and Monitoring of Earthen Dam Seepage. Massachusetts Institute of Technology: Department of Earth, Atmospheric and Planetary Sciences, Earth Resources Laboratory.

Nelson, D., McManus, B., Urbanski, S., Herndon, S., Zahniser, M., 2004. High precision measurements of atmospheric nitrous oxide and methane using thermoelectrically cooled mid-infrared quantum cascade lasers and detectors. Spectrochim. Acta, Part A, Mol. Biomol. Spectrosc. 60, 3325–3335.

Norman, S., Barker, D., Jones, J., 1992. Evaluation of optical fiber distributed temperature sensing for underground mass transit systems. In: Proceeding of IEE Colloquium on Fibre Optics Sensor Technology, 29 May, pp. 2/1–2/3.

Nostrand, R., Cook, K., 1966. Interpretation of Resistivity Data, Geological Survey Professional Paper 499. US Government Printing Office.

Pierce, K., Liechty, D., Rittgers, J., 2012. Geophysical Investigations Electrical Resistivity Surveys Santee Basin Aquifer Recharge Study. U.S. Department of the Interior Bureau of Reclamation Technical Service Center, Technical Memorandum No. TM-86-68330-2012-23.

Poczesny, T., Prokopczuk, K., Domanski, A., 2012. Comparison of macro-bend seismic optical fiber accelerometer and ferrule-top cantilever fiber sensor for vibration monitoring. In: Berghmans, F., Mignani, A., Moor, P. (Eds.), Proc. SPIE, vol. 8439, Brussels, Belgium, 16 April, pp. 84392N-1–84392-N-8.

Sabatier, J., Ning, X., 2001. An investigation of acoustic-to-seismic coupling to detect buried antitank landmines. IEEE Trans. Geosci. Remote Sens. 39 (6), 1146–1154.

Samouelian, A., Cousin, I., Tabbagh, A., 2005. Electrical resistivity survey in soil science: a review. Soil Tillage Res. 83, 173–193.

Sato, T., Honda, R., Shibata, S., 2001. Ground strain measuring system using optical fiber sensors. In: Proc. SPIE, vol. 4328, Newport Beach, US, August 6, pp. 35–46.

Schroeder, R., 2002. The present and future of fiber optic sensors for the oilfield service industry: where is there a role. In: Proceeding of 15th Optical Fiber Sensors Conference Technical Digest. Portland, US, 10 May, pp. 39–42.

Shan, S., Wang, L., 2007. Leakage detection of oil pipeline using distributed fiber optic sensor. In: Du, S., Leng, J., Asundi, A. (Eds.), Proc. SPIE, vol. 6423, Harbin, China, 18 July, 64231O.

Shi, Y., Feng, H., An, Y., 2014. Research on wavelet analysis for pipeline pre-warning system based on phase-sensitive optical time domain reflectometry. In: Proceeding of IEEE/ASME International Conference on Advanced Intelligent Mechatronics. Besancon, France, 8–11 July, pp. 1177–1182.

Shindo, Y., Yposhikawa, T., 2001. Earthquake observation on the seafloor by the fiber-optic accelerometer. In: Proceeding of the 4th Pacific Rim Conference on Lasers and Electro-Optics. Chiba, Japan, 15–19 July, pp. I-492–I-493.

Soto, G., Fontbona, J., Cortez, R., Mujica, L., 2016. An online two-stage adaptive algorithm for strain profile estimation from noisy and abruptly changing BOTDR data and application to underground mines, Measurement, to be published, pp. 17.

Sreechaka, G., Bordakov, G., 2008. Reflected wave spectrum analysis for estimation of Earth subsurface fluid properties. In: Imbriale, W. (Ed.), Proceeding of International Symposium on Antennas and Propagation Society. San Diego, US, 5–11 July, pp. 1–4.

Sun, L., Li, Y., 2010. Acoustic emission sound source localization for crack in the pipeline. In: Cao, H., Zhu, X. (Eds.), Proceeding of Control and Decision Conference (CCDC). Xuzhou, China, 26–28 May, pp. 4298–4301.

Sun, Z., Qian, J., 2010. Research of the array model and algorithm in acoustic emission. In: Mahadevan, V., Zhou, J. (Eds.), Proceeding of the 2nd International Conference on Computer Engineering and Technology (ICCET). Chengdu, China, 16–18 April. pp. V2-714–V2-719.

Sun, W., Xu, Q., Zhang, H., 2012. Research on detection and visualization of underground pipelines. In: Gahegan, M. (Ed.), Proceeding of 2nd International Conference on Remote Sensing, Environment and Transportation Engineering (RSETE). Nanjing, China, 1–3 June, pp. 1–4.

Sun, X., Lee, W., Hou, Y., 2014. Underground power cable detection and inspection technology based on magnetic field sensing at ground surface level. IEEE Trans. Magn. 50 (7), 6200605.

Ulrych, T., Sacchi, M., 2005. Information-Based Inversion and Processing with Applications. Elsevier, New York.

Wang, Y., 2013. Study on Numerical Simulation and Inversion of Electrical Resistivity Method for Heavy Metal Contaminated Sites. China University of Mining and Technology, Beijing.

Wang, W., Shan, X., 2010. Dynamic fuzzy clustering for infrasound as a precursor of earthquakes. In: Tan, Z., Wan, Y., Xiang, Z. (Eds.), Proceeding of 3rd International Congress on Image and Signal Processing (CISP). Yantai, China, 16–18 Oct., pp. 3582–3586.

Wang, Y., Li, F., Xiao, H., 2006. Unattended ground sensor system based on fiber optic disk accelerometer. In: Ye, C. (Ed.), Proceeding of Optics Valley of China International Symposium on Optoelectronics. Wuhan, China, 12 Nov., pp. 33–36.

Wang, G., Dong, J., Diao, Y., 2011. Characterizing the infiltration process in low-permeable bedrock by high-density electrical resistivity tomography. In: Proceeding of International Symposium on Water Resource and Environmental Protection (ISWREP). Xi'an, China, 20–22 May, pp. 1972–1974.

Watanabe, T., Takahashi, H., 2004. Seismological monitoring on the 2003 Tokachi-oki earthquake derived from permanent OBSs and land-based observation. In: Proceeding of IEEE Oceans MTTS. Kobe, Japan, 9–12 Nov., pp. 1961–1968.

Watts, A., 1997. Gravity anomalies and magmatism along the western continental margin of the British Isles. J. Geol. Soc. 154, 523–529.

Wnuk, V., Mendez, A., 2005. Process for mounting and packaging of fiber Bragg grating strain sensors for use in harsh environment applications. In: Udd, E. (Ed.), Proc. SPIE, vol. 5758, San Diego, US, 6 March, pp. 46–53.

Xia, T., Zhou, B., 2008. A coal mine security monitoring system based on multiplexed fibre Bragg grating sensors and coherence-multiplexing technique. In: Golovchenko, E., He, S. (Eds.), Proceeding of Optical Fiber Communication & Optoelectronic Exposition & Conference. Shanghai, China, 30 Oct.–Nov., pp. 1–3.

Xiang, X., Tu, P., Zhao, J., 2011. Application of fiber Bragg grating sensor in temperature monitoring of power cable joints. In: Wang, R., Shi, P. (Eds.), Proceeding of International Conference on Electronics, Communications and Control (ICECC). Ningbo, China, 9–11 Sept., pp. 755–757.

Yang, Y., Liu, X., 2013. The analysis and design of optic fiber methane sensor based on the spectrum absorption type. In: Li, Z. (Ed.), Proceeding of 3rd International Conference on Consumer Electronics, Communications and Networks. Xianning, China, 20–22 Nov., pp. 658–660.

Yang, R., Ma, T., 2006. A study on the applications of acoustic emission technique. J. North Univ. China (Natural Sci. Edn.) 27 (5), 456–461.

Yang, L., Zhu, Z., 2010. Design of distributed fiber optical temperature measurement system based on Raman scattering. In: Hong, W., Yang, G., Gu, K. (Eds.), Proceedings of International Symposium on Signals, Systems and Electronics. Nanjing, China, 17–20 Sept., pp. 1–4.

Yao, L., Cao, P., Song, K., 2010. A software design based on distributed architecture for seismic exploration system. In: Luo, Q. (Ed.), Proceeding of Second IITA International Conference on Geoscience and Remote Sensing. Qingdao, China, 28–31 Aug., pp. 567–570.

Yilmaz, O., 2001. Seismic Data Analysis: Processing, Inversion and Interpretation of Seismic Data. Society of Exploration Geophysicists, Tulsa.

Yordkayhun, S., Suwan, J., 2012. A university-developed seismic source for shallow seismic surveys. J. Appl. Geophys. 82, 110–118.

Zeng, N., Shi, C., Zhang, M., 2004. A 3-component fiber-optic accelerometer for well logging. Opt. Commun. 234, 153–162.

Zhang, L., 2012. Research on Concrete Damage Detection by Using Acoustic Emission Technology. Dalian Maritime University, Dalian, China.

Zhang, Y., Li, S., 2005. Fiber Bragg grating sensors for seismic wave detection. In: Proc. SPIE, vol. 5855, Bruges, Belgium, 23 May, pp. 1008–1011.

Zhang, W., Li, X., 2009. Underwater fiber laser geophone: theory and experiment. In: Proceeding of Asia Communications and Photonics Conference and Exhibition. Shanghai, China, 2–6 Nov., pp. 1–2.

Zhang, D., Shi, B., Cui, H., Xu, H., 2004. Improvement of spatial resolution of Brillouin optical time domain reflectometer using spectral decomposition. Opt. Appl. 34 (2), 291–301.

Zhang, Y., Ning, J., Yang, S., 2009. Fiber-optic sensors for the exploration of oil and gas. In: Proceeding of 14th Opto-Electronics and Communications Conference. Hong Kong, China, 13–17 July, pp. 1–2.

Zhang, B., Yi, C., Xie, G., 2011. The Method for Seismic Data Processing, second edn. Petroleum Industry Press, Beijing (Chapter 4).

Zhong, M., Long, Y., Zhang, W., 2009. Multi-fractal analysis of the explosion seismic signal based on seismic exploration. In: Feng, J., Min, J. (Eds.), Proceeding of 1st International Conference on Information Science and Engineering. Nanjing, China, 26–28 Dec., pp. 600–603.

Zhou, B., Chen, Z., 2014. Active fiber gas sensor for methane detecting based on a laser heated fiber Bragg grating. IEEE Photonics Technol. Lett. 26 (11), 1069–1072.

Zhou, B., Chen, Z., Zhang, Y., 2014. Active fiber gas sensor for methane detecting based on a laser heated fiber Bragg grating. IEEE Photonics Technol. Lett. 26 (11), 1069–1072.

Zhu, K., 1996. Analysis of response of the electromagnetic induction for detection of buried objects. In: Proceeding of International Conference on Geoscience and Remote Sensing Symposium, vol. 4, Lincoln, US, 27–31 May, pp. 2041–2043.

Zhu, X., 2014. Study on Characteristics and Detecting Technology of Infrasonic Signal Processed by Rock Fracture. Chengdu University of Technology, Chengdu, China.

Zhu, R., Nan, S., Gao, Q., 2011. Application of distributed optical fiber sensor technology based on BOTDR in similar model test of backfill mining. Proc. Earth Planet. Sci. 2, 34–39.

Zou, Z., Zhou, H., 2011. Crustal and upper-mantle seismic reflectors beneath the Three Gorges Reservoir region. J. Earth Sci. 22 (2), 205–213.

SUBCHAPTER 2.2

GPR Technologies for Underground Sensing

Tian Xia
University of Vermont, Vermont, USA

2.2.1 INTRODUCTION TO GROUND PENETRATING RADAR

Ground penetrating radar (GPR) is a nondestructive evaluation (NDE) technology for detecting, locating, and inspecting objects or structures buried underneath ground surface. It is also applicable to detect objects behind walls or other dielectric barriers. Comparing with other nondestructive evaluation technologies, such as chain drag, half-cell, acoustic and hammer test, GPR is superior for its easy deployment, high efficiency, and imaging capabilities.

In its operation, the GPR antenna radiates a wideband electromagnetic (EM) wave to penetrate through the ground surface. When hitting the boundaries of layers or objects of different electrical properties, EM wave is reflected back and received by the GPR receiving antenna. By analyzing the reflection signal, including the amplitude and phase parameters, the reflectors' features can be characterized. If an area scan is performed, 2D or 3D subsurface image can be obtained, which essentially maps the impedance contrast of the subsurface structure. The impedance contrast is primarily determined by the structure's electrical properties, including permittivity (or dielectric constant), conductivity, and permeability.

GPR applications span a very wide scope, ranging from archaeological investigation (Goodman et al., 1995; Gracia et al., 2000), building condition evaluation (Orlando and Slob, 2009), mine detection (Ho et al., 2004), geophysical exploration (Greaves et al., 1996), and pipes detection (Zeng and McMechan, 1997) to surveillance sensing (Hunt et al., 2001), buried victim search on the rescuing site (Jaedicke, 2003), etc. In recent decades, GPR's unique nondestructive sensing and imaging capability has received great attention by various transportation agencies. GPR has been utilized in transportation infrastructure surveys, specifically for evaluating roadway, bridge deck, and railroad structural reliability and structural defects (Benedetto et al., 2012a; 2012b; Al-Qadi and Lahouar, 2005; Xu et al., 2012; Zhang et al., 2014, 2015a, 2015b), such as debond and delamination, missing rebar, rebar corrosion, cavity, void, fouled railroad ballast, draining problem, high degree moisture, etc. These defects are mainly the results of material deterioration, which in turn cause the formation of additional interfaces or dielectric discontinuities. As the changes of material's EM properties affect EM wave's propagation velocity and amplitude attenuation, through GPR signal measurement and analysis, the features of subsurface structure can be characterized.

2.2.2 OPERATING MECHANISM OF GPR

Fig. 2.2.1A is a diagram illustrating GPR operating mechanism. The GPR transmitting antenna radiates an EM wave to the structure under inspection. At the interface of two different layers, reflection and refraction occur, where part of the signal penetrates through the layer and propagates continuously, while the other part is reflected back and captured by the receiving antenna. Fig. 2.2.1B shows an A-scan waveform obtained from an impulse GPR. The waveform contains multiple pulse components: A1 is the pulse reflected from the ground surface, A2 and A3 are pulses reflected from the interfaces between layer 1 and layer 2 as well as layer 2 and base, respectively. The phase of the pulse component is determined by the layer thickness and EM signal velocity, while the amplitude is determined by the reflection coefficient at the layer interface, and the attenuation factor of the layer. All these effects are ultimately determined by the dielectric material's electromagnetic properties, specifically the permittivity, the conductivity, and the permeability. Note, on the A-scan waveform, there exits another pulse A0, which is produced by the direction coupling between the transmitting antenna and the receiving antenna. As a large amplitude A0 might degrade other reflection signal components receiving and identification, it should be alleviated as much as possible.

In order to leverage GPR sensing efficiency and effectiveness, as summarized by D.J. Daniel in Daniels (2005), there are strict requirements to GPR functions:

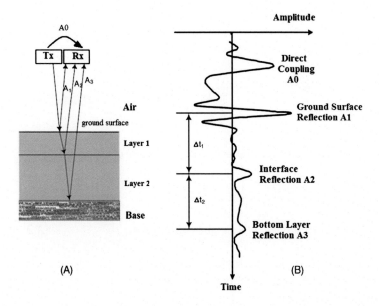

Figure 2.2.1 (A) GPR scanning a multi-layer structure; (B) A sample A-scan waveform

- Efficient coupling of the transmitted electromagnetic energy into the ground
- Adequate penetration of the energy through the ground, relative to target depth
- Reception from buried objects or other dielectric discontinuities of a sufficiently large reflected signal for detection at or above the ground surface
- Adequate bandwidth in the detected signal, relative to the desired resolution and noise levels
- Adequate signal-to-clutter ratio

2.2.2.1 GPR Signal Propagation in Dielectric Materials

The GPR signal reflection and refraction at the layers interface can be interpreted utilizing Snell's law. While for analyzing EM signal propagations inside the dielectric material, Maxwell's equations set the theoretical foundations.

Fig. 2.2.2 illustrates EM signal reflection and refraction at the interface of two materials, where S1 and S2 are the incident signal and the reflection signal in material 1, respectively, and S3 is the refracted signal penetrating into material 2.

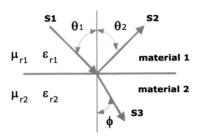

Figure 2.2.2 EM signal reflection and refraction at the interface of two materials

According to Snell's law, one has

$$K_1 * \sin\theta_1 = K_2 * \sin\theta_2 = K_3 * \sin\varphi \qquad (2.2.1)$$

where θ_1, θ_2 and φ are angles measured from the normal of the boundary. K_1, K_2 and K_3 are the wave numbers in each material respectively, which can be characterized as

$$K_r = \omega\sqrt{\mu_r\varepsilon_r}/c \qquad (2.2.2)$$

where ω is the angular frequency, μ_r and ε_r are the relative permeability and permittivity, and c is the speed of light in air. As the incident signal and the reflection signal are in the same media, one has $\theta_1 = \theta_2$. Whereas for the refraction signal angle φ, its relationship with θ_1 can be characterized by the relative properties of two interfacing materials. For nonmagnetic materials, μ_r equals 1, hence

$$\frac{\sin\theta_1}{\sin\varphi} = \frac{\sqrt{\varepsilon_{r_2}}}{\sqrt{\varepsilon_{r_1}}}. \qquad (2.2.3)$$

For propagating inside the medium, the GPR signal can be modeled as a one-dimensional plane wave. Assuming the propagation is along the z-direction, the electrical field (E) equation can be expressed as

$$\frac{\partial E}{\partial Z^2} = \mu\varepsilon\frac{\delta^2 E}{\delta t^2}, \qquad (2.2.4)$$

where the wave's phase velocity is

$$v_r = \frac{1}{\sqrt{\mu\varepsilon}} = \frac{1}{\sqrt{\mu_0\mu_r\varepsilon_0\varepsilon_r}} \qquad (2.2.5)$$

in which μ_0 and ε_0 are the permeability and permittivity of free space.

In a nonmagnetic material, as $\mu_r = 1$, the EM wave phase velocity equals

$$v_r = \frac{1}{\sqrt{\mu\varepsilon}} = \frac{c}{\sqrt{\varepsilon_r}}. \tag{2.2.6}$$

The intrinsic impedance equals the ratio of the transverse components of the electrical field and the magnetic field,

$$Z_r = \frac{E}{H} = \sqrt{\frac{j\omega\mu}{\sigma + j\omega\varepsilon}}. \tag{2.2.7}$$

For a nonconducting medium whose conductivity σ is zero, the intrinsic impedance is

$$Z_r = \sqrt{\frac{\mu}{\varepsilon}}. \tag{2.2.8}$$

The reflection coefficient at the interface between material 1 and material 2 equals

$$\Gamma_{12} = \frac{Z_2 - Z_1}{Z_2 + Z_1} = \frac{\sqrt{\varepsilon_{r_1}} - \sqrt{\varepsilon_{r_2}}}{\sqrt{\varepsilon_{r_1}} + \sqrt{\varepsilon_{r_2}}}, \tag{2.2.9}$$

Γ_{12} specifies the amplitude ratio of the reflected waves and the incident wave, ε_{r_1} and ε_{r_2} are the relative dielectric constants of material 1 and material 2. As disclosed by Eq. (2.2.9), the value of the reflection coefficient Γ_{12} is determined by the dielectric constants of two interfacing materials. When $\varepsilon_{r_1} > \varepsilon_{r_2}$, such as when EM wave propagates from air into a dielectric material, like soil, sand, etc., the reflection coefficient has a positive value. When the incident wave propagates from a low permittivity material to a high permittivity material, or from a high impedance material to a low impedance material, the reflection coefficient will have a negative value, which means the reflection signal changes the polarity. Eq. (2.2.9) inspires a simple, yet practical approach to measure the permittivity of material 2 if the permittivity of material 1 is known a priori. In Bertrand et al. (2006), the dielectric constant measurement of a concrete layer is demonstrated. In the measurement, a large metal plate is placed on the layer surface. Due to high conductivity and nearly zero impedance of the metal plate, the incident EM wave is fully reflected. The reflection signal's amplitude is the same as that of the incident signal, and is recorded as A_{pl}, while its polarity is reversed. Then the metal plate is removed, and the amplitude of the new

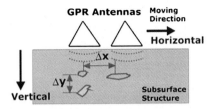

Figure 2.2.3 GPR scanning diagram

reflection wave is measured as A_c. The reflection coefficient equals

$$\Gamma_{12} = -\frac{A_c}{A_{pl}}. \tag{2.2.10}$$

Since material 1 is air whose relative dielectric constant value is 1, the dielectric constant of the concrete layer can then be calculated as

$$\varepsilon_c = \left[\frac{1 + A_c/A_{pl}}{1 - A_c/A_{pl}}\right]^2. \tag{2.2.11}$$

2.2.2.2 GPR Sensing Resolution

Range Resolution

The range resolution characterizes GPR's ability to distinguish two objects that are closely located at different depths in the vertical direction. For an impulse radar, the range resolution is determined by the pulse width and EM signal's travel velocity.

Fig. 2.2.3 illustrates the GPR scan operation. In a standard measurement, the subsurface object's depth is calculated by measuring EM wave traveling time between the antenna and the object. Assuming y is the vertical depth, v is the wave traveling speed, and t is the time-of-flight,

$$y = \frac{1}{2}v * t. \tag{2.2.12}$$

The factor of $1/2$ accounts for the round trip.

If the wave's traveling times for two objects are t_1 and t_2, respectively, then their distance can be calculated as

$$\Delta y = \frac{1}{2}v * (t_1 - t_2) = \frac{1}{2}v * \Delta t. \tag{2.2.13}$$

For an impulse GPR to distinguish two objects buried in a media of permittivity ε_r, it requires the pulse width τ to be shorter than $\Delta t/2$ in order not to mingle two reflection pulses. Hence the range resolution equals

$$r_y = \frac{1}{2}\frac{c}{\sqrt{\varepsilon_r}} * \tau. \qquad (2.2.14)$$

In a pulse compression system, the range resolution can be related to the signal bandwidth BW as

$$r_y = \frac{c}{2\sqrt{\varepsilon_r}} * \frac{1}{BW}. \qquad (2.2.15)$$

Eq. (2.2.15) reveals that by leveraging GPR signal bandwidth, a higher range resolution can be obtained.

Cross-Range Resolution

Cross-range resolution specifies GPR's ability to differentiate two objects in horizontal direction at the same depth level. Primarily, the cross-range resolution is determined by multiple factors: (a) GPR pulse repetition frequency (PRF). For GPR scan in the horizontal direction, within a certain spatial distance, more pulse radiations leads to a finer sensing resolution; (b) Antenna directivity. A higher directivity antenna has a more focused radiation beam, which is beneficial for detecting objects of small feature sizes; (c) The distance to the objects from the antennas. Usually increasing the distance spreads the antenna beam and footprint, which in turn reduces the ability to distinguish closely located objects.

In Cabrera (2007), an empirical formula is provided to characterize the cross-range resolution:

$$r_x = \frac{C}{4*f*\sqrt{\varepsilon_r}} + \frac{D}{\sqrt{\varepsilon_r + 1}}, \qquad (2.2.16)$$

where C is the speed of light in air, f is the antenna's center frequency, D is the distance between objects and the antennas.

Note that Eqs. (2.2.15) and (2.2.16) only characterize the range resolution and the cross-range resolution approximately. There are other factors that are hard to be quantized in a closed form. For instance, strong interference signals in the sensing environment can smear GPR signal transmission and receiving to degrade GPR sensing resolution. As the interference signals have complex and varying properties, their effects on GPR operations

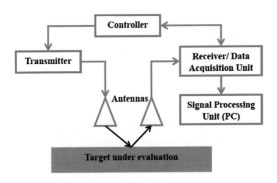

Figure 2.2.4 Impulse GPR system diagram

are generally nonlinear and are difficult to be quantitatively analyzed exactly.

2.2.3 GPR SYSTEM DESIGN

GPR is an ultrawide bandwidth (UWB) radio frequency system. Depending on radar signal generation, there are two types of GPR implementation, continuous wave (CW) GPR and impulse GPR. CW GPR typically radiates sinewaves of many frequency tones in sequence, while the impulse GPR transmits and receives narrow width pulse signals directly. Due to its design simplicity, impulse GPR is more widely designed and employed. As shown in Fig. 2.2.4, a typical impulse GPR consists of several functional elements: (1) transceiver electronics, (2) UWB antennas, (3) data acquisition unit, (4) radar signal processing unit, and (5) the digital controller to coordinate the operations of all functional elements.

In the following subsections, some selected functional components in an impulse GPR will be introduced.

2.2.3.1 Pulse Generator

The pulse generator plays a critical role in determining GPR performance, including sensing depth, resolution, etc. The key specifications of the GPR pulse signal include pulse width, pulse amplitude, and tail ripple or ringing. From the above analysis, it is known that the pulse width is a determining factor of sensing resolution while the pulse amplitude determines the penetrating depth. A narrow pulse usually facilitates small size feature detection, and a high amplitude can tolerate a large propagation attenuation, leading to a deep penetrating depth. Whereas it is not feasible to increase the amplitude of pulse signal arbitrarily high due to several constraints: A high amplitude pulse generation typically results in high complexity circuit design and high power consumption. In addition, the Federal Communications Commission (FCC) has a strict regulation (FCC 02-48) on UWB system power radiation, which sets a limit of GPR pulse amplitude.

GPR pulse generation circuits can be designed using various devices, such as tunnel diode (Yu et al., 1994), oscillator (Xu et al., 1996; Lemaire and Xia, 2009), avalanche transistor (Krishnaswamy et al., 2007), or step recovery diode (SRD) (Han and Nguyen, 2002; Xu et al., 2012), etc. Among different designs, the SRD based pulse generator is very popular due to SRD's superior ability to sharpen pulse signal transition, and its easiness to use. Unlike the regular P–N junction diode, the SRD has an intrinsic layer that makes it a P–i–N junction device with unique dynamic characteristics. When the SRD is forward biased, it behaves like a normal diode to conduct the current. In the meantime, a charge is accumulated in the intrinsic layer. When the bias voltage changes the polarity by switching from a positive value to a negative value, the SRD becomes reverse biased. Due to the charge stored in the intrinsic layer, the diode resistance remains low for a short duration, in which the anode current does not cease but changes the direction, and the charge in the intrinsic layer diminishes gradually. When the intrinsic layer charge is completely exhausted, the diode resistance increases sharply to a very high value and the SRD enters the cut-off state. Such behavior has been utilized in various pulse generator circuits to produce very narrow pulses.

Fig. 2.2.5 shows the schematic of a pulse generator designed in Xia et al. (2012), which consists of three functional elements, including a conditioning circuit, a Gaussian pulse generator, and a pulse shaping filter. The conditioning circuit converts a single polarity square wave pulse into a dual-polarity one, which, as a stimulus signal, is fed to the SRD and the shorted transmission line based Gaussian pulse generator. To leverage impedance matching and alleviate internal reflection, a power attenuator is inserted in the midst. The attenuator essentially improves the signal ratio between the incident signal and the reflection signal: When the reflection signal due to the impedance mismatch propagates back from the pulse generator to the stimulus source node and then returns to the pulse generator, its amplitude is attenuated at least twice by the attenuator, whereas the incident signal is only attenuated once.

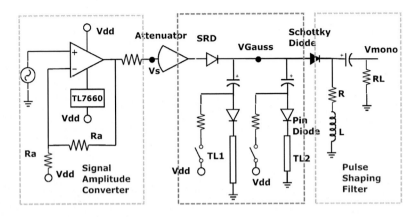

Figure 2.2.5 Schematic of a Gaussian monocycle pulse generator (Xia et al., 2012)

In the Gaussian pulse generator unit, a step recovery diode (SRD) sharpens the transition edge of the incident square wave, which then connects to short-circuited microstrip lines of varying lengths through the PIN diodes and DC decoupling capacitors. The PIN diode is controlled by a digital switch. When the switch for a specific microstrip line turns on, the PIN diode is forward biased, and the edge sharpened step signal gets connected to the microstrip line. The shorted-circuit nature of the microstrip line results in the step signal reflection with an opposite phase. At the SRD output, the incident step signal and the reflection signal add together and produce a Gaussian pulse whose pulse width is proportional to the signal propagation delay along the microstrip line, which in turn is determined by the length of the microstrip line. By using PIN diodes to select microstrip lines of different lengths, the Gaussian pulse width is tunable. The produced Gaussian pulse connects to a Schottky diode, which acts as a half-wave rectifier to conduct the positive pulses while eliminates the negative ones. The pulse-shaping unit, composed of a shunt resistor, an inductor, and a series-connected capacitor, plays two roles. On the one hand, it is a high-pass filter to eliminate low-frequency ripples. On the other hand, it is a differentiator to the input Gaussian pulse signal to produce a monocycle pulse. The amplitude of the monocycle pulse is proportional to the slope of the Gaussian pulse. Fig. 2.2.6 measures the Gaussian monocycle pulse waveforms. As shown, the pulse signal width is 1 ns and the pulse amplitude is 10 V, and the pulse signal to ringing is higher than 20 dB.

In De Angelis et al. (2009), a different high voltage pulse generator circuit utilizing an avalanche transistor is developed. Fig. 2.2.7 depicts the

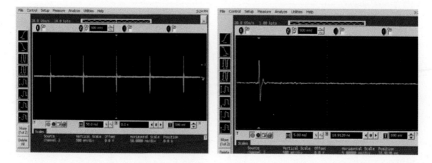

Figure 2.2.6 Pulse signal waveforms measurement upon 20 dB attenuation: (A) multiple pulses; (B) single pulse (Li and Wan, 2010)

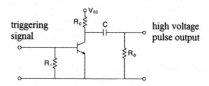

Figure 2.2.7 Pulse generator circuit utilizing the avalanche transistor (De Angelis et al., 2009)

circuit schematic, where the circuit power supply voltage is 300 V. In the operation, when the transistor base terminal voltage is zero, the transistor turns off, the capacitor C is charged by the power supply. When the transistor's base is driven by a trigger signal, the transistor turns on and enters the avalanche mode, where an abrupt voltage drop is produced between the collector and emitter junction. The capacitor is then discharged. The discharging current path is formed by C, Ro, and the transistor. As the collector and emitter junction resistor is very small in the avalanche mode, the discharging time is very short. As a result, a very narrow negative pulse is produced. In Fig. 2.2.8, a 65 V pulse is measured, whose rise time is 0.6 ns, the fall time is 0.8 ns, and the pulse width is 1.3 ns.

2.2.3.2 GPR Antenna

Antenna is another critical functional unit in GPR system. GPR antenna design is different from antennas in other remote sensing systems. First, the GPR antenna's operating bandwidth is ultra–wide, which requires intensive design considerations to achieve good impedance matching and high radiation efficiency across the whole spectrum. Second, the GPR antenna is typically mounted near the ground surface. Therefore the ground

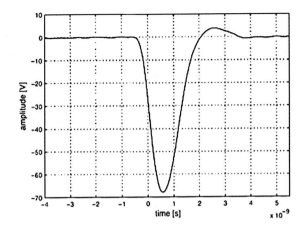

Figure 2.2.8 Measured high voltage pulse waveform (De Angelis et al., 2009)

surface is in the near-field region of the antenna to affect antenna's radiation characteristics. Moreover, as the transmitting and receiving antennas are closely located together, a strong direct coupling exists between them, which requires special electrical and mechanical measures to isolate the two antennas.

In a GPR system, three types of antenna are primarily designed, which are the element antenna, the frequency independent antenna, and TEM horn antenna (Daniels, 2005).

Element Antenna

The element antenna, such as dipole antenna and bow–tie antenna, is a nondispersive antenna with relatively simple structure. The main limitation of element antennas is the moderately narrow bandwidth. In order to extend the operating bandwidth, various techniques have been investigated. Among them, a widely adopted approach is the resistive loading, i.e., the end loading, the distributed loading, or the tapered resistive loading. The resistive loading can reduce the reflection at the aperture. As a result, less signal distortion and lower level ringing effect can be achieved across a wide band spectrum, whereas the price of resistive loading is the reduced radiation gain. Fig. 2.2.9 depicts an end resistive loading bow–tie antenna, whose key structural parameters are the element length and the angle. The bow–tie antenna is primarily designed and utilized in the ground coupled GPR system, where the signal propagation loss is smaller than the air coupled GPR signal.

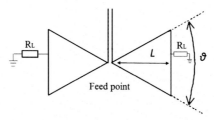

Figure 2.2.9 An end loading bow-tie antenna

Figure 2.2.10 A logarithmic spiral antenna (Thaysen et al., 2001)

Frequency Independent Antenna

The frequency independent antenna has an advantage over the element antenna for its functional parameters, including radiation pattern, polarization and impedance, are nearly constant over a wide operating bandwidth. One frequency independent antenna commonly employed in GPR application is the spiral antenna. Fig. 2.2.10 is a photo of a logarithmic spiral antenna designed in Thaysen et al. (2001). The measurements show that it maintains stable specifications in a very wide spectrum, from 0.35 to 4.5 GHz. In spite of its frequency independent properties, the spiral antenna does have a critical drawback: its response time to the transient electric field is long. When used in the impulse GPR, a large ringing effect will be produced to degrade sensing accuracy and sensing resolution.

TEM Horn Antenna

TEM horn antenna is mainly used in air coupled GPR. The structure of a sample TEM horn antenna (Ahmed et al., 2016) is shown in Fig. 2.2.11. Comparing with bow-tie antenna and spiral antenna, TEM horn antenna's response time to the input or output field is very short, which is effective to reduce the ringing effect in impulse GPR. Additional, TEM antenna has a relatively narrow beam width which facilitates a high directivity and a high gain over a wide spectrum. To design high performance TEM horn antenna

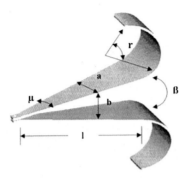

Figure 2.2.11 A TEM horn antenna (Ahmed et al., 2016)

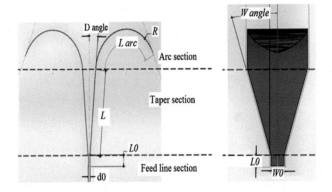

Figure 2.2.12 TEM horn antenna structural model (Ahmed et al., 2016)

for GPR application, the key considerations are to limit the amount of propagation mode and to leverage impedance matching.

To achieve wide band impedance matching, there are two main structural points needing intensive considerations: one is the feed port, and the other is the interface at the aperture. For the air coupled GPR antenna, the characteristic impedance at aperture is 377 Ω, while the feed line impedance is 50 Ω. Design measures need to gradually accomplish impedance transition so as to minimize antenna's internal reflections.

In Ahmed et al. (2016), an UWB TEM antenna is designed. As shown in Fig. 2.2.12, the antenna consists of three structural sections: feedline, waveguide taper segment, and a rounded shaped aperture. In the design, the feedline and the taper section are approximately modeled as a series of N parallel–plate transmission line segments. Each segment consists of two metal plates that are separated by a dielectric media of varying widths.

The impedance of the feedline is characterized in Eq. (2.2.17), where Z_{in}, Z_{out} are the feedline input and output terminal impedances, respectively, and l is the feedline length:

$$Z_{in} = Z_0 \frac{Z_{out} + jZ_0 \tan(\beta l)}{Z_0 + jZ_{out} \tan(\beta l)}. \tag{2.2.17}$$

Z_0 is characteristic impedance and β is the wave number of the transmission line.

The feedline width and feedline height are labeled as w and d, respectively. The electromagnetic simulations reveal that when $l = 6$ mm, $d = 3$ mm, and $w = 12$ mm, the feedline S11 parameter is below -10 dB across the frequency band ranging from 600 MHz to 6 GHz.

For the waveguide taper section, it can be modeled as a staircase structure consisting of N segments. Each segment can be assumed homogeneous when the segment length is small in comparison with the signal wavelength. To minimize the discontinuity effect between the adjacent segments, a large N value can be selected in the analysis for segmenting the taper section and reducing the length of each segment so as to smooth out the structure transition and leverage modeling accuracy. The input impedance Z_{in} of each segment can be calculated using Eq. (2.2.18), where β_i, Z_{0_i}, and l_i are the wave number, characteristic impedance, and length of the ith segment, respectively. $Z_{in_{i+1}}$ is the input impedance of the $(i+1)$th stage loading the ith segment. The 0th segment is the touching point between the feedline and the taper section. The Nth segment is the end of the taper section connecting the aperture arc:

$$Z_{in_i} = Z_{0_i} \frac{Z_{in_{i+1}} + jZ_{0_i} \tan(\beta_i l_i)}{Z_{0_i} + jZ_{in_{i+1}} \tan(\beta_i l_i)}, \quad \forall i = 0, \ldots, (N-1). \tag{2.2.18}$$

An analytical model based on Eq. (2.2.18) can be used to identify suitable values for D_angle and W_angle, depicted in Fig. 2.2.12. In the model, the loading impedance of the last stage Z_L equals the free space characteristic impedance (377 Ω). After performing the parametric analysis, it is obtained that when D_angle equals 5.5°, W_angle equals 13°, and the taper section length is 180 mm, optimum S11 result is obtained across the wide spectrum.

The shape and length of the arc section also stand as critical design variables. The antenna's lower end frequency is designed to 600 MHz whose wave length is 500 mm. The total length of the antenna is set to be half of

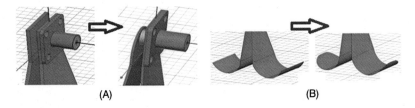

(A)   (B)

Figure 2.2.13 Rounding the edges at the aperture and the feed point to smooth out signal flow (Ahmed et al., 2016)

that wave length. Since the taper section length is 180 mm, the arc section length is thus set to 70 mm. The EM simulation validates that such configurations lead to optimum impedance matching spanning a wide bandwidth.

To further improve antenna performance, extra structure optimization measures are also taken to smooth out EM signal flow and to alleviate EM signal internal reflection. Fig. 2.2.13 shows antenna's aperture edges and the feed point which are rounded up.

2.2.4 GPR IMAGE PROCESSING

GPR signal processing focuses on two aspects, one is the desired reflection signal enhancement and the other is feature extraction. GPR signal is usually mixed with strong noise and interference, which can significantly degrade sensing quality. There are various noise and interference sources. For instance, the mechanical vibration of antennas can cause reflection signal time skew; the ground surface reflection and direct coupling between the transmitting and receiving antennas can produce strong clutters; the in-band radio frequency interference signal existing in the sensing environment can smear GPR reflection signals. To alleviate these effects, multiple steps of GPR signal processing are typically performed. In the following subsections, some commonly applied signal processing techniques are described.

2.2.4.1 Vibration Effect Correction

In the test, the mechanical vibrations of GPR antennas may cause spurious level misalignments among different scan waveforms. For instance, Fig. 2.2.14A shows a group of A-scan waveforms. The highlighted black dots reveal that different traces are misaligned due to the vibration effect. To fix the problem, level tracking and level correction processing steps are

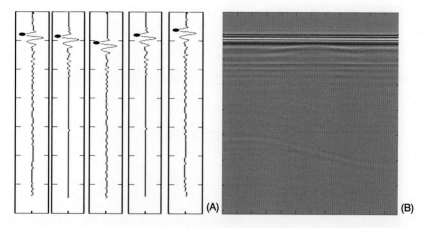

Figure 2.2.14 (A) Misaligned A-scan traces; (B) B-scan image after vibration effect correction

needed. For level tracking, a constant peak level existing in every trace is selected and identified. Level correction algorithm then selects the peak point time index in the first trace as the reference, and adjusts the time indexes of all other traces by the relative offsets. Upon completing level tracking and level correction, time zero line is identified. Usually the antennas installation plane is selected as the time zero reference plane. In a GPR system, one usually places the transmitting and receiving antennas in close proximity, which results in a strong EM coupling between them. In the received signal waveform, the first peak pulse is usually attributed to the direct coupling. As its amplitude is large, it can be easily recognized and conveniently used as the reference point. Fig. 2.2.14B plots a B-scan image after vibration effect correction, where the top flat line is constructed with the reference pulses in all A-scan waveforms.

2.2.4.2 Radio-Frequency Interference Reduction

The GPR data collection may be subject to RF interference (RFI) in the test environment. If the RF interference signal is out of GPR frequency band, then an appropriate filter, such as a high pass filter, a bandpass filter, etc., can be employed for interference removal. However, as GPR's frequency band is very wide, there may exist significant in-band RF interferences. To remove them, a different filtering approach is needed. For transient and random RFI signal removal, median filtering has proved to be effective. In Xu et al. (2012), a 3 × 3 windowing median filtering is

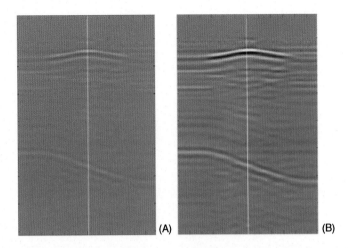

(A) (B)

Figure 2.2.15 B-scan images before and after median filtering

demonstrated for GPR B-scan image data resampling, where 9 data samples in the window are first sorted according to their amplitudes. Then, the median value is selected to replace the center pixel in the resampling window. Fig. 2.2.15 shows the B-scan images before and after median filtering. As can be seen, the features in the B-scan image after median filtering are enhanced, which proves the effectiveness of median filtering for improving image quality.

2.2.4.3 Clutter Removal

GPR data are also prone to contamination by various systematic interferences, such as transmitting and receiving antennas direct coupling, ground surface reflection, etc., which are principal causes of GPR image clutter. If clutter signals are strong, they can mask the real reflection signals from the scattering object under inspection. To remove clutters, a commonly adopted signal processing approach in practice is averaging and subtraction. For averaging, in a raw GPR B-scan image, a number of A-scan traces are selected to calculate an average trace, which is then used as the reference to be subtracted from all other A-scan waveforms. The rationale of such signal processing approach is that when the clutter signals are stationary and constant in every A-scan waveform, the ensemble averaging calculation can significantly alleviate the random noise components; then by performing the reference wave subtraction, the deterministic and constant clutter components can be canceled out (Xu et al., 2012).

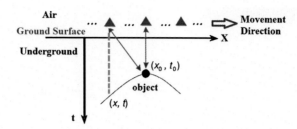

Figure 2.2.16 GPR scan diagram for demonstrating hyperbola curve fitting

2.2.4.4 Feature Extraction

Image feature extraction is an important procedure to characterize subsurface object structural parameters, e.g., size, shape, burying depth, structural conditions, etc. In GPR B-scan images, one commonly detected pattern is a hyperbola curve, which is produced due to the distance change between the underground object and GPR antennas. As Fig. 2.2.16 shows, in the scanning, when GPR antennas are approaching and leaving the object, their mutual distance changes from high to low then to high. When the antennas are located right above the object, the minimum distance is achieved. As radar signal propagation attenuation is proportional to the traveling distance, the reflection signal's amplitude varies from low to high then to low following the distance changing pattern. As a result, a hyberbola curve is produced.

The hyperbola curve can be constructed for subsurface object feature analysis. For instance, the span of the hyperbola curve specifies the size of the object; the coordinates of the hyperbola curve vertex indicate the object's burying depth and position; the magnitudes of pixels on the hyperbola curve specify reflection signal strength, etc. To construct the hyperbola curve, curve fitting algorithm (Xu et al., 2012) can be utilized in accordance with the standard hyperbola's characteristic equation,

$$\frac{t^2}{a^2} - \frac{x^2}{b^2} = 1, \tag{2.2.19}$$

where the parameter values a and b are calculated and feature the characteristics of the scatter.

The hyperbola curve fitting has been utilized in many GPR applications (Mertens et al., 2016; He et al., 2009). In Xu et al. (2012), an experiment is performed to detect two rebars buried in a concrete slab. Fig. 2.2.17 illustrates the concrete slab and the corresponding GPR B-scan image, where

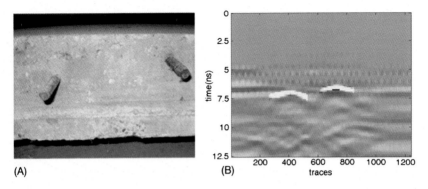

(A)   (B)

Figure 2.2.17 (A) A concrete slab with two rebars; (B) Hyberbola curve fitting of rebar patterns

two hyberbola curves are highlighted, specifying the image patterns of two rebars.

2.2.4.5 Statistical Analysis for Singular Feature Detection

For GPR sensing, its primary mission is to detect the sporadically distributed objects and structural defects buried underneath the ground surface. As the sensing data volume is generally huge, i.e., 10~100 GB or even larger, searching for and locating the sparse objects of interest is not trivial, and typically requires extensive computing power and time. Moreover, for sophisticated feature characterizations, many complicate signal processing steps have to be taken. It is very difficult, if not impossible, to process the whole big data set with completely identical processing procedures. Therefore, finding how to effectively and efficiently search and identify the singular features of interest is of paramount importance and value.

To tackle such issues, statistical signal processing method can be utilized. In Zhang et al. (2015a, 2015b), an entropy based statistical analysis algorithm is developed and applied for rebar detection. To identify a rebar in a GPR image, the traditional curve fitting approach is to model and search for rebar's signature hyperbolic pattern. Although effective, its computation time is long; hence it is only applicable for small volume GPR data processing. Moreover, the curve fitting method is only operative when object's characteristic pattern is a priori known and well defined. Alternatively, in Soldovieri et al. (2011), an approximate linear scattering model is proposed to reconstruct rebar reflection signal utilizing the sparse nature of scatters. In the processing, the double integral and minimization algorithm based on loops of matrix multiplication is implemented. The approach has the same limitation of requiring intensive computation resources and is not suitable for processing big data set directly.

In Zhang et al. (2015a, 2015b), a two-dimensional (2D) entropy analysis algorithm is developed whose objective is to narrow down the data scope or reducing data volume in searching regions of interest. With data volume reduction, more sophisticated post-processing methods, such as curve fitting, spectrum analysis, etc., become feasible.

In information theory, by evaluating the uncertainty associated with a random variable, entropy quantifies the expected value of information contained in a message. For GPR data processing, entropy value is calculated to evaluate subregion data singularity. In particular, a high entropy value specifies a high degree data similarity from the background, whereas a low entropy value indicates a low similarity, or a high degree of singularity from the background. Therefore through entropy analysis, the special regions of interest can be detected.

Assuming that the received GPR reflection signal is $Y(t)$, it can be modeled as:

$$Y(t) = D(t) + S(t) \tag{2.2.20}$$

where $D(t)$ represents the reflection signal of interest, $S(t)$ is the interference and noise. In entropy calculation, the first step is power normalization which is performed as

$$y_i(t) = \frac{\|Y_i(t)\|^2}{\sum_{i=1}^{M} \|Y_i(t)\|^2} \tag{2.2.21}$$

where $y_i(t)$ is the normalized signal, i is the trace index, M is the total number of traces included, and t is the time index. Upon power normalization, a generalized Renyi's entropy (Zhi and Chin, 2006) is computed to assess data singularity:

$$E_\alpha(t) = \frac{1}{1-\alpha} \log_e \left\langle \sum_{i=1}^{M} [y_i(t)]^\alpha \right\rangle \tag{2.2.22}$$

where $E_\alpha(t)$ is the entropy quantification and α is the entropy order. When $\alpha = 1$, Eq. (2.2.22) transforms to the basic Shannon entropy. In the demonstrated GPR signal processing, α equals 3, leading to optimal processing result.

The figures below illustrate a GPR scan test and the test results. Fig. 2.2.18A shows the test setup where a rebar is suspended in air.

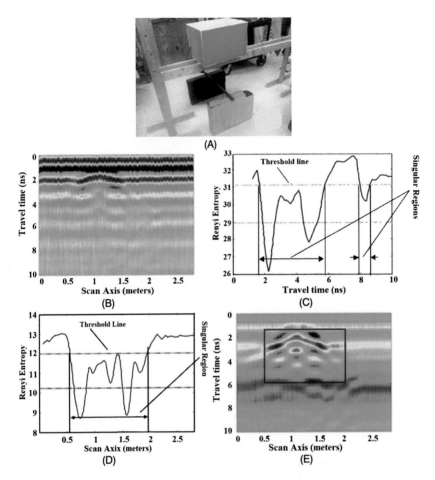

Figure 2.2.18 Experimental results of entropy analysis: (A) rebar in air test setup; (B) raw B-scan image; (C) entropy curve calculated for traces along Y-axis; (D) entropy curve calculated for traces along X-axis; (E) the final B-scan image with rebar region detected

Fig. 2.2.18B is the raw B-scan image. Figs. 2.2.18C and 2.2.18D are Renyi's entropy curves calculated along y-axis and x-axis traces in the B-scan image, respectively. Utilizing OTSU thresholding technique, the entropy curves are segmented into subregions, where the subregions having lower entropy values contain singular data sets. Then applying additional data processing steps, such as short time Fourier transform (Zhang et al., 2015a, 2015b), refined singular region in B-scan image is correctly identified, shown in Fig. 2.2.18E. In a separate test in Zhang et al. (2015a, 2015b), GPR is demonstrated to scan a big structure designed to emulate the railroad foundation. Utilizing entropy analysis, the burying objects are correctly detected. Moreover, the study reports achieving 95% data reduction rate.

Other GPR Design Technologies

In the literature there are other GPR design technologies. Based on radar signal waveform generation, an alternative approach is the continuous wave (CW) GPR. Unlike the impulse GPR that transmits and receives short pulse signals directly, the CW GPR radiates multiple individual frequency tones sequentially. Through measuring the reflection signal's magnitude and phase responses corresponding to each frequency tone, the scatter's spectrum response can be characterized. By controlling the number of frequency tones and the frequency step size, the bandwidth of CW signals can be manipulated. In addition, by assembling the magnitude and phase responses of all frequency tones, and performing the inverse Fourier transform, the scatter's impulse response can be synthesized. Comparing with the impulse GPR, the CW GPR has its unique advantageous: First, as the continuous wave of each frequency tone is emitted and received sequentially. The CW GPR essentially operates in the narrow band mode, which relaxes the requirements for the high cost high speed analog to digital converter (ADC) for data acquisition. In addition, it facilitates high quality narrow band signal filtering, which in turn improves system dynamic range and signal-to-noise ratio (SNR). However, the CW GPR has one important drawback: as each individual frequency tone must be generated, transmitted, and received sequentially, the CW GPR operation speed is lower than the impulse GPR. For certain applications that requires fast speed scan, such as roadway, bridge deck, and railroad structure inspection, the applicability of CW GPR is limited. To overcome such limitations, there are extensive researches investigating various solutions. Some published approaches include multitone signal generation, compressive sensing, etc. (Gurbuz et al., 2009; Zhang et al., 2015c; Metwally et al., 2015).

Moreover, many commercial GPR systems are single channel systems that are equipped with a single transmitting antenna and a single receiving antenna for scan operation. As the antenna footpint is small (focused beam with dth is designed for the consideration of high directivity and penetrating depth), the sensing area coverage is limited. For inspecting roads of multiple lanes or surving underground utilities in a large area, the single channel GPR is time consumping and is of low efficiency. To improve sensing efficiency, multichannel GPR systems are being investigated and designed. For instance, in Eriksen et al. (2004), Francese et al. (2009), GPR systems are designed equipped with multiple transmitting and receiving antennas. Different channel antennas are mounted and operated in parallel in transverse direction so as to achieve wide coverage. Their operations are coordinated by a central digital controller.

REFERENCES

Ahmed, A., Zhang, Y., Burns, D., Huston, D., Xia, T., 2016. Design of UWB antenna for air-coupled impulse ground-penetrating radar. IEEE Geosci. Remote Sens. Lett. 13 (1), 92–96.

Al-Qadi, I.L., Lahouar, S., 2005. Measuring layer thicknesses with GPR–theory to practice. Constr. Build. Mater. 19 (10), 763–772.

Benedetto, A., Benedetto, F., Tosti, F., 2012a. GPR applications for geotechnical stability of transportation infrastructures. Nondestruct. Test. Eval. 27 (3), 253–262.

Benedetto, A., Manacorda, G., Simi, A., Tosti, F., 2012b. Novel perspectives in bridges inspection using GPR. Nondestruct. Test. Eval. 27 (3), 239–251.

Bertrand, C., Michalk, B.K., Xing, H., Li, J., Liu, R.C., Oshinski, E., Claros, G.J., 2006. New method for pavement dielectric constant measurement using ground-penetrating radar. In: Transportation Research Board 85th Annual Meeting (No. 06-0280).

Cabrera, R.A., 2007. GPR antenna resolution. http://geoscanners.es/appnotes/antres.pdf.

Daniels, D.J., 2005. Ground Penetrating Radar. John Wiley & Sons, Inc.

De Angelis, A., Dionigi, M., Moschitta, A., Carbone, P., 2009. A low-cost ultra-wideband indoor ranging system. IEEE Trans. Instrum. Meas. 58 (12), 3935–3942.

Eriksen, A., Gascoyne, J., Al-Nuaimy, W., 2004. Improved productivity and reliability of ballast inspection using road-rail multi-channel GPR. Railway Eng., 6–7.

Francese, R.G., Finzi, E., Morelli, G., 2009. 3-D high-resolution multi-channel radar investigation of a Roman village in Northern Italy. J. Appl. Geophys. 67 (1), 44–51.

Goodman, D., Nishimura, Y., Rogers, J.D., 1995. GPR time slices in archaeological prospection. Archaeolog. Prosp. 2, 85–90.

Gracia, V.P., Canas, J.A., Pujades, L.G., Clapés, J., Caselles, O., García, F., Osorio, R., 2000. GPR survey to confirm the location of ancient structures under the Valencian Cathedral (Spain). J. Appl. Geophys. 43 (2), 167–174.

Greaves, R.J., Lesmes, D.P., Lee, J.M., Toksöz, M.N., 1996. Velocity variations and water content estimated from multi-offset, ground-penetrating radar. Geophysics 61 (3), 683–695.

Gurbuz, A.C., McClellan, J.H., Scott, W.R., 2009. A compressive sensing data acquisition and imaging method for stepped frequency GPRs. IEEE Trans. Signal Process. 57 (7), 2640–2650.

Han, J., Nguyen, C., 2002. A new ultra-wideband, ultra-short monocycle pulse generator with reduced ringing. IEEE Microw. Wirel. Compon. Lett. 12 (6), 206–208.

He, X.Q., Zhu, Z.Q., Liu, Q.Y., Lu, G.Y., 2009. Review of GPR rebar detection. In: PIERS Proceedings. March, pp. 804–813.

Ho, K.C., Collins, L.M., Huettel, L.G., Gader, P.D., 2004. Discrimination mode processing for EMI and GPR sensors for hand-held land mine detection. IEEE Trans. Geosci. Remote Sens. 42 (1), 249–263.

Hunt, A., Tillery, C., Wild, N., 2001. Through-the-wall surveillance technologies. Correct. Today 63 (4), 132–133.

Jaedicke, C., 2003. Snow mass quantification and avalanche victim search by ground penetrating radar. Surv. Geophys. 24 (5–6), 431–445.

Krishnaswamy, P., Kuthi, A., Vernier, P.T., Gundersen, M.A., 2007. Compact subnanosecond pulse generator using avalanche transistors for cell electroperturbation studies. IEEE Trans. Dielectr. Electr. Insul. 14 (4), 873.

Lemaire, O., Xia, T., 2009. Design of a monolithic width programmable Gaussian monocycle pulse generator for ultrawideband radar in CMOS technology. In: Circuits and Systems and TAISA Conference, 2009. Joint IEEE North-East Workshop on NEWCAS-TAISA'09. IEEE, pp. 1–4.

Mertens, L., Persico, R., Matera, L., Lambot, S., 2016. Automated detection of reflection hyperbolas in complex GPR images with no a priori knowledge on the medium. IEEE Trans. Geosci. Remote Sens. 54 (1), 580–596.

Metwally, M., L'Esperance, N., Xia, T., 2015. Compressive sampling coupled OFDM technique for testing continuous wave radar. J. Electron. Test. 31 (1), 75–83.

Orlando, L., Slob, E., 2009. Using multicomponent GPR to monitor cracks in a historical building. J. Appl. Geophys. 67 (4), 327–334.

Soldovieri, F., Solimene, R., Monte, L.L., Bavusi, M., Loperte, A., 2011. Sparse reconstruction from GPR data with applications to rebar detection. IEEE Trans. Instrum. Meas. 60 (3), 1070–1079.

Thaysen, J., Jakobsen, K.B., Appel-Hansen, J., 2001. A logarithmic spiral antenna for 0.4 to 3.8 GHz. Appl. Microwave Wirel. 13 (2), 32–45.

Xia, T., Venkatachalam, A.S., Huston, D., 2012. A high-performance low-ringing ultrawideband monocycle pulse generator. IEEE Trans. Instrum. Meas. 61 (1), 261–266.

Xu, L., Szipöcs, R., Spielmann, C., Krausz, F., 1996. Ultrabroadband ring oscillator for sub-10-fs pulse generation. Opt. Lett. 21 (16), 1259–1261.

Xu, X., Xia, T., Venkatachalam, A., Huston, D., 2012. Development of high-speed ultrawideband ground-penetrating radar for rebar detection. J. Eng. Mech. 139 (3), 272–285.

Yu, R.Y., Konishi, Y., Allen, S.T., Reddy, M., Rodwell, M.J., 1994. A traveling-wave resonant tunnel diode pulse generator. IEEE Microw. Guided Wave Lett. 4 (7), 220–222.

Zeng, X., McMechan, G.A., 1997. GPR characterization of buried tanks and pipes. Geophysics 62 (3), 797–806.

Zhang, Y., Venkatachalam, A.S., Xia, T., Xie, Y., Wang, G., 2014. Data analysis technique to leverage ground penetrating radar ballast inspection performance. In: 2014 IEEE Radar Conference. IEEE, pp. 0463–0468.

Zhang, Y., Burns, D., Huston, D., Xia, T., 2015a. Sand moisture assessment using instantaneous phase information in ground penetrating radar data. In: SPIE Smart Structures and Materials+ Nondestructive Evaluation and Health Monitoring. International Society for Optics and Photonics, p. 943726.

Zhang, Y., Candra, P., Wang, G., Xia, T., 2015b. 2-D entropy and short-time Fourier transform to leverage GPR data analysis efficiency. IEEE Trans. Instrum. Meas. 64 (1), 103–111.

Zhang, Y., Wang, G., Xia, T., 2015c. Compressive orthogonal frequency division multiplexing waveform based ground penetrating radar. In: 2015 IEEE Radar Conference (RadarCon). May. IEEE, pp. 0684–0689.

Zhi, W., Chin, F., 2006. Entropy-based time window for artifact removal in UWB imaging of breast cancer detection. IEEE Signal Process. Lett. 13 (10), 585–588.

CHAPTER 3

Geotechnical Underground Sensing and Monitoring

Magued Iskander[#]
New York University, New York, USA

In construction underground, where the engineer deals with materials having properties that vary not only in space but in time, details of construction have significant influence on the behavior of the structure and of the surrounding soil.

Ralph Peck

3.1 INTRODUCTION

Field monitoring of geomechanics (FMGM) began approximately 75 years ago by the late Ralph Peck, a geotechnical engineering legend, and long time associate of Karl Terzaghi, the father of geotechnical engineering. Peck was the first to recognize that sensors were needed because conditions underground cannot be prescribed, or defined, as well as in dealing with other materials, so equations don't always predict the behavior of site conditions, accurately. As a result, designers tend to be conservative. At the same time, some designs fail to meet the desired level of performance.

A classic example to justify the need for use of instrumentation in geotechnical engineering is the necessity to determine the loads in the struts supporting the braced excavation shown in Fig. 3.1, in order to size them properly. Strut loads are of course a function of the earth pressure. Theories of earth pressure predict vastly different strut load values depending on weather the wall is moving away from the soil (active condition), moving towards the soil (passive condition), or remains stationary (at rest condition). In reality strut loads depend not only on the soil conditions and dimensions of the excavation. They also depend on strut location,

---

[#] Professor & Chair, Civil & Urban Engineering Department, Tandon School of Engineering (Formerly Polytechnic Institute/University), Six Metrotech Center, Brooklyn NY 11201.

Underground Sensing.
DOI: http://dx.doi.org/10.1016/B978-0-12-803139-1.00003-5
Copyright © 2018 Elsevier Inc. All rights reserved.

Figure 3.1 Braced cut excavation (photo credit Deep Excavation, LLC)

construction sequence, workmanship, time between excavation and strut installation, among other factors. If a strut is slightly longer than the excavation, conditions approach the passive condition. Similarly, if the strut is slightly shorter than the width of the excavation, loads approach the active condition. Thus, workmanship has a very large impact on the loads in the struts, and instrumentation can be employed to determine the actual strut loads and corresponding earth pressure.

Use of instrumentation during the design phase helps define actual site conditions. During construction, instrumentation is commonly used to monitor the effects of construction, prevent failure, protect third party property, enhance public relations, and to avoid legal liability. Often instrumentation is also employed for fact finding after undesirable performance or failure occurs.

The *Observational Design Method* employs instrumentation to determine actual site conditions, and modify the design according to the observed measurements to meet a pre-prescribed performance level. Design thresholds are established, and the design is modified according to the measured site behavior. For example, the observational method can be employed in the braced excavation described previously in order to determine if a third level of bracing is required, or may be omitted, based on the observed lateral ground deformations and the loads measured in the struts.

This chapter introduces the most commonly used sensors employed for monitoring of underground geomechanics. The chapter begins with measurement of strain, since strain is often employed for monitoring a variety of common geotechnical processes such as pressure, load, displacement, and

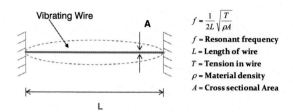

Figure 3.2 Operating principal of vibrating wire strain gage

tilt. The chapter ends with a discussion of common measurement errors and typical sensor specifications.

3.2 MONITORING STRAIN

Strain is one of the most important quantities measured in engineering. For an axial member, strain is typically defined as the change in length of a member divided by the length of the original member. However, in a continuum such as soil one must consider the presence of three orthogonal strains, and the corresponding shear strains. In any case, knowledge of strain permits converting it to load, pressure, deformation, tilt, or torque, depending on the geometric configuration of the strain measurement.

Although there are many ways to monitor strain, only three conventional ways are commonly used to measure strain in geotechnical engineering practice. The most common approaches employ vibrating wire gauges, or electrical resistance (Foil) gages. In the last decade, fiber-optic sensors have increasingly been employed for measuring strain; and they are expected to gain in popularity with time for field monitoring of geotechnical phenomena.

3.2.1 Vibrating Wire (VW) Strain Gages

Vibrating wire strain gage technology tends to be used for field applications, since it is one of the most robust and durable sensing technologies, with some vibrating wire devices being in service for over three decades.

3.2.1.1 Operating Principle of VW Gages

VW strain gages operate on the principle that a tensioned wire, when plucked, vibrates at a frequency that is proportional to the strain in the wire (Fig. 3.2). In that respect they are similar to a guitar, and other musical string instruments. They depend on measuring the resonant frequency of the wire. The relationship between the resonant frequency (f) and the tension in the wire (T) depends on the length (L), cross-sectional area (A), and material density of the wire (ρ), and is expressed as

$$f = \frac{1}{2L}\sqrt{\frac{T}{\rho A}}. \tag{3.1}$$

By substituting tension as equal to the multiple of the strain (ε), cross-sectional area (A) and Young's modulus (E) (*i.e.*, $T = \varepsilon AE$), and rearranging the equation, strain can be expressed as

$$\varepsilon = f^2 \frac{4\rho L^2}{E}. \tag{3.2}$$

It is convenient for strain gage manufacturers to lump all constant terms into one constant referred to as the *Gage Factor (GF)*, as follows:

$$GF = \frac{4\rho L^2}{E}, \tag{3.3}$$

$$\varepsilon = GFf^2. \tag{3.4}$$

Thus strain is related directly to the square of the frequency. Finally, because the sensor must contain some built-in tension to function properly, it is essential to subtract the value of that tension from the measured value in order to obtain the strain that occurs in the strain gage. This is done by subtracting the initial frequency (f_0) measured by the gage, as follows:

$$\varepsilon = GF(f^2 - f_0^2). \tag{3.5}$$

3.2.1.2 Commercial Vibrating Wire Strain Gages

A typical vibrating wire strain gage consists of a metal tube in which the vibrating wire is inserted. The wire is free to move within the tube, and the tube's main function is to protect the wire. The wire is generally attached to end caps that also prescribe the length of the gage. End caps are attached to the structure being monitored, in order to permit monitoring the local strain at that location.

In most instances, the metal tube is surrounded by a small magnetic coil that is used for (a) exciting the wire, and (b) for reading the resonant frequency of the vibrating wire. First, a voltage pulse is supplied to the coil that creates magnetic attraction which causes the wire to vibrate. Next, the coil becomes a listening device, because wire vibrations cause an alternating voltage to be induced in the plucking coil. The frequency of the induced electric current is the natural frequency of the vibrating wire. Finally, the voltage signal is transmitted along the signal cable to a frequency counter. This process takes approximately 100–200 milliseconds, making it inappropriate for dynamic applications. For dynamic applications, two coils are employed in what is referred to as an auto–resonant arrangement. One coil is continuously exciting the wire and the other is continuously detecting the frequency. The use of two coils introduces two difficulties. First, determining the natural resonant frequency of an auto–resonant gage requires some signal conditioning and is more difficult than when using a single coil. Second, continuous excitation of the gage reduces its life due to fatigue. Thus the vast majority of vibrating wire strain gages employ a single coil.

Vibrating wire strain gages are typically packaged for one of the following installation methods (Fig. 3.3):

- **Spot weldable strain gages** are supplied with flanges at each end that can be welded to the structure. The strain in the structure travels through the flange into the vibrating wire, where it can be measured. These gages are commonly used to monitor bridges, piles, tunnels, etc. Spot welding is perhaps the easiest method to install gages on metallic members, as long as the welding does not affect the heat treatment of the member.
- **Surface mounted strain gages** are supplied with end blocks that can be attached to the structure. Gages are installed by drilling holes into a structure, and attaching the end blocks to the structure with bolts or screws. Loading the structure results in a change in the distance between the two end blocks, which results in a change in the tension of the measurement wire. These gages are typically shipped mounted on a transport plate, so the gage is not damaged in transit. An installation tool is commonly provided to facilitate locating the holes perfectly. The screws used to attach to a transport plate are typically used for mounting the gage to the blocks attached to the structure.
- **Sister bars** are strain gages mounted on a piece of reinforcement rebar with a development length on each side of the gage. They are tied alongside an existing rebar prior to casting a concrete structure so that they become embedded in concrete after the concrete cures. The development length is on the order of 40–60 diameters, which permits the gage to experience the same strain like the concrete structure. The sister

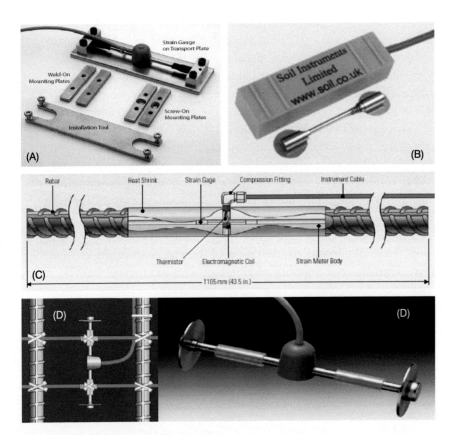

Figure 3.3 Common configurations of VW strain gages (A) Surface mounted, (B) Weldable, (C) Sister bar, and (D) Embedment type

bar configuration permits easy installation of a strain gage under typical field conditions. Thus, Sister Bars are commonly employed for monitoring load transfer in cast-in-place concrete members, such as drilled shafts, slurry walls, and tunnel linings.

• *Embedment gages* have two flat discs at the end. They are intended to be attached to the reinforcement cage inside a concrete beam or column. After concrete is poured, embedment gages become monolithic with the structure. Strain in the concrete causes the two flanges (discs) to move relative to each other, thus increasing or decreasing the tension in the measurement wire. These gages are used for concrete structures, but are less popular than Sister Bars due to the difficulty in keeping them aligned during the concrete casting process.

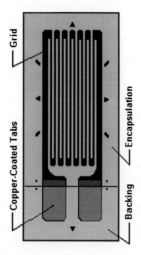

Figure 3.4 Typical electric resistance (a.k.a. Foil) strain gage (photo credit Vishay Micro Measurements, Inc.)

3.2.2 Foil Strain Gages

3.2.2.1 Operating Principle of Foil Gages

Foil strain gages, depend on measuring the change in resistance in a wire, and they tend to be more commonly used for indoor applications. In principle, a foil strain gage consists of a small diameter wire that is attached to a backing material, and the wire is looped back and forth to create a longer wire. The resistance of any wire (R) is a function of its resistivity "ρ", multiplied by its length "L", divided by its cross–sectional area "a" (*i.e.*, $R = \rho L/a$). Thus when a wire is stretched, its resistance increases because (i) the wire becomes longer, and (ii) the cross-sectional area becomes smaller. The longer the wire, the bigger the change in resistance; however, there are practical limitations on how long the wire can be made, so the wire is looped back and forth, to form a strain gage. At each end of the wire a solder tab is attached. The solder tab permits connecting a lead wire to a foil strain gage (Fig. 3.4).

Present day foil strain gages are made by etching a metallic compound on a thin backing material. The gage is attached to the member where strain is being measured using an adhesive glue, or an epoxy compound. The adhesive and backing material together serve to transfer the measured strain from the instrumented member to the sensing "wire," while electrically isolating the metal "wire" from the member on which strain is measured.

3.2.2.2 Commercial Foil Strain Gages

Commercial strain gages come in several hundred variations, in order to accommodate a plethora of usage conditions. Appropriate selection of a proper strain gage is essential to optimizing of gage performance, increasing accuracy and reliability of the installation, facilitating easy installation, and minimizing the total cost of the installation. The following is a brief explanation of the most important selection decisions:

Gage Series

Gage series refers to the combination of materials employed to manufacture the gage's (i) sensing element (*i.e.*, "wire") and (ii) backing material. The most commonly used alloy is *Constantan* (45% Nickel and 55% Copper). Other specialized alloys that are sometimes employed include annealed Constantan for high-elongation applications; Karma, a nickel–chromium alloy used for high-performance gages; platinum; platinum–iridium; nickel, among others.[1] Two backing materials are in use. The most common backing material is a polymeric material known as *Polyamide*. Fiberglass is sometimes employed because it is stiffer, and can therefore prevent signal attenuation in the backing material. However, polyamide is preferred because it is easy to handle while fiberglass is fragile. The surface of the strain gage is often covered with a layer of polyamide to protect the gage from damage during handling (*i.e.*, encapsulated). In civil engineering, encapsulated Polyimide Constantan is the most widely used gage type for laboratory stress analysis.

Self-Temperature Compensation

The strain gage "wire" should have the same expansion coefficient of the material that the gage is mounted on. Otherwise, temperature variation may result in apparent strain, which is known as thermal strain. Thus gauges are designed such that the temperature coefficient of their resistive elements is adjusted to match that of the measured member. The self-temperature compensation (STC) number describes the expansion coefficient of a strain gage, and each STC number is intended for use on a different material. Self-temperature compensation is only possible for certain alloys, but fortunately it is possible for Constantan, the most commonly used alloy.

---

[1] For further details, refer to Micro Measurements Tech Note TN 505-4.

Gage Pattern

Foil strain gages are supplied as either uniaxial gages that measure strain in one direction only, or *Strain Rosettes* that measure strain in two or three axes. The direction of the major principal stress is known for many civil engineering applications, because it coincides with the direction of gravity, thus making bi-axial (Tee) strain rosettes more common for monitoring of geomechanics than strain rosettes having three strain gages.

Gage Length

In the United States, gage length is typically expressed in mils (1 mil = 0.001 in or 0.0254 mm). Gages are available in a variety of lengths, but 5–12 mm are most common. Longer gages are easier to install; however, smaller gages are often used for a variety of reasons. First, space may simply not be available in a surprisingly large number of applications to install a large gage. Second, many structural applications involve a large strain gradient and use of a large gage will average the strain occurring beneath it. Finally, larger gages generate more heat, which may be difficult to dissipate. Excessive heat may change the properties of the backing material, the adhesive material, or the test part. Heat generation is one of the most challenging issues for measurements conducted on polymeric members, since the modulus of thermoplastics is temperature dependent.

Gage Resistance

Strain gages are commonly available in 120 Ω or 350 Ω. Specialized 1000 Ω gages are available in limited sizes, especially for use on polymeric materials. Higher resistance gages are preferable because they (a) reduce the heat generation, (b) improve signal-to-noise ratio due to random resistance change, and (c) reduce the effect of lead-wire resistance changes due to temperature fluctuations.

Options

A number of options are available to facilitate with installations, especially in laboratory environments. These options include preattached lead wires, gauge encapsulations, and preplaced solder dots.

3.2.2.3 Surface Preparation for Foil Strain Gages

All strain gages must be installed on a clean, dry, adequately prepared surface. There are a number of surface preparation steps that are uniquely required for foil gages. These include: (1) surface degreasing to remove oils, organic contaminants, and chemical residues; (2) surface abrasion, since each adhesive requires a certain degree of roughness to work properly; (3) surface conditioning which involves washing the test surface with a mild acidic solution, in order to remove any surface oxidation; and (4) neutralizing the surface using a mild alkaline solution in order to neutralize the conditioner and provide optimum pH for the strain gage adhesive. It is important to note that surface conditioning and neutralizing will have to be repeated if the prepared surface is not used within 30–45 minutes depending on the test material.

3.2.2.4 Bonding of Foil Strain Gages

A thin uniform layer of adhesive is employed to bond the strain gage to the test surface. The adhesive must allow the gage to undergo the same strain experienced by the member being instrumented. Nearly all adhesives have a smaller modulus of elasticity than the test member, so it is important to keep the glue line as thin as possible in order to provide perfect coupling of the strain phenomenon. A second reason for keeping the glue line thin is that the adhesive insulates the strain gage and reduces its ability to dissipate heat. Excessive heat can damage the gage or introduce errors.

Two classes of adhesives are commonly employed to attach foil strain gauges to the test surface. The first is a *cyanoacrylate*-type glue similar to the consumer products *Super Glue* and *Krazy Glue*. Glue-based adhesives depend on absorbing moisture from the atmosphere to initiate the cure reaction. Alternatively, an addition-cure type epoxy consisting of a resin and a hardener that are mixed to start the curing process can be used. Glue-based adhesives are easier to use and less expensive than epoxy, but may degrade with time and humidity. Epoxy based adhesives allow for higher temperature ranges, have high resistance to moisture, and thus tend to be used for permanent installations. In either case, a 5–20 psi pressure has to be applied on the surface during the curing process. Glue can typically cure in a few minutes at room temperature. Epoxy, on the other hand, may require several hours at an elevated temperature to cure properly. A variety of methods can be used to apply the required stress. For glue-based adhesives thumb pressure will typically suffice. For epoxy-based adhesives sand bags are typically employed to apply pressure to horizontal surfaces. For vertical surfaces, a variety of contraptions employing Bungee cords or reactions from inclined beams with hanging weights can be used. Elevation

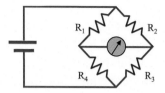

Figure 3.5 Wheatstone bridge (Note R_1, R_2, R_3, and R_4 are four resistances or strain gages corresponding to the strains ε_1, ε_2, ε_3, and ε_4, respectively)

of temperature necessary for epoxy-type adhesives to properly cure can be achieved using heat lamps.

3.2.2.5 Attaching Lead-wires and Protection of Foil Strain Gages

After a foil strain gage is attached to the test surface, lead-wires must be attached to the gage. Because the gage is thin and fragile, a controlled temperature soldering iron must be used in order not to burn the gage. After lead-wires are attached, a variety of methods are employed to protect the strain gage from moisture and/or physical damage. A commonly used material is acid-free liquid-thinned polyurethane that can be brushed on the gage and solder joints. The material dries completely at room temperature. There are also a variety of acid-free silicon adhesive sealants that are often employed for moisture and physical protection. These one-part moisture-cure materials cure at room temperature in an environment of 30–80% relative humidity.

3.2.2.6 Wheatstone Bridge Circuit

The change in resistance with strain is very small, to the extent that it is not practical to measure. To overcome this difficulty, a *Wheatstone bridge* circuit is employed. A Wheatstone bridge consists of 4 equal resistors as shown in Fig. 3.5. One, two, or four resistors are replaced by foil strain gages in what is known as quarter, half, and full bridge, respectively. The 4 resistors form 4 nodes at the intersection of each 2 gages. Direct current (DC) power is supplied to diagonal nodes and the voltage change in the other two gages is recorded. The relationship between the output voltage (V_o) and input voltage (V_i) is given by the following equation:

$$\frac{V_o}{V_i} = \frac{F}{4}(\varepsilon_1 - \varepsilon_2 + \varepsilon_3 - \varepsilon_4) \tag{3.6}$$

where ε_1, ε_2, ε_3, and ε_4 are the strains corresponding to the 4 arms of the bridge and F is the gage factor. Gage factors are supplied by the manufacturer and typically range between 2 and 2.2. It is useful to note that opposite arms of the bridge are additive and adjacent arms of the bridge are subtractive. These properties can be taken advantage of to increase the output of the bridge, for example, by installing gages on a beam where one side is in compression and the second in tension and connecting them next to each other.

3.2.2.7 *Optimizing the Excitation of Foil Strain Gages*

Voltage application across the Wheatstone bridge leads to power loss in each arm. This loss is dissipated in the form of heat. The amount of heat generated increases with the size of the gage grid area, and decreases as the resistance of the gage increases. A third factor that affects the temperature of the installation is the heat sink capacity of the test surface. Some test materials such as aluminum and copper have a high capacity to dissipate heat. Alternatively, polymers have a very poor ability to dissipate the generated heat, with steel being somewhat in between.

There is a linear relationship between the readout voltage and the applied excitation voltage. An increase in the applied voltage is desirable in order to expand the range of the output voltage. However, a strain gage installation must also control the level of voltage excitation in order to control the generated heat. Failure to control the gage temperature may lead to zero (no-load) instability, and or development of hot spots due to imperfections or voids in the glue line. At extreme cases gages may de-bond from the test surface if the temperature becomes intolerably hot.

The best method to determine the optimal bridge excitation level is experimentally. The bridge excitation is gradually increased under zero-load conditions and the output is converted to engineering units. The standard deviation of the voltage fluctuation over a fixed period of time is computed, and the process is repeated until definite zero instability is observed. The excitation should then be reduced until a zero reading becomes stable again.

For an equal-arm bridge arrangement, where the voltage across the active arm is one-half of the bridge voltage, the bridge excitation voltage (V_o) can be calculated[2] as a function of the gage resistance (R_G) in ohms,

---

[2] Vishay Micro Measurements Tech Note 502.

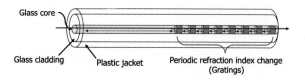

Glass core

Glass cladding   Plastic jacket   Periodic refraction index change
(Gratings)

Figure 3.6 Schematic diagram of fiber optic with Bragg grating

grid power density (P_G) in watts/in^2, and grid area (A_G) in in^2:

$$V_o = 2\sqrt{R_G P_G A_G}. \tag{3.7}$$

Strain gage vendors publish easy-to-use charts based on Eq. (3.7) that relate the excitation voltage to the grid area and the heat sink capacity of
the backing material. These charts are the most common way employed to
select an initial excitation voltage that may be adjusted up or down experimentally.

3.2.3 Fiber-Optic Strain Gages

An optical fiber is a plastic fiber that has the ability to guide light along
its axis. The three parts of an optical fiber are (1) core, (2) cladding, and
(3) protective coating, or buffer. Optical fibers are classified as multimode,
having a core large enough to propagate more than one mode of light; and
single-mode, through which only one mode of light propagates. Fiber-
optic strain gages are made of single mode fiber. To use a fiber optic as a
strain gage a property of light such as phase or frequency of a propagating
light is correlated to an externally induced strain. There are many types
of fiber-optic strain gages but Fiber Brag Grating (*a.k.a.* FBG) is the most
widely used type for monitoring of geomechanics.

Bragg grating is a set of closely spaced refractive index changes in the
fiber. FBG strain gages employ optical communication fiber that is modified by introducing small periodic variations in the refractive index of the
core of the fiber (Fig. 3.6). Refraction index changes are made by exposing
the fiber to special lighting in a fixed pattern.

In an FBG strain gage, the fiber itself acts as the sensing medium. The
gage is bonded to the area where strain is measured. When a light beam
travels through an FBG gage, light is refracted, and some of it is reflected,
while the rest is transmitted (Fig. 3.7). The intensity of the reflected and
transmitted light depends on the FBG pitch. The gage responds faithfully

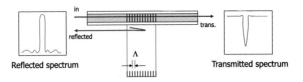

Figure 3.7 Operating principal of FBG strain gage

to changes in strain, with strain leading to change in refracted and reflected light.

A single sensing fiber can be several kilometers long, and can contain as many as 100 FBG sensors. Sensors typically use light from the infrared region of the spectrum. Each FBG strain gauge is designed to reflect a slightly different color of light. The precise light reflected from each FBG sensor depends on the period of the grating of that sensor. Gratings with large periods reflect redder, or longer wavelengths of light, while gratings with small periods reflect bluer, or shorter wavelengths of light. A light is shined down the sensing fiber and the color of the laser light is gradually varied, with each grating in turn reflecting its characteristic color back towards the light source.

Fiber-optic gages offer a number of advantages, including (1) long term stability, (2) small size, (3) electrical immunity, and that (4) many sensors can be multiplexed in one fiber. Sensors are, however, (i) fragile and (ii) the technology remains complex. Nevertheless, the technology is gaining in popularity.

3.2.4 Installation of Strain Gages

The installation of all strain gages generally follows similar steps. The installation surface must be clean, dry, and free of grease. On metal structures, all rust must be removed with a sander to create a flat surface. Next, the exact location where the strain is measured should be marked with a sharp tool. Foil strain gages require the use of a number of additional steps including surface abrasion, conditioning, and neutralization that have been presented above.

Some consideration during installation of strain gages must be given to the orientation or the gages. Strain gages should be positioned parallel to the axis of interest. Failure to do so results in measuring the cosine component of the load. The magnitude of the error is small for small deviations, but increases as the deviation increases. Another important consideration

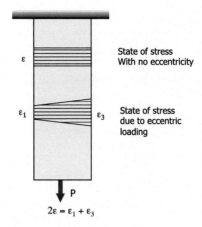

State of stress
With no eccentricity

State of stress
due to eccentric
loading

$2\varepsilon = \varepsilon_1 + \varepsilon_3$

Figure 3.8 Effect of eccentricity on measured strain

is load eccentricity, which is common in most structures. If the structure is subjected to eccentric loading, it introduces bending, so spot measurement of strain may not necessarily capture the correct average applied strain (Fig. 3.8). Bending will increase the strain on one side of the neutral axis and decrease it on the other side. Axial strain can be isolated from bending strain by installing gauges on the opposite sides of the test member, and averaging the results.

3.3 MONITORING LOAD

In geotechnical engineering load cells are very commonly used in load tests and proof tests for measuring loads applied to piles, drilled shafts, tie-backs, rock-bolts, cranes, and sheet pile walls. Load cells can also be used for monitoring loads applied to columns, beams, cables, and material testing frames.

3.3.1 Electric Load Cells

In general, loads can be determined using one of two principal ways; either by comparing masses like in a balance or by measuring the deformation or strain resulting from the application of the load. A third method is to monitor the pressure developed in a sealed container due to the application of the load (*e.g.*, hydraulic jacks). Today, loads are typically measured taking advantage of Hook's law, by measuring the strain occurring in a member of known modulus of elasticity (E) and dimensions (Fig. 3.9). This is com-

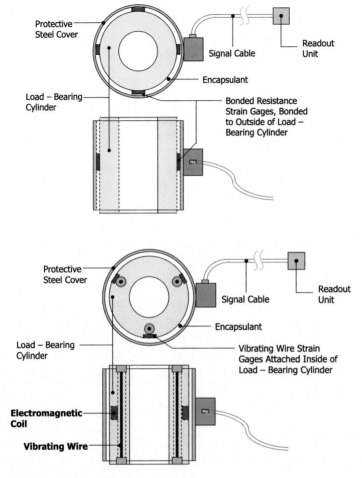

Figure 3.9 Schematic of typical donut load cell design with (A) 4 foil strain gages – top, and (B) 3 vibrating wire strain gages-bottom

monly done for a wide range of consumer and commercial applications ranging from bathroom scales to grocery checkout counters. Thus monitoring of load is simply an extension of the above discussion on strain.

Load cells take many shapes; however, the most common shapes employed in civil engineering are the donut, S-beam, and universal load cells. Donut load cells are popular because many of the encountered loads are axial, and a donut cell permits passing the load through the center of the load cell to avoid eccentric loading of the load cell. Regardless of the design, 100% of the measured force must travel through the load cell. Thus

Figure 3.10 Donut load cell with six vibrating wire strain gages

the load cell must be strong enough to resist the stresses caused by the measured force, and remain in the elastic range. In donut load cells, compressive strain is typically measured using foil strain gages for indoor and dynamic applications. Alternatively, vibrating wire strain gaged load cells are typically employed for long term monitoring of outdoor applications (Fig. 3.10). In either case, strain is converted to an electrical signal that is calibrated to the applied load.

Many electric resistance load cells depend on measurement of shear such as the universal load cell, and the S-shaped load cell. A universal load cell consists of a metal disk, with a recess on one face and a protrusion on the opposite face. Eight holes are typically drilled perpendicular to the cross-section creating 8 ribs that can be instrumented (Fig. 3.11). A tension Collar is typically supplied on the side of the recess. This arrangement permits compressive or tensile load to travel through the edge, through the ribs to the protrusion. Load is monitored my measuring shear strain in the ribs, where shear is constant. The universal load cell design is one of most popular designs, because it provides 16 surfaces (on both sides of the ribs) for installing foil strain gages. The accuracy of universal load cells increases with the increase in the number of strain gages employed. Universal load cells are available in a wide range of capacities, and many are supplied in a number of different accuracies depending on the number of strain gages employed.

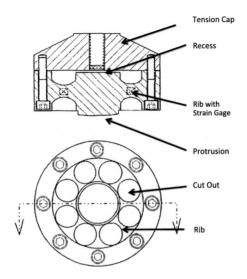

Figure 3.11 Cross section of universal load cell (photo credit Interface, Inc.)

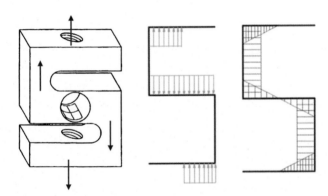

Figure 3.12 Operating principal of S-shaped load cell: loading diagram (LHS), shear–force diagram (center), bending moment diagram (RHS)

The S-shaped load cell also works in shear as shown in Fig. 3.12. Load travels from one end to another through the S-shape. A cutout hole is drilled in the middle beam where shear is constant. Foil strain gages are installed in the cutout hole, and are connected to a Wheatstone bridge circuit.

Load can never be applied in a perfectly concentric manner, especially in field applications. Thus good load cell design must include multiple strain gage measurements in order to average out stress concentrations caused by

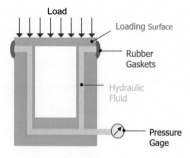

Figure 3.13 Schematic of hydraulic load cell.

Figure 3.14 Weigh in motion load cells in service

eccentric loading. Good foil strain gaged load cells can have as many as 24 foil strain gages per bridge. While good vibrating wire strain gaged load cells can have as many as 6 vibrating wire strain gages (Fig. 3.10). As the number of strain gages increases, the specified accuracy of the load cell increases. However, caution must be exercised to prevent excessive eccentric loading of any load cells since all load cells are designed to carry load along their primary axis only.

3.3.2 Hydraulic Load Cells

Hydraulic load cells measure the pressure generated in a fluid filled chamber, instead of measuring strain (Fig. 3.13). The principal is relatively old. For many decades before load electric load cells became popular and affordable, loads were measured in load tests by monitoring the pressure in the hydraulic actuator and correcting for friction losses in the actuator bushing through calibration; or at times by simply reducing the derived load by 10–15% to account for friction in the bushings. Hydraulic load cells can be used traffic engineering applications where load cells are built into weigh in motion stations to measure the weight of traffic (Fig. 3.14). In that environment loads are dynamic and VW technology is too slow for monitoring

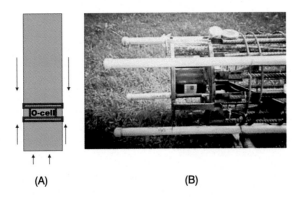

(A)   (B)

Figure 3.15 Osterberg load cell: (A) principal of operation (LHS), and (B) as installed on a drilled shaft cage (RHS)

traffic, and the environment is too harsh for foil strain gages to survive. Hydraulic load cells can also accommodate eccentric loading better than strain gage based load cells.

3.3.3 Osterberg Load Cells

An Osterberg cell is similar to a hydraulic load cell. The term Osterberg refers to the inventor the late Prof. Jorj Osterberg. An O-Cell consists of a calibrated hydraulic jack that is used to apply the load on a drilled shaft foundation at a point somewhere along the shaft (Fig. 3.15). Therefore, the top of the shaft is pushed against the bottom of the shaft, without the need for a reaction weight or piles as is commonly done, which greatly reduces the cost of setting up a load test. Usually, one half of the shaft fails before the other half, unless the engineers have been exceptionally careful and lucky. As a result an Osterberg test typically yields a lower estimate of capacity than a conventional test. Multiple O-cells have been employed to test the load transfer in specific soil layers.

In addition to measuring the pressure in the O-Cell and converting it to load, the expansion of the load cell, and the displacement of the shaft top are also monitored. This information is used to plot two load settlement curves representing the load tests on the top and bottom portions of the shaft. Both load settlement curves are combined to derive a load settlement curves that represents the behavior of the shaft had it been loaded, conventionally, at the top. After the test is complete, the O-cell is grouted with a cementitious grout and the shaft is typically used as a production shaft.

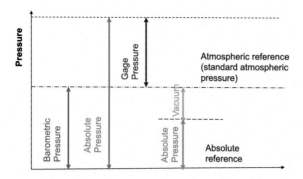

Figure 3.16 Pressure terminology

3.4 MONITORING PRESSURE

Geotechnical engineers are concerned with monitoring two types of pressure, piezometric pressure and total stress (*a.k.a.* total earth pressure). Most geotechnical phenomena depend on the so-called principle of effective stress, which is defined as the average stress carried by the soil skeleton. The effective stress (σ') acting on a soil is calculated as the difference between two parameters, total stress (σ) and pore water pressure (u), *as follows:*

$$\sigma' = \sigma - u. \tag{3.8}$$

Effective stress is a concept, much like love or patriotism, in that it cannot be directly measured. Therefore, the effective stress is measured by measuring total stress and the pore water pressure (*a.k.a.* piezometric pressure). Finally, it is important to recall that pressure is equal in all directions while stress is directional.

3.4.1 Monitoring of Piezometric Pressure
3.4.1.1 Pressure Terminology

The *barometric (a.k.a. atmospheric) pressure* is equal to the weight of the overlying air and it is nominally considered to be 14.7 psi or 101 kPa. Barometric pressure decreases with altitude, and it is subject to fluctuation due to weather patterns. Barometric pressure represents zero gauge pressure, or the measurement datum. Positive gauge pressure is simply referred to as gauge pressure, while negative values of gauge pressure are referred to as *vacuum* (Fig. 3.16). Absolute pressure is equal to the sum of the atmospheric pressure and the gauge pressure (positive or negative). Absolute pressure is typically employed in calculations, because it negates the effect of weather changes on measured piezometric pressures, which can be important in some projects.

3.4.1.2 Piezometric Measurements

When dealing with piezometric (water) pressure it is important to recall Bernoulli's equation which states that the total head (h) is equal to the sum of the pressure head (p/g = pressure/fluid density), elevation head (Z), and velocity head ($v^2/2g$ where, g is the gravitational acceleration), as follows:

$$h = \frac{p}{\gamma} + Z + \frac{v^2}{2g}. \tag{3.9}$$

The velocity head is very small in nearly all geotechnical applications, and is neglected. Flow can only take place between two points if there is a difference in their total heads, with flow emanating from the region with higher total head towards the region with lower total head.

In geotechnical engineering practice, piezometric measurements help engineers to control placement of fills, predict slope stability, deal with lateral earth pressures, deal with uplift pressures and buoyancy, and monitor seepage and verify models of flow. Piezometers can also be used in environmental engineering practice to monitor surface water runoff, water levels at contaminated sites, and to find the rate and direction of movement of contaminant plumes. In hydrology, piezometers are used to predict volume of water in an aquifer and its recharge rate, monitor tidal effects on coastal soils, and encroachment of salt water into fresh water aquifers.

In geotechnical engineering, the simplest method for monitoring piezometric pressure is to employ an observation well, or an open standpipe piezometer. An observation well permits monitoring the natural ground water table (level). It consists of a boring that is lined with a slotted pipe, with the annular gap between the pipe and the natural soil packed with filter material (typically sand). A seal made of Bentonite clay is provided at the top to prevent surface water from affecting the measurement (Fig. 3.17).

Total head does not always increase linearly with depth, for example, a consolidating clay layer sandwiched between two sandy layers may experience a higher head at its center. An open standpipe piezometer is employed to measure the total head at a specific point. It is similar to an observation well except that slotted pipe and surrounding filter soil is provided in a short segment of the borehole (on the order of 3 ft., 0.9 m) and the rest

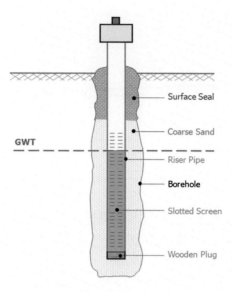

Figure 3.17 Typical observation well setup

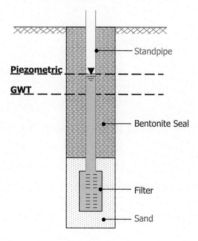

Figure 3.18 Typical standpipe piezometer setup

of the borehole is filled with seal material typically made of Bentonite clay (Fig. 3.18).

A water level indicator can be employed with both observation wells and open standpipe piezometers to conveniently determine the water level in the standpipe. A water level indicator consists of a tape measure fitted with a battery-operated probe at its end. The probe contains 2 electrodes

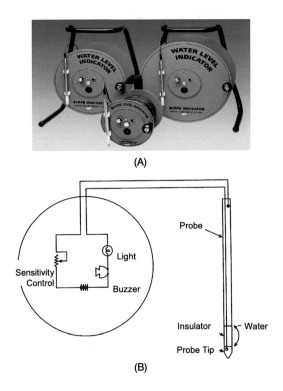

Figure 3.19 Water level indicator: (A) device (top) and (B) circuit (bottom)

(Fig. 3.19). When the probe reaches ground water, the water completes the circuit between the electrodes and the probe buzzes and/or flashes. This process is simple, but labor intensive. Therefore pressure transducers are commonly employed for long-term monitoring projects.

3.4.1.3 Piezometric Pressure Transducers

A large variety of pressure transducers (*i.e.*, sensors) exist; they typically depend on having a sensing face and a way of measuring its deformation. Sensing face deformation can be measured using a foil strain gage, a vibrating wire strain gage, inductance, piezoelectric properties, and ultrasonic proximity sensors. Pressure measurement is a highly developed field with geotechnical measurements representing a tiny fraction of the business volume. A wide range of transducers is made to satisfy a variety of industrial and engineering needs. Piezometric pressure sensors are distinguished from

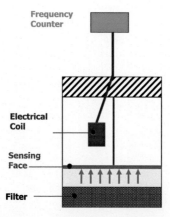

Figure 3.20 Schematic diagram of vibrating wire (VW) piezometer

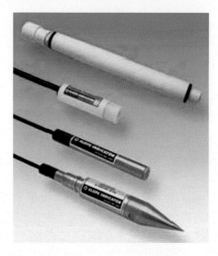

Figure 3.21 Photographs of piezometers (photo credit Slope Indicator Inc.)

nearly all other pressure sensors in that they are typically fitted with a filter that separates earth pressure from pore water pressure (Fig. 3.20).

In geotechnical engineering, the most commonly employed piezometer design involves a vibrating wire strain gage employed to monitor the deformation of a sensing face under piezometric pressure (Fig. 3.21). Vibrating wire gages are preferred due to their long-term stability, and the technology's ability to tolerate wet wiring. Hermetically sealed foil gages are sometimes used with built in data acquisition systems. Early vibrating wire piezometers have been employed for more than three decades to monitor

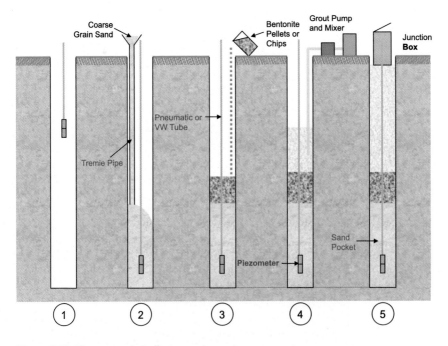

Figure 3.22 Piezometer installation steps

piezometric pressures in several dams, which is a testament to their durability. As a result, VW piezometers are one of the most commonly deployed sensors in practice.

Vibrating wire piezometers are typically installed in a borehole (Fig. 3.22). The measurement location is typically filled with coarse sand, which acts as a filter. A Bentonite seal is often employed to hydraulically separate the measurement location from the rest of the borehole. Multiple piezometers can be installed, at the same boring, in this manner.

A number of variations on the classical installation method described above are sometimes employed. The oldest is to employ a so-called *Push-In* piezometer, which consists of a conical device resembling a cone penetrometer (CPT) and the measurement port is located on the side (Fig. 3.21). The device can be pushed down, in some soils, from the ground surface using a CPT rig, thus eliminating the need for a borehole. Alternatively, a Push-In piezometer is sometimes employed to reduce the response time of the piezometer system due to the formation of low permeability smear zone along the perimeter and bottom of the borehole due to the circulating auger. The smear zone around a pushed one-inch diameter piezometer

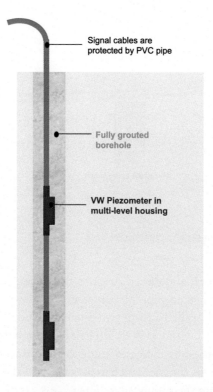

Signal cables are
protected by PVC pipe

Fully grouted
borehole

VW Piezometer in
multi-level housing

Figure 3.23 Schematic of fully grouted borehole with multiple VW piezometer

is thinner than that at the bottom or side of a four-inch borehole. So the device is pushed beneath the bottom of the boring to the depth of interest.

A relatively new way to install multiple piezometers in the same borehole has emerged in the last decade. The procedure employs a fully grouted borehole (Fig. 3.23). The main idea is that the grout permits the transfer of hydraulic pressure laterally, over a distance measured in inches, but not vertically, over a distance measured in feet. The grout employed has a hydraulic conductivity on the order of 1×10^{-4} to 1×10^{-5} cm/s. The procedure is possible in certain soils only and cannot be universally applied.

3.4.1.4 Pneumatic Piezometers

Pneumatic piezometers have been employed to measure piezometric pressure for many decades, and may now be considered near the end of the product cycle. They are still employed when concerns over the durability of the electronic components is questionable due to environmental factors

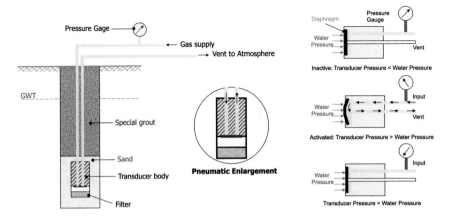

Figure 3.24 Schematic of pneumatic piezometer

such as temperature or pH, and direct access to the measurement location is not possible through a stand pipe piezometer. The device consists of a rubber diaphragm that is pushed against two pneumatic ports by the piezometric pressure (Fig. 3.24). Compressed nitrogen gas is sent down a tube to one of the ports. As the pressure increases inside the transducer housing, the pressure of the gas exceeds the piezometric pressure and the diaphragm is forced outward away from the supply vent. The excess gas escapes from the return vent. Manual adjustments to the supply pressure are performed until it is just equal to the piezometric pressure and the value is noted. The technology remains popular in environments of high corrosion or temperature, but is difficult to automate.

3.4.1.5 Piezometric Time Lag

A finite time is required for the value of water pressure in the measuring system to reach the value of water pressure in the surrounding soil. This equalization process depends on soil permeability, k, shape factor of piezometric tip, F, volume factor of the measuring system, V, presence of air or gases, and volume change required to actuate the sensor unit.

The time lag to reach a specific equalization ratio was developed by Hvorslev (1951). An equalization ratio of 90% represents the measurement system reading 90% of the actual piezometric pressure in the ground. Hvorslev's theory predicts that 5.8 days are needed to reach an equalization ratio of 90% for the standpipe piezometer shown in Fig. 3.25. The effect of time lag is small in many long-term measurements. Nevertheless, time

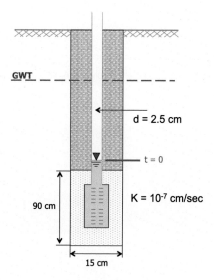

GWT

d = 2.5 cm

t = 0

K = 10⁻⁷ cm/sec

90 cm

15 cm

Figure 3.25 Example used to compute Hvorslev's time lag

lag can mask important phenomena when rapid change in pore pressures occur in low permeability soils.

3.4.2 Monitoring of Total Stress (Total Earth Pressure)

In most projects it is possible to infer the effective stress by measuring the piezometric pressure and guessing the total stress, knowing the unit weights of the soil layers involved. Nevertheless, earth pressure cells are employed to verify design assumptions where the magnitude of earth pressure is uncertain. For example, total stress cells are commonly employed to determine stresses acting on deep basements, slurry walls, and tunnels, especially when new construction activities may redistribute stresses acting on existing structures.

Total stress cells typically consist of a fluid filled pressure chamber that is attached to a pressure transducer, which is used to monitor pressure fluctuations caused by changes in total stress. The cell measures the combined effect of effective stress and pore water pressure. The most common design consists of two steel plates welded together, around the periphery. The plates are separated by a narrow gap filled with fluid, and any pressure on the external plates is transferred to the fluid, which activates the pressure transducer (Fig. 3.26). A vibrating wire pressure gauge is typically employed due to the technology's long-term stability and ability to tolerate

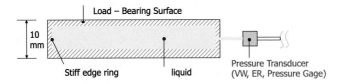

Figure 3.26 Schematic of total stress cell

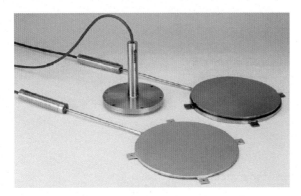

Figure 3.27 Total stress cells

wet wiring. However, electric resistance pressure transducers can also be used in situations where dynamic pressures are monitored.

Total stress cells are designed to either be embedded in the ground or to be installed flush with a structural member (Fig. 3.27). In either case, care has to be employed to minimize redistribution of earth pressure around the cell. In cases where the cell is attached against a structure, cells are typically mounted with mortar in a recess such that the sensitive face is flush with the structure. Cells can also be attached to the formwork before concrete is cast. Special total stress cells are designed for mounting in diaphragm walls. These are attached to a plate of equal diameter, and a pneumatic jack in between the cell and the plate. The cell is mounted on the reinforcing steel cage. When the steel cage is in position, the jack is activated and locked, forcing the cell into contact with the soil, and reacting against the reaction plate. The jack remains activated until the concrete hardens.

In cases where total stress cells are embedded in the ground, they are either installed in pockets below a structure, or in boreholes. Cells are placed in excavated pockets and covered in hand-compacted fill before normal fill and compaction operations can resume. The excavated soil should be used and the original water content should be maintained when possible.

Figure 3.28 Avongard gage

Pockets are typically installed one meter deep, and cables are installed in trenches. If a borehole is employed, it may be difficult to employ the parent soil as a fill material, so clean sand is often employed as a fill material. In either, case it is important that the compacted fill soil has the same stiffness like the parent soil in order to prevent redistribution of earth pressure by arching. Finally, care must be exercised to ensure that the cell is oriented to capture the stress of interest.

Arching can redistribute stresses in the ground and render the total stress measurement in error. If multiple total stress cells are employed, it is important that cells are installed in a manner such that they do not influence the stress being measured by each other. Also it is important to avoid areas where local conditions may not be representative of the global conditions, such as near corners. For this reason finite element analysis is sometimes employed to determine optimal cell layout.

3.5 MONITORING DEFORMATION

Monitoring of deformation is one of the oldest and most common applications in the field of instrumentation. Historically, displacement has been monitored optically using surveying methods. However, increasingly other electrical and at times manual methods are more commonplace. The following is a summary of the most common methods employed for monitoring deformation of soils and structures in the context of geomechanics.

3.5.1 Manual Methods

There is a wide range of manual and mechanical devices that can be adapted for monitoring displacement including dial gages, Vernier calipers, and depth indicators. Perhaps the most commonly used such device is the Avongard gage, which is very commonly used to monitor displacement across cracks and joints. An Avongard gage consists of two plastic plates,

Figure 3.29 Moire tell tale gage

each attached to on one side of the crack being monitored. A grid printed on the backing plate, and a cross hair is printed on the front plate. Initially, the crosshair is centered in the middle of the grid. Displacement can then be monitored by tracking the movement of the cross hair relative to the grid (Fig. 3.28). It is, however, important to note that all crack measurements experience daily and annual thermal cycles and that many apparent joint movements are simply a result of thermal cycles. A recent variation on the Avongard Gage is the Moire tell tales, which employs two sets of patterns made of concentric circles printed on each plate. Initially the circles are centered (Fig. 3.29). If movement occurs the interaction of the two patterns results in the generation of a Moire pattern, which can be converted into precise movement.

3.5.2 Linear Potentiometers

The simplest and cheapest way to measure displacement electrically in geomechanics applications is using a linear potentiometer. A linear potentiometer is simply a variable resistance. The resistor typically consists of a ceramic material that has an electric wire wound around it. A slider contact permits transferring movement between the measurement location and a location in the middle of the resistor (Fig. 3.30). A constant voltage is supplied across the entire resistor and voltage drop between the contact and the reference end is measured. Displacement is then deduced as the length of the resistance element multiplied by the ratio of measured voltage divided by the supply voltage. Linear potentiometers are inexpensive; but more importantly their output is sufficiently large and therefore does not require signal conditioning. As a result, linear potentiometers are a common component in many displacement monitoring systems. Rotary resistive potentiometers and optical linear and rotary encoders, which fol-

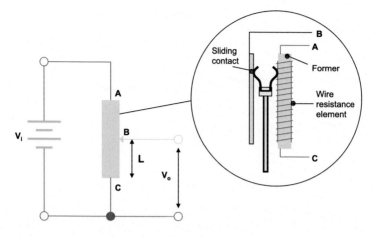

Figure 3.30 Operating principal of linear potentiometer

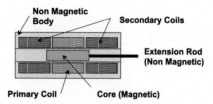

Figure 3.31 Schematic diagram of LVDT in null position

low similar principles, are commonly employed in mechanical engineering, but less common for monitoring of geomechanics.

A *soil strain meter* that can be employed to monitor soil movements is typically built using a sliding potentiometer and a rod mounted between two anchors. Relative movement between anchors causes a change in the output of the potentiometer. The initial reading of the strain meter is used as a datum. Subsequent readings are compared to the datum to calculate the magnitude and rate of movement.

3.5.3 LVDT

A *linear variable differential transformer (LVDT)* is an electromechanical trans-ducer that converts motion of an object to an electrical signal. LVDTs are available in a variety of ranges, but most are 1/4 to 8 inches (6–200 mm). An LVDT consists of 3 independent coils placed in series between two nonmagnetic concentric stainless steel tubes (Fig. 3.31). The primary coil,

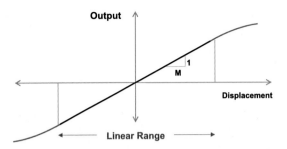

Figure 3.32 Nonlinearity of typical LVDT output

which is located in the middle, is excited with a high frequency alternating current on the order of 3000 Hz. A core made of mild steel is coupled mechanically to the moving object and is placed inside the smaller tube, while the outside tube is fixed to the reference datum. If the core is located midway between the two secondary coils it induces equal magnetic flux in the two secondary coils and the voltages induced in the secondary windings cancel each other. When the core moves away from the null position differential AC current is induced in the two secondary coils, and the magnitude of the difference in current can be calibrated to the distance traveled by the core. The AC current is converted to DC current using a demodulator.

LVDTs became popular because AC signals are more immune to electrical current interference than DC signals. However, at the present time most LVDTS are packaged as direct current differential transformer (DCDT). A DCDT consists of (1) a modulator that converts DC current to AC current (2) an LVDT, and (3) a demodulator that converts AC current to DC current. A DCDT effectively operates as a DC device. Both LVDTs and DCDTs have a linear range over which they are expected to operate and a large nonlinear range over which readings do not fit the linear calibration constant, so care must be exercised in the placement of the core to ensure that the readings are in the linear calibration range (Fig. 3.32).

LVDTs and DCDTs are commonly used in a wide variety of applications. They are environmentally robust and tend to be long lasting because of their friction free operation. Additionally, they also provide an absolute as opposed to incremental output and are insensitive to misaligned loading. They are primarily used in laboratory and short-term field applications, such as load tests on piles. They are also employed for displacement measurement of embedded concrete anchors, deformation and creep of

Figure 3.33 VW joint meter employed to monitor an expansion joint

concrete walls and settlement of bedrock. LVDTs offer a good dynamic response and can be used for dynamic measurements of fatigue.

3.5.4 Vibrating Wire Joint Meters

VW joint meters (*a.k.a.* Crack Gages) consist of a spring attached in series to a vibrating wire strain gage. The purpose of the spring is to extend the range of the strain gage from the micron range up to 100 mm. The load in the spring can be linearly related to its elongation by the spring constant. Load is of course directly related to strain. Therefore, the displacement of the spring can be linearly calibrated to the measured strain. Thus one end of the spring is attached to a VW strain gage and the other end is attached to a sliding mechanism (Fig. 3.33). The strain gage and the sliding end are connected to expandable or groutable anchors that are installed at opposite sides of the joint being monitored.

3.5.5 Rod Extensometers

Rod extensometers are commonly employed for monitoring settlement or heave in excavations, near foundations and embankments, or above tunnels, and other underground openings. Rod extensometers measures distance between the top of the borehole and a number of anchors embedded in borehole. The anchors are attached to the soil using either (1) groutable anchors, (2) hooks known as Borros points, (3) snap in anchors, or (4) hydraulic anchors. Each anchor is attached to a rod that transfers its movement to the top of the borehole (Fig. 3.34). The head of the borehole is typically configured with a number of linear potentiometers, VW crack meters, or LVDTs to monitor the movement of the rods. Either the top of the borehole or the bottom most anchor is assumed stationary, and the information is combined to provide the displacement of each anchor. Rods are either made of stainless steel or fiberglass. Fiberglass requires about half the labor involved to install stainless rods, but have a much higher coeffi-

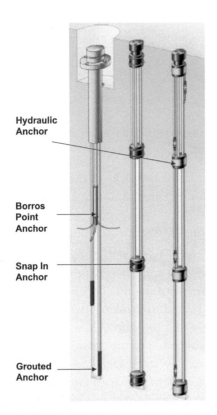

Figure 3.34 Typical configuration of rod extensometer

cient of expansion. Therefore fiberglass is preferred in environments where temperature is expected to remain constant, while stainless is preferred in situations where ground temperature is expected to change, such as near excavations.

3.5.6 Probe Extensometers

A number of devices involve lowering a probe into a borehole to track the location of pre-placed markers using magnetic or inductance properties of the markers. The markers are typically made of low-grade magnetic stainless steel rings, brass rings, or magnets. The markers are installed in the borehole using a variety of patented ways. For example, the Sondex method involves installing stainless steel rings on a corrugated pipe that is installed outside of an inclinometer casing (Fig. 3.35). The Increx system utilizes brass rings positioned at one-meter intervals along inclinometer casing. The magnetic

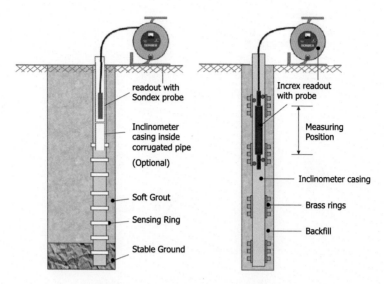

Figure 3.35 Typical configuration of probe extensometers (A) Sondex (LHS) and Increx (RHS)

Figure 3.36 Photo of spider magnet and access tube

probe involves spider magnets that are installed around a small diameter access tube (Fig. 3.36). All methods involve conducting an initial survey of the installed markers and comparing subsequent surveys to the initial one.

3.5.7 Slope Extensometers

A number of approaches can be employed to locate the depth of a slip surface and its rate of movement. Shear strips are used to locate slip surfaces. They consist of a parallel electrical circuit made up of resistors mounted on a brittle backing strip. The resistors are typically spaced two to three feet apart. The circuit is placed in a borehole that is drilled perpendicular to

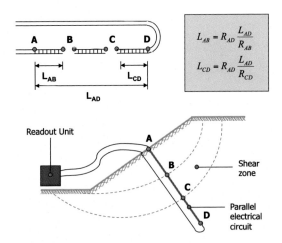

Figure 3.37 Schematic diagram of a shear strip

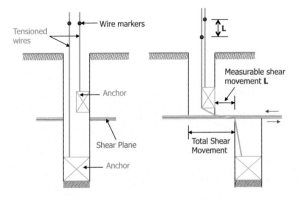

Figure 3.38 Schematic diagram of slope extensometer

the slope (Fig. 3.37). The resistance of the entire circuit is measured at the beginning and end of the circuit. The method involves locating the position of the slip circle using the resistance measured after slope movements. Location of up to two breaks can be determined by measuring resistances at top and bottom of the shear strip.

Slope extensometers are used to determine the rate of movement of a slip surface. Anchors are installed above and below a slip surface. Tensioned wires are attached to each marker (Fig. 3.38). Initial shear deformation will not cause an equivalent movement in the wire markers, but wire markers

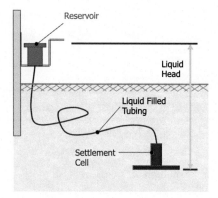

Figure 3.39 Schematic diagram of settlement cell

will change position after the borehole has been separated completely, and the rate of movement will correspond to movement of the wire markers.

3.5.8 Liquid Level Gages

A common liquid level can be used as a datum to measure differential settlement. The technique has been used since ancient times to level buildings. A settlement cell consists of a liquid reservoir, tubing, and a pressure transducer that monitors the settlement (Fig. 3.39). The transducer measures the pressure created by the column of liquid in the tubing. As the transducer settles with the surrounding ground, the height of the liquid column is increased and the settlement cell measures higher pressure. Settlement is calculated by converting the change in pressure to liquid head.

Settlement cells are supplied in two types, vented and non-vented. The tubing of non-vented settlement cells consist of supply and return lines, in addition to the electric wiring. The device measures the combined effect of piezometric changes and barometric pressure changes, so barometric pressure must be independently measured and mathematically subtracted from the piezometric pressure. The tubing of vented settlement cells adds a vent line, and the sensor measures differential pressure so it automatically corrects for barometric pressure changes.

A multipoint liquid level system can also be used to monitor settlement over a wide area. The system consists of a distribution manifold that is able to establish a common liquid level at multiple measurement locations with submillimeter accuracy (Fig. 3.40). Displacement is measured against the common liquid level using a system made of an LVDT and a floating disk

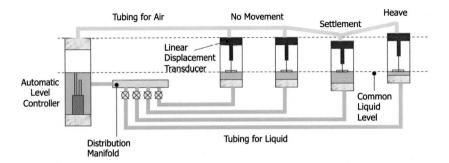

Figure 3.40 Schematic diagram of multipoint liquid level system

at each measurement location. The system has been used for monitoring differential settlement in structures and controlling compensation grouting in sensitive projects.

3.5.9 Optical Methods

Recent availability of robotic total stations has made it possible to employ classical surveying methods to automatically monitor movement of soils and structures at a very low cost per target. Automatic motorized total stations (AMTS) are employed together with optical prisms that are installed at each monitored location. The AMTS can be located as far as 500 m from the prisms, but practical considerations can limit this distance considerably. The robotic total station is able to monitor the displacement of each prism in three orthogonal coordinates, with some total stations able to achieve sub-millimeter accuracy. Automated readings are set at user-defined intervals, typically on the order of 1–3 hours. Accuracy increases as the number of control points increases, and a minimum of 5 is recommended.

LIDAR which refers to light detection and ranging laser has also been recently introduced to the field of monitoring geomechanics. LIDAR employs laser scanners that are able to achieve very dense cloud maps. To date the technique is less competitive than robotic total stations in terms of cost and accuracy, but it provides for vastly denser cloud point maps and its cost is dropping rapidly.

3.6 MONITORING TILT

The words tiltmeter, inclinometer, and clinometer all refer to devices that measure angular displacement with respect to the vertical gravity vector.

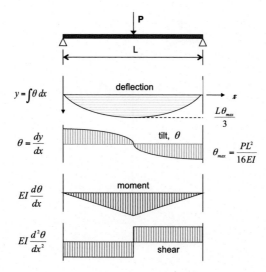

Figure 3.41 Relationship between tilt and deflection, bending, shear

Tilt is one of the most versatile measurements in engineering because integrated tilt yields deflection or displacement, while differentiating tilt yields moment, and double differentiation yields shear (Fig. 3.41).

Tilt is employed for monitoring behavior of structures under load such as rotation of retaining walls, piers, and piles. Tilt is also commonly used for documenting the effects of nearby excavations and providing an early warning of potential damage.

3.6.1 Measurement of Tilt

A variety of sensors are employed for measuring tilt, which are presented next.

3.6.1.1 Electrolytic Tilt Sensors

An electrolytic tilt sensor consists of a glass vial, similar to a carpenter's level. The vial contains a conductive liquid and has three electrodes that are connected in a Wheatstone circuit arrangement (Fig. 3.42). When the device tilts, the bubble in the vial moves, which increases the resistance between two pairs electrodes and decreases the resistance between the opposite pair. The circuit is similar to that used to monitor strain in electric resistance strain gages. Sensors are supplied with a constant voltage and voltage change is correlated with tilt.

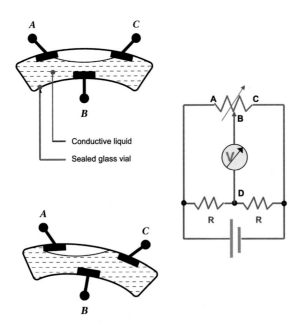

Figure 3.42 Schematic diagram of electrolytic tilt sensor (photo credit Slope Indicator, Inc.)

Electrolytic tilt sensors are very sensitive with sensitivities down to 1 arc-second, but their range is typically limited to less than ±10°. This makes them suitable for structural applications requiring high resolution. Electrolytic tilt sensors are robust because they do not have any moving parts.

3.6.1.2 Accelerometric Tilt Sensor

Accelerometric tiltmeters employ a force balance servo accelerometer to measure inclination. The device consists of a steel pendulous mass suspended over a precise null detector. The mass is also surrounded by pairs of torque coils (Fig. 3.43). When the device tilts, the pendulous mass moves, and the null detector activates a feedback control circuit that excites the torque coils to push the mass back to the null position. Electricity passing through a coil has a magnetic effect that is used to push the pendulum to the null position. The current in the coil is regulated by means of a servo amplifier, hence the name. The power required to push the pendulum to the null position is calibrated against tilt. Force balance servo accelerometers are commonly available in bi-axial and submersible models. They possess a wide operating range up to ±50°.

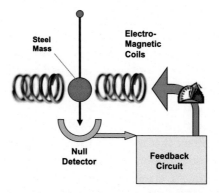

Figure 3.43 Operating principal of force balance servo accelerometer

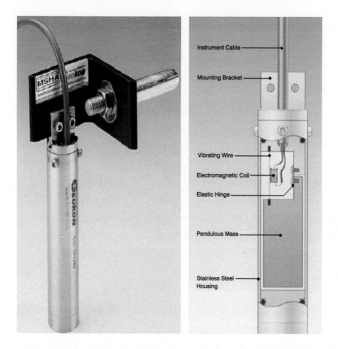

Figure 3.44 Vibrating wire tilt sensor (photo credit Geokon, Inc.)

3.6.1.3 Vibrating Wire Tilt Sensors

A vibrating wire tiltmeter consists of a heavy pendulous mass that can swing under the force of gravity. Attached to the mass is a tensioned vibrating wire strain gage (Fig. 3.44). When the mass swings it changes the tension in the strain gage. The strain gage is read as discussed above.

3.6.1.4 MEMS Based Tilt Sensors

MEMS based tilt sensors employ micro-electromechanical systems (MEMS) technology, where sensors are embedded in a semiconducting chip. MEMS sensors typically consist of a microprocessor packaged together with several components that can sense the environment. MEMS sensor components are approximately 1 to 100 microns in size, with the sensor being generally less than a millimeter in size. Tilt is typically measured by means of a solid-state accelerometer that measures the effect of gravity on a tiny mass suspended in an elastic support structure. When the device tilts, the mass moves slightly, causing a change in the capacitance between it and the supporting structure. The tilt angle is calculated from the measured capacitances. MEMS are inexpensive, shock tolerant, have good frequency response, require low power, produce little noise, and are thus becoming increasingly popular in many applications involving tilt and acceleration monitoring.

3.6.2 Tilt Beams

Because structures do not deform uniformly, it is difficult to compute lateral sway (displacement) by integrating tilt. A common solution is to attach a tilt sensor to a beam. The beam can bridge over nonuniform tilt, providing the average tilt between its two ends (Fig. 3.45).

Another popular application for beam sensors is for monitoring settlement of active train tracks (Fig. 3.46). Beam sensors permit monitoring of settlement with very high frequency, as often as multiple times per minute, thus permitting setting up of an alarm system in case the tracks are subject to settlement or heave resulting from construction activities.

3.6.3 Inclinometers

The inclinometer is one of the oldest and most versatile devices employed for monitoring geomechanics. At the present time, inclinometers are available in a number of types, including traversing inclinometers, in-place inclinometers, and shape accelerometer arrays.

3.6.3.1 Traversing Inclinometers

The setup for a traversing inclinometer includes (1) casing, (2) probe, (3) data logger, and (4) software. Casing is typically made of polymer and is permanently affixed in a borehole with grout. The key characteristic of inclinometer casing is the presence of 4 precisely-cut longitudinal

Figure 3.45 Beam sensor (photo credit GZA Geoenvironmental of NY)

Figure 3.46 Beam sensor employed to monitor settlement of rail track (photo credit RST Instruments)

grooves along the length of the casing (Fig. 3.47). The probe of a traversing inclinometer consists of a biaxial accelerometer mounted inside an approximately 0.5 m long metal tube. The tube is fitted with two sets of guide wheels, which permit lowering the probe inside the casing with the wheels traveling in the longitudinal grooves. This permits the probe to profile the shape of the casing in a consistent manner with time. Casing is installed in a borehole that passes through suspected zones of movement. Casing comes in several diameters (*e.g.*, 85, 70, and 48 mm) to accommodate different casing curvatures caused by installation of the casing in materials with varying stiffness ranging from soft soils to concrete. The casing is typically

Figure 3.47 Typical traversing inclinometer apparatus: (A) probe and casing (LHS) (photo credit Gouda Geo) and (B) traversing probe inclinometer (RHS) (photo credit Geokon, Inc.)

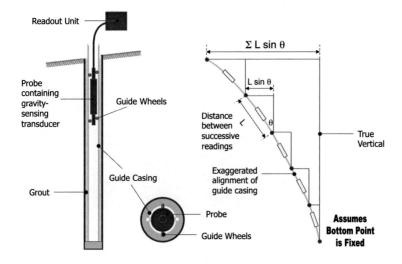

Figure 3.48 Principal of operation of traversing inclinometer

delivered in sections, approximately 3 m (10 ft. long), which are provided with watertight coupling mechanism that permits the grooves to continue uninterrupted from one section to the next.

The probe is used to profile the inclinations along the inclinometer casing as it travels in it. An important assumption is that the lowest portion of the inclinometer is stable, which can be achieved by embedding the casing in rock, or having it extend well below the region of interest (Fig. 3.48). Bottom fixity permits accurate computation of the lateral translation of the inclinometer from the measured inclinations. The grooves in the casing are typically oriented perpendicular and parallel to the anticipated movement. The probe is lowered along one set of grooves and readings are collected as it travels down and up the casing, thus providing two sets of readings. Data is collected using a data logger for the inclinations along two perpendicular planes. Thus typically, a second set of readings is also taken using the second set of grooves, and the average of 4 sets of readings is computed.

An initial survey is used to profile the inclinations along the installed casing, which are typically never straight. The initial inclinations are subtracted from all subsequent readings, to provide the inclinations of interest. Data is collected manually at internals that permit profiling the movements of interest (Fig. 3.49). The data is processed using specialized software that performs the requisite trigonometric calculations, averages the results, and registers readings from different dates to each other.

3.6.3.2 In-place Inclinometers

An in-place inclinometer is similar in principal to a traversing inclinometer, except that it consists of a series of beam sensors, 3–10 ft. long, which are daisy chained to one another. Each beam sensor consists of a single or biaxial tilt meter. The sensors are installed in conventional inclinometer casing, and connected to a data logger. Tilt is typically measured with either vibrating wire or MEMS tilt sensors. A key design detail is that the connection between individual beam sensors must permit free rotation in both axes, such that the beam sensor does not deform and remains straight (Fig. 3.50). The main advantage of the in-place system is that it permits continuous remote logging of the data, at the expense of a lower resolution, since the sensors are spaced at a greater distance than the typical reading frequency of the traversing systems.

3.6.3.3 Shape Accelerometer Arrays (SAA)

A shape accelerometer array consists of compact array of MEMS sensors that measures 3D shape and vibration. The device is typically supplied as a water proof "hose" consisting of 30 measurement links that rolls up for easy shipping and storage (Fig. 3.51). Each link consists of an accelerometer capable of resolving tilt in two orthogonal directions. The main advantage of the system is that it eliminates the need for casing to line the borehole. A preliminary survey is used to profile the initial inclinations and these values are subtracted from all subsequent readings, to provide the inclinations

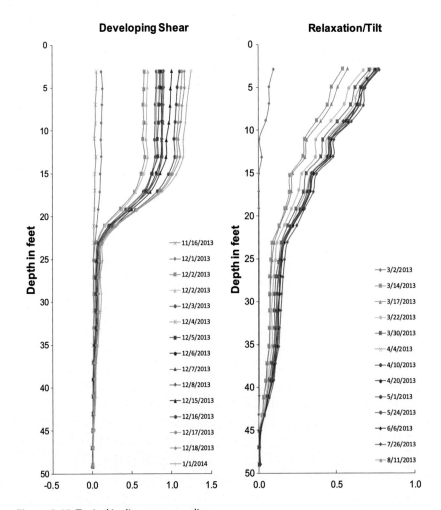

Figure 3.49 Typical inclinometer readings

of interest. Data is processed using specialized software that performs the requisite trigonometric calculations.

3.7 MONITORING VIBRATION

Failure due to vibration is a common phenomenon. Design of civil infrastructure should account for different vibrations, such as earthquakes, wind loads, tornados, hurricanes, machine vibrations, and at times vibrations due to blasting activities and/or explosions.

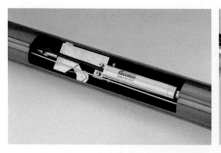

Figure 3.50 Typical in-place inclinometer (photo credit Geokon, Inc.)

Figure 3.51 Typical shape accelerometer array (photo credit Measurand, Inc.)

Liquefaction is another important phenomenon in which the strength and stiffness of the soil is reduced due to vibration. Vibration increases the pore water pressure in the ground, resulting in a reduction of the effective stress. As the effective stress (total stress minus pore water pressure) approaches zero, granular soils lose their stiffness, since stiffness of granular soils is directly related to the ambient effective stress. Liquefaction due to earthquakes has been responsible for significant damage all over the world.

Vibration monitoring began nearly 90 years ago. The first applications were related to mining and quarry blasting. Today, vibration monitoring is employed for a variety of applications including monitoring earthquakes, machines, and winds, among others. Monitoring of earthquake vibrations is employed to ensure structural integrity and to predict liquefaction of

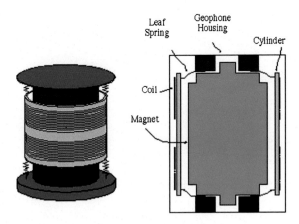

Figure 3.52 Operating principal of geophone

soils. Machine vibration studies are common for ensuring human comfort and structural integrity. Similarly, wind vibrations affect human comfort and structural integrity. In construction, vibration monitoring is often employed to prevent structural damage, limit cosmetic damage, and generate data that can resolve legal claims.

Vibrations can be classified into periodic or transient vibrations. Periodic vibrations are typically modeled using oscillatory simple harmonic motion consisting of an elastic member that stores potential energy and a mass that stores kinetic energy. Vibration is the transfer of kinetic to potential energy and vice versa. Damping is introduced to dissipate energy, thus permitting motion to come to an end. Simple harmonic motion can be defined in terms of (i) *frequency*, which is the number of cycles occurring per second in units of Hz, (ii) *period*, which is the time required for a motion to repeat itself, and (iii) *amplitude*, which is the distance from the mean position to the peak displacement.

3.7.1 Sensors for Monitoring Vibration

Monitoring of vibrations is commonly done using geophones and accelerometers. However, microphones and proximity sensors can also be used. The following is a brief summary of vibration monitoring sensors:

3.7.1.1 Geophones

Geophones are employed to monitor velocity. They consist of a magnetic core, surrounded by an electric coil (Fig. 3.52). The magnetic core is at-

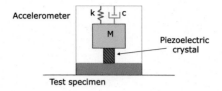

Figure 3.53 Operating principal of an accelerometer

tached to the housing and moves with it. The coil is attached to the housing using delicate leaf springs. As the ground vibrates, the housing moves but the coil tends to stay stationary. Movement of the magnet within the coil induces an electrical current that is calibrated to the velocity of vibration.

An important feature of geophones is that they can only monitor frequencies above their natural frequency, up to a specified spurious frequency. A natural frequency on the order of 10 Hz, and a spurious frequency on the order of 250 Hz are common. This frequency range precludes using geophones for certain applications, such as monitoring of pile driving. Nevertheless, geophones are one of the most common sensors employed for monitoring blasting and construction activities.

3.7.1.2 Accelerometers

Accelerometers are used to measure and record acceleration. A wide variety of accelerometers is available; however, in geotechnical engineering, piezoelectric accelerometers are the most commonly used type. Naturally occurring or manufactured piezoelectric quartz generates a transient electric charge under load or pressure. The signal dissipates quickly with time, even if the load or pressure is maintained. A piezoelectric geophone consists of a piezoelectric crystal supporting a small mass. The properties of the accelerometers are adjusted using a spring and a damping device as shown schematically in Fig. 3.53.

An important feature of accelerometers is that they can only monitor accelerations occurring at frequencies below their natural frequency. A natural frequency on the order of 5,000–20,000 Hz is common. Accelerometers are therefore more versatile than geophones, and are often used to measure velocity, by integrating the acceleration-time record.

In recent years MEMS-based accelerometers have been introduced. As discussed earlier, these employ micro-electromechanical systems (MEMS) technology, where sensors are embedded in a semiconducting chip. The device measures the effect of gravity on a tiny mass suspended in an elastic support structure. When the mass moves it causes a change in the capacitance between it and the supporting structure that can be correlated to acceleration.

3.7.1.3 Microphones

Two types of microphones may be encountered in geotechnical applications. The first is *linear overpressure microphones*, which can be used to monitor the pressure wave that typically occurs subsequent to blasting. These microphones amplify the lower frequencies in the 2–200 Hz range, which are common in blasting, but are unsuitable for monitoring sound. The second type is the *sound level microphone* that can be used to monitor the sound level for investigating sounds causing human discomfort.

3.7.1.4 Proximity Sensors

Proximity sensors are ultrasonic sensors that detect the presence of an object in front of the sensor. Ultrasonic sensors are increasingly being used in consumer products such as to turn the water on or off in faucets mounted in public restrooms. A combination of two ultrasonic sensors can be used to measure velocity, knowing the distance between the two sensors. This is commonly done to monitor the velocity and efficiency of pile hammers.

3.7.2 Installation of Geophones and Accelerometers

Geophones and accelerometers should be installed on the foundation of the structure or if access to the foundation is not possible they should be placed as close as possible. Geophones can also be buried in soil within 10 ft. of the structure, but less than 10% of the distance from the source of vibration, relative to the structure. If the vibration level is less than 0.2 g, geophones can be simply placed at the location of interest. At vibration levels between 0.2 and 1.0 g geophones can be anchored (*i.e.*, spiked) into the ground, glued into concrete, or placed under a 15 lb sand bag. Geophones and accelerometers should not be mounted on carpeted or wooden floors, thick grass and loose gravel, paneled or sheet rock walls, in order to minimize the likelihood of recording false positives. Installing geophones and accelerometers in highly trafficked areas should obviously be avoided.

Geophones must be installed level. Tri-axial geophones and accelerometers have unique orientations that must be followed. In particular vertical geophones must be oriented vertically with the top pointing against gravity.

For permanent installation, cables should ideally be routed in conduits, for among other reasons preventing trip hazard. This also helps reduce electromagnetic interference (EMI).

Data acquisition units are typically set to sample at least 10 times the anticipated frequency. A frequency of 1024 Hz is common for monitoring construction activities. Units can be set to record waveforms (single shot, continuous mode, or manually). However, a more common practice is to record the peak particle velocity (PPV) over a set period (say, 5 minutes). Units are typically set to monitor vibration at specified periods, but record data that exceeds a certain vibration threshold, only. A histogram of PPV plotted against time is generated. A trigger in the range of 0.25–0.5 in/s is common. A combination of waveform and histograms are also possible.

Most sites encounter a variety of vibration sources unrelated to the activities being monitored. Therefore, it is crucially important to establish a baseline of the ambient vibration level, prior to vibration monitoring.

Because vibration monitoring is often involved in resolving legal disputes, it is important to keep professional field notes, including logging all construction activities, serial numbers and calibration constants of all equipment, photographs of the installation and construction activities. The exact location of each sensor should be known. Modern equipment permits using a GPS for determining the exact location of each sensor.

3.8 COMMON MEASUREMENT ERRORS

3.8.1 Notation

Some of the largest failures in civil engineering have taken place due to not communicating the sign notation employed in measurements or calculations. For example, tension is considered positive in classical mechanics but is considered negative in geotechnical engineering. Many engineering organizations are growing through mergers and acquisitions, thus consolidating units with different cultural and notational standards, resulting in multiple sign notations employed within the same organization. Similarly, many medium engineering organizations operate globally in markets with divers units and notational standards. It is therefore incumbent on all engineers and scientists to note the units and sign notation employed, on every measurement and calculation sheet.

3.8.2 Conformance

Conformance is the effect of the presence of the measurement instrument on the parameter being measured. If a sensor alters a measured value, it has poor conformance. The classic example of a sensor subject to poor conformance is earth pressure cells. An earth pressure cell should have the same deformation characteristics as the soil it replaces, but it is very difficult to meet this criterion adequately, so earth pressure cells may redistribute the stresses that they are supposed to measure due to their high stiffness. Conformance errors tend to be larger in laboratory measurements than in field measurements, because the soil samples are small. For example, an LVDT rod may be too heavy for measuring surface deformation, especially in elevated-gravity centrifuge tests. Sometimes the process of measurement itself causes an error. Another example of conformance is measurement of pore pressure in soil samples. A pore pressure transducer has a diaphragm that must flex in order for it to measure the pressure. If the diaphragm flexes too much, it may reduce the pressure being measured, at least temporarily, especially in laboratory experiments.

3.8.3 Electric Noise

Noise is measurement variation caused by external factors. The most common sources of noise include interference from electric currents and radio frequency. Electric circuits having inadequate ground circuits, or having more than one ground, resulting in *ground loops*, often cause interference. Thus 60 Hz electrical frequency is the most common type of noise encountered.

Extensive noise in a system may mask small changes in the source data (Fig. 3.54). Two techniques can be employed to reduce or eliminate noise, one is filtration, and the other is oversampling. For example, a 10 Hz analog filter can be used to reduce 60 Hz noise for quasistatic measurements. Similarly, a 100 Hz filter can be used to reduce 60 Hz noise for dynamic measurements. Alternatively, if a signal is oversampled, and then averaged, noise can be effectively reduced by averaging every ten or hundred readings. For example, the data shown in Fig. 3.54 is from a foil strain gage installed on a sister bar in a drilled shaft. Shaft head displacement is plotted against the load in the sister bar. It is evident that electric noise has masked the true phenomenon, if the measured data is observed. The real data was recovered by averaging the measured data.

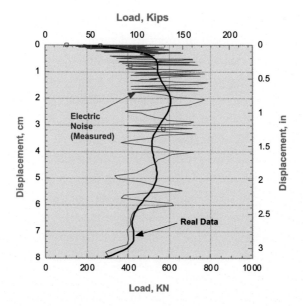

Figure 3.54 Typical electrical noise

Figure 3.55 Typical electrical drift

3.8.4 Drift

Drift is the change in readings, despite stable experimental conditions (Fig. 3.55). For example, a load cell that yields changing readings under constant load is said to be drifting. Obviously, stability is desirable. Drift is often caused by the sensor being too hot due to the use of an excessive excitation voltage. Proper sensor selection and following manufacturer specifications can reduce noise.

3.8.5 Signal Aliasing

All phenomena must be sampled in order to determine the value of the quantity being monitored. Signal aliasing is a phenomenon where a cyclical phenomenon becomes distorted (*i.e.*, aliased), as a result of being sampled infrequently. Many civil engineering measurements encounter cyclical frequencies such as daily and annual thermal and sunlight cycles. Preventing signal aliasing is particularly important when measuring tilt or piezometric pressure in coastal areas. Theoretically, a sampling frequency equal to 4 multiples of the physical frequency is necessary. In practice, sampling rate should be approximately 10 times that of the frequency of the physical phenomenon of interest. It is sometimes possible to reduce the required sampling frequency by taking advantage of other available information about the behavior. For example, temperature is cyclical. By sampling mornings and nights, its effect can be bounded.

The data shown in Fig. 3.56A depicts measurement of tilt vs. time. Building tilt is nearly always affected by the daily and seasonal thermal and sunlight cycles. When tilt is sampled a few times per week, it is evident that it is absolutely hopeless to understand the data, mainly due to the paucity of the data points. In that case, tilt is captured correctly if it is sampled 4 times per day (Fig. 3.56B).

3.8.6 Bias (Systematic) Errors

Bias is the difference between the average value of the measurements and the true value (Fig. 3.57). Bias errors tend to have a fixed value. They occur consistently every time measurement is repeated, and are therefore not susceptible to statistical analysis. A classic example of bias error is the use an 11.9″ ruler to measure distance, thinking it is a 12″ ruler. The measurement always contains a fixed 0.1″ bias error for every foot (12″) measured. User bias is another type of bias error that often makes the user unable to see the truth for what it is. Bias errors are very common in measurements and in society at large, due to the fact that the truth is difficult to know, and that the measurement environment is suboptimal for comparison. Because they are consistent and repeatable, bias errors will not reduce by repeating the experiment. They can only be studied through calibrations (*i.e.*, comparisons with different sensors).

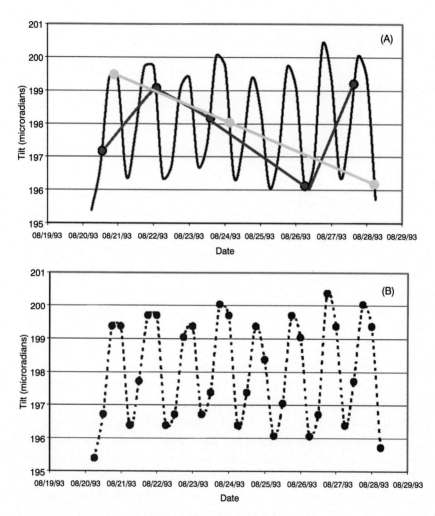

Figure 3.56 Typical signal aliasing due to improper sampling frequency (top). Correct phenomenon captured when sampling 4 times per cycle

3.8.7 Precision (Random) Errors

Precision is the difference between a single measured value and mean value of the measurements. Precision is sometimes expressed using statistical parameters such as the standard deviation, or a multiple of it. Because it tends to be random in nature, it can be studied through statistical analyses and is reduced when an experiment is repeated. In many respects, precision is similar to *repeatability*, or ability of a transducer to reproduce output reading

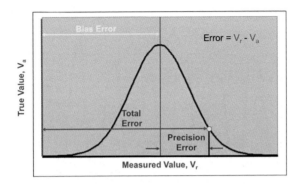

Figure 3.57 Total, bias, and precision errors

when the same measured value is applied. Precision errors are also reduced by the correct choice of measurement instrument, use of filters, recording environmental factors and application of correction factors to account for them.

3.8.8 Sampling Errors

Sampling error is the variability in measured parameter due to variations between observers. Sampling errors were very common when analog read-out units were employed. Nevertheless, the use of digital equipment does not totally eliminate sampling errors. For example, different readers may provide a different warm-up time for the same device resulting in a variation in the read out value, based on the observer. Sampling errors may be random or systematic in nature. Employing automated sampling using a data logger or a data acquisition system reduces sampling errors. Employing multiple sensors can also reduce random sampling errors.

3.8.9 Gross Errors

Gross errors, miss-readings, computational errors, and blunders are unfortunate but common. These errors can be reduced by duplicate readings, automation of calculations, and employing reviewers and checkers.

3.9 SENSOR SPECIFICATIONS

The following definitions are provided to aid readers with sensor selection. Most sensors have performance specifications expressed in terms of range, accuracy, precision, linearity, resolution, etc. These terms are described next.

3.9.1 Range

Range defines the operating limits of the device in which it can properly function. It is typically specified as the lowest and highest input values within which the sensor is capable of measuring, with the specified accuracy, precision, etc. For example, an LVDT can have a range of $\pm 0.5''$. The LVDT will continue to read outside the specified range but the calibration will be widely off. In other cases, such as in the case of a load cell, the sensor may be damaged if it is loaded at a level well above its range.

3.9.2 Sensitivity

Sensitivity is the ratio of change in transducer output to a change in the value of the measured quantity. For example, a $\pm 10,000$ lb load cell, having an output of ± 5 V, possesses a sensitivity $= 0.5$ mV/lb.

3.9.3 Resolution

Sensor resolution is the smallest change in the mechanical input that produces a detectable change in the output signal. For example, images taken with a 20 MP camera have more resolution than ones with a 5 MP camera. The first has 20 million pixels, and each of these pixels has a depth in terms of the color range. Most sensors are connected to a digital data acquisition system. So resolution is often related to the smallest detectable change that the combined sensor-logger system can register. High resolution is desirable, but does not imply in by itself high accuracy or precision. It just means that the system has many measurement divisions. For example, a camera can be high resolution but have poor optics or poor accuracy in terms of capturing color.

3.9.4 Linearity

Linearity is the widest gap between a straight line fit and the true calibration curve (Fig. 3.58). Linearity is important since linear calibration constants are often preferred, and employed.

3.9.5 Hysteresis

Hysteresis is a measure of the widest gap between the loading and unloading curves (Fig. 3.59). It is commonly caused by energy losses, especially due to

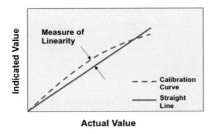

Figure 3.58 Nonlinearity

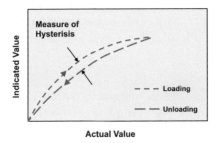

Figure 3.59 Hysteresis

friction. It is typically measured by determining the maximum difference between output readings for the same point, one point obtained while increasing from zero to full scale and the other while decreasing from full scale to zero.

3.9.6 Precision (Repeatability)

Repeatability is the ability of a sensor to reproduce output readings when the same measured value is applied to it consecutively, under the same conditions (Fig. 3.60). Repeatability is typically expressed as a percentage of full-scale output.

3.9.7 Accuracy

Accuracy is closeness to the true value (Fig. 3.60). In instrumentation, accuracy is the difference between a measurement mean value and the true value. It is not the same as uncertainty; you can have an accurate measurement with a large uncertainty in it. Many instrumentation vendors refer to maximum error as accuracy, defined as the combined error of nonlinearity, repeatability, and hysteresis expressed as a percent of full-scale output.

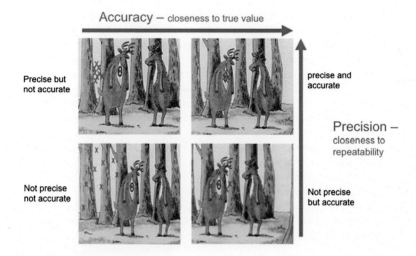

Figure 3.60 Accuracy vs. precision

However, few report the probability that the sensor will exceed the reported accuracy threshold. In any case, an accuracy of ±1% is a reasonable specification to aim for in geotechnical and structural instrumentation.

3.10 CLOSING COMMENT

A large number of sensors have been presented in this chapter along with a number of common errors. These sensors cover over 90% of commonly employed instruments for monitoring of geomechanics. The availability of complex or advanced tools, by itself, is not a reason for employing these tools. Sensors should only be employed to help answer specific questions. Ralph Peck's, the pioneering legend's of field monitoring in geomechanics, advice *"If you can do it with a ruler, do it with a ruler"* remains relevant.

FURTHER READING

ASCE Task Committee Guidelines for Instrumentation & Measurements for Monitoring Dam Performance.

Bassett, R., 2012. A Guide to Field Instrumentation in Geotechnics. Spon Press.

Dunnicliff, J., 1988. Geotechnical Instrumentation for Monitoring Field Performance. Wiley.

Hanna, T.H. (1985) Field Instrumentation in Geotechnical Engineering, Trans Tech Publications (out of print & difficult to find).

Hvorslev, J., 1951. Time Lag and Soil Permeability in Ground-Water Observations, Waterways Experiment Station Corps of Engineers. U.S. Army, Vicksburg, Mississippi.

Hvorslev, J., 1976. The Changeable Interaction Between Soils and Pressure Cells; Tests and Reviews at the Waterways Experiment Station. Final Technical Report S-76-7, U.S. Army Engineer Waterways Experiment: Station Soils and Pavements Laboratory, Vicksburg, MS.

Glisic, B., Inaudi, D., 2007. Fibre Optic Methods for Structural Health Monitoring. Wiley.

Vishay, 2010. Micro Measurements Technical Note TN 502: Optimizing Strain Gage Excitation Levels.

Vishay, 2014a. Micro Measurements Technical Note TN 504: Strain Gage Thermal Output and Gage Factor Variation with Temperature.

Vishay, 2014b. Micro Measurements Technical Note TN 505-4: Strain Gage Selection: Criteria, Procedures, Recommendations.

Siskind, D., Stagg, M., Kopp, J., Dowding, C., 1989. Structure Response and Damage Produced by Ground Vibration from Surface Mine Blasting. Report of Investigations 8507. US Bureau of Mines.

CHAPTER 4

Environmental Underground Sensing and Monitoring

Tissa H. Illangasekare*, Qi Han*, Anura P. Jayasumana[†]
*Colorado School of Mines, Golden, CO, USA
[†]Colorado State University, Fort Collins, CO, USA

4.1 INTRODUCTION

Protection of the environment for the general well-being of humans, animals, and plant life and sustainability of water for potable use, food production, protection of human and ecological health and quality of life, and energy development will continually challenge environmental scientists and engineers for many decades to come. Global warming and its detrimental impacts on climate will further exacerbate these problems, adding more complexities and challenges. Subsurface, which is an important component of the earth system, is central to meet both these goals of environmental protection and water sustainability. Meeting the goals of environmental protection and water sustainability requires the management of the subsurface as a critical component of the earth system. Throughout the world, groundwater as a central source of water supply is threatened from overdraft and contamination due to agricultural and industrial activities. The processes occurring in the subsurface are also central to greenhouse gas (GHG) loading to the atmosphere that contributes to global warming leading to climate change. The primary greenhouse gases such as methane and carbon dioxide are emitted naturally or as a result of engineered activities such as leakage from geologically sequestrated carbon dioxide and hydraulic fracturing. Monitoring of the subsurface will be of critical importance in managing the subsurface to address existing and emerging problems in the earth, water, and environment.

Traditional subsurface water monitoring technologies primarily rely on the use of monitoring wells or bore holes that are used to grab water samples (Loden et al., 2009; Porta et al., 2009). Samples are collected periodically and analyzed on-site using hand held instruments or at off-site laboratories. This approach of monitoring based on manual sampling is la-

Underground Sensing.
DOI: http://dx.doi.org/10.1016/B978-0-12-803139-1.00004-7
Copyright © 2018 Elsevier Inc. All rights reserved.

bor intensive and costly. In monitoring large systems using sampling from sparsely-distributed wells, the sampling frequency may be quite large— typically weekly or even monthly. This approach of sparse spatial and temporal sampling will result in misinterpretation and inaccurate assessment of the behavior of the system. In some cases, this low-frequency sampling may even completely fail in monitoring highly transient events. The sampling may disrupt the flow field thus biasing the data.

Monitoring subsurface gas and vapor migration produces additional challenges. In the case of either CO_2 or methane, the leakage occurs deep in the formation as a result of a well failure and preferential pathway development through natural faults or induced fractures (Plampin et al., 2014). When the gas or fluids in the formation migrate, the resulting plume when reaching the shallow subsurface will impact large areas up to several kilometers. As the location of the source will be not known, and the contamination area is large, deciding where to sample is not feasible. Also, any sampling above the land surface to determine atmospheric loading results in additional challenges as the readings are affected by the atmospheric climate conditions, which makes it hard to interpret the leakage signal to assess the atmospheric loading of the gases. In the case of methane leaking from production and distribution infrastructure (e.g., buried pipelines), it becomes difficult to decide in advance on where the monitoring and sampling need to be done.

The limitations and challenges in long-term monitoring of large geologic systems using conventional methods require the development of innovative technologies. Such underground monitoring technologies should use low-cost and low-energy sensors that can be deployed over large areas, thus drastically reducing the need to collect and analyze samples in the laboratory. In this chapter we explore such environmental monitoring technologies specifically focusing on the underground where other challenges exist due to limitations of transmitting data wirelessly. First, a literature review of the state of environmental monitoring using sensors is provided. This is then followed by a general introduction of a wireless sensor network (WSN) based real-time monitoring system, which uses closed-loop data acquisition, inversion, and simulation that are suitable for subsurface monitoring and decision making. Last, we describe the potential use of this WSN technology for traditional problems in subsurface remediation as wells as emerging problems associated with greenhouse gas loading to atmosphere that contributes to global warming.

4.2 OVERVIEW OF CONVENTIONAL AND TRANSITIONAL ENVIRONMENTAL SENSORS

Use of sensors for monitoring the environment has become more prevalent during the last several decades. Applications include monitoring of environmental phenomena such as weather and storms, volcanoes, air quality, agriculture systems, forests, and ecological systems. Environmental sensors such as biological or chemical sensors have recently become more and more accurate and small, making their use very practical in remote areas where human manual sampling is impossible, and for microscopic applications (Wolfbeis, 2004). These sensors are widely used for infrastructure and environmental monitoring. Most environmental monitoring applications require sensors that can measure parameters that define air quality and dissolved constituents in water. Sensors for monitoring water quality variables such as pH, conductivity, dissolved oxygen, and turbidity are commercially available (e.g., Global Water Instruments). However, they are generally expensive for large scale distributed applications at a unit cost ranging between $500–1000. In 2003, the EPA reviewed emerging sensor technologies for long-term groundwater monitoring of volatile organic compounds, VOCs (EPA, 2004). The report states that "*the process of collecting ground water samples is time consuming and labor intensive, and analysis of the samples is generally conducted in off-site laboratories. Due to the required labor and analysis, ground water monitoring is relatively expensive. In addition, due to the collection, preparation, and transportation of samples from monitoring wells to distant laboratories, ground water monitoring is prone to errors. The large number of samples collected and analyzed each year, the relatively high cost of collecting and analyzing each sample, and the potential for errors warrant research into new technologies for facilitating ground water monitoring.*" A National Research Council (NRC, 2013) study highlighted the potential of in situ sensor technology for the detection of groundwater contaminants of interest, such as low concentrations of chlorinated solvents. In 2001, the Army conducted a study to validate sensors specifically used for detection of VOCs incorporated into a Penetrometer System (Davis et al., 2001). At the time when that report was written, no rapid on-site method to investigate the extent of groundwater and soil contamination for VOCs was available. The 2003 EPA study identified the key requirements for effective sensor-based long-term groundwater monitoring such as sensitivity, accuracy, precision, reversibility, speed, durability, reliability, simplicity, selectivity, affordability, and acceptability. This report summarized the capabilities and the technologies behind a number of sensors that were able to detect a variety of chlorinated compounds (Table 4.1).

Table 4.1 State of VOC sensors

Sensor description	Developer	Measurements
Chemoresistors	Sandia National Laboratory (Ho, 2003)	VOCs, BTEX (benzene, toluene, ethylbenzene, and xylenes)
Quartz Crystal Microbalances	Nomadics, Inc.	Chlorinated compounds
Ion Mobility Spectrometry	Boise State University (Kanu et al., 2008)	VOCs and BTEX below the maximum contaminant level (MCL)
Resonance Enhanced Multiphoton Ionization fiber optics	Univ. of South Carolina (Chinni et al., 2004)	BTEX compounds in the ~1 μg/L range
Bio-Optoelectronic Sensor System (BOSS)	Georgia Tech (EPA, 2004)	Detection capabilities in the range of approximately 50 μg/L to 100 μg/L
Mid–Infrared Fiber-optic Sensors	Georgia Tech (EPA, 2004)	VOCs with detection limits of ~100 μg/L

EPA (2004) report also provided approximate cost per unit. Summarizing the technology status, the report concluded that the manufacturing cost might be around $100; that does not represent the cost of a fully-functional instrument. A fully developed sensor to measure two or three parameters was estimated to cost around $7,500. An Environmental Security Technology Certification Program (ESTCP) funded project reported the testing of a set of sensors for monitoring VOCs (Lieberman, 2007). The sensors tested included XSD-MIP systems (halogen-specific detector – membrane interface probe) and laser-induced fluorescence, based on the rapid optical screening tool (ROST). The costs of these systems ranged in the thousands of dollars.

Fluorescence-based probes have also been used in groundwater contamination analysis by coupling them to a direct-push method (Lieberman, 1998). In such a system, the sensor can be pushed into the ground while taking continuous measurements, permitting real-time data gathering. In-situ chemical sensing techniques have been developed by Sandia National Laboratories to detect volatile organic contaminants with an electrochemical sensor (Ho, 2003). Other sensors, such as polymer-absorption, metal-oxide-semiconductor, fiber-optic, and mass sensors might also be used for such an application. Further testing and verification remains to be done with these sensors on the field scale to assess their real capabilities and effectiveness.

4.3 WIRELESS SENSOR NETWORKS FOR ENVIRONMENTAL SENSING APPLICATIONS

4.3.1 Background and Current State-of-the-Art

Wireless sensor networking has emerged as an alternate technology that facilitates bridging the physical environmental sensing and the Internet based communication and computing infrastructure. It provides dramatic improvements over traditional data acquisition technologies, and offers the potential to significantly increase the time and space granularities of data acquisition, and ability to connect seamlessly to the Internet or cloud-based computing where models can be coupled directly.

A wireless sensor is a small device that contains at a minimum a simple processor, a wireless transceiver, and digital and analog interfaces to connect appropriate sensors. Advances in many technologies, such as sensors for field installation, power harnessing, sources, distributed communication, and cloud based data storage are synergistically merging in wireless sensor network (WSN) applications. WSNs are a key component of the emerging Internet of Things (IoT) paradigm. With sensors deployed for underground plume tracking, the plume can be envisioned as a "thing" being monitored, tracked, modeled, and learned from using the Internet of Things (IoT) infrastructure. Wireless sensor nodes with significant processing and communication capabilities are already commercially available from multiple vendors. Multiple standards for short and long-distance wireless communications with sensors provide a broad selection in terms of distance, cost, frequency band, technology, power consumption, etc., for communication with sensors under different deployment conditions (Frenzel, 2016). Standards such as IEEE802.15.4 (IEEE, 2006) for wireless communication and 6LoPAN (Montenegro et al., 2007) for IPv6 based addressing have resulted in a plethora of devices that can seamlessly communicate with each other, thus significantly easing their field deployments. The programmability, processing capability, and storage capacity of nodes as well as the connectivity to the Internet make novel intelligent and adaptive techniques possible for large-scale environmental monitoring.

WSNs connecting the spatially distributed sensors that make in situ environmental measurements, e.g., dissolved contaminant concentration, gas concentration, temperature, pressure, and water quality parameters such as pH are at the core of this monitoring paradigm. Ad hoc networks of sensor nodes will not only deliver the measurements to appropriate points of interest (such as computing models or visual displays) in real-time, but also allow the operators or models to change the sensing modalities such as sampling frequency, noise filtering, and trigger mechanisms for initiating sampling dynamically.

WSNs have already been deployed worldwide for monitoring of infrastructures (bridges, tunnels, buildings, etc.) and environmental and natural phenomena such as weather and storms, volcanoes, air quality, landslides, crops, forests, and ecological systems (e.g., Culler et al. (2004); Haenggi (2005); Werner-Allen et al. (2006); Martinez et al. (2009)). While environmental monitoring has been considered one of the most important applications since the field's inception, wireless sensing for the underground monitoring has been limited with only a few developments. For instance, a network of 30 wireless nodes was deployed to study the ecosystem of the soil (Musaloiu et al., 2006). Project "Suelo" (Ramanathan et al., 2009) used a wireless sensor system for soil monitoring and was deployed in three places: the Ganges delta in Bangladesh, a forest in the James Reserve, and the junction between the Merced and San Joaquin rivers in California. PipeNet is a wireless network designed to detect leaks and other anomalies in municipal water systems (Stoianov et al., 2007). PipeNet monitors pressure, acoustic and hydraulic velocity values from pipelines, sewers, and storm drains. SASA (Li and Liu, 2012) is a system consisting of 27 Mica2 motes deployed in an underground mine, where its scalability and reliability have been evaluated.

In addition to these applications and deployments, there exist several studies focusing on the communication aspects of underground sensing. Examples include the viability of wireless communication in a subsurface environment (Stuntebeck et al., 2009) and effect of soil properties on radio signal propagation is in subsurface communication environments (Yoon et al., 2012). Advanced channel models have been developed to characterize the underground wireless channel (Vuran and Silva, 2009) in wireless underground sensor networks (WUSNs). Another class of work aims to support underground sensing via different communication methods. For instance, MINERS (Markham and Trigoni, 2012) have built magnetic transceiver nodes for underground sensor networks. These nodes can transmit data through rock and soil at higher data rates than those achieved previously.

4.3.2 Recent Advances in WSN Hardware Suitable for Underground Environmental Applications

Many challenges exist to develop sensors capable of meeting the monitoring demands of the underground environments, including: long-term sensor calibration, improved resolution and accuracy, and ability to quantify complicated chemical compounds that reside in soil and groundwater (Trubilowicz et al., 2009; Porta et al., 2009; Ramanathan et al., 2009). Although early work pointed to the need for network protocols which route information efficiently between network nodes and a central computer (Akyildiz et al., 2002), recent research has made significant progress in this regard. A class of such protocols are based on using geographic coordinates (Karp and Kung, 2000), while another class uses connectivity based coordinates such as virtual coordinates and topology coordinates (TCs) (Dhanapala and Jayasumana, 2011, 2014). Although obtaining geographic coordinates of sensor nodes automatically is not an easy task in dense sensor networks, for subsurface monitoring applications such as those discussed here, it is possible to program the coordinates of the sensors using a GPS unit at the deployment time. On the other hand, connectivity based coordinate systems (Dhanapala and Jayasumana, 2011; Jayasumana et al. 2016) are also likely to perform well in this context when no mobile nodes are involved. While a critical aspect of early WSN protocols was to ensure data reliability while minimizing energy consumption to maximize the battery life (Polastre et al., 2005), there are energy harvesting techniques (Priya and Inman, 2009) such as solar cells that can also fulfill the needs of this application. Reliable sensors and robust network and communication protocols mean fewer trips to the field to maintain this sensing technology. Network access also means that not only the individual sensors, but also the state of the network nodes can be monitored remotely. Adherence to standards provides interoperability, and protocols stacks provide seamless communication among different devices.

Assimilating WSN data into accepted models of solute transport, or even transient flow models, is itself a formidable task. Trial-and-error model calibration is the manual adjustment of model parameters until a reasonable match is obtained between model predictions and field observations. Automatic (i.e., computer-driven) inversion of flow and transport models is gaining popularity though trial-and-error calibration is still often seen in practice (Carrera et al., 2005; Poeter and Hill, 1997). The use of WSNs is different from traditional data collection for subsurface monitoring in the following ways:

- WSNs sample, process and relay data in real-time. Typical hydrological data collection may require days of manual field sampling and collection. In some cases, it may be weeks or years before data is post-processed and analyzed;
- Wireless communication may have down times due to interference sources or link failure. In fact, there may be environments where wireless transmission is not possible, for example, when the area being monitored has obstructions and interfering sources. Data may be missing for periods of time when the network conditions deteriorate. A cumbersome option is to store data on a storage device and then manually or opportunistically collect the data from local storage if some delay is tolerable;
- In situ sensors that are used for element or compound quantification are currently less accurate than analyzing grab samples using laboratory equipment; however, in situ sensing equipment can sample the environment frequently, depending on energy resource constraints. The cost savings realized by eliminating the manual collection process may be invested in better and stable sensors as well; and
- Data processing by a WSN can refer to any number of data reduction, amalgamation, inter/extrapolation and correction techniques depending on what the node software is designed to perform.

In the rest of this chapter, we discuss several environmental applications that can benefit from the user of the wireless sensor network technologies.

4.4 FUNDAMENTALS OF WSN SUPPORTING ENVIRONMENTAL APPLICATIONS: ADVANCES AND OPEN ISSUES

4.4.1 Sensor Network Deployment

When WSNs are being used for long-term environmental monitoring applications, the locations of sensor nodes are often predetermined strategically according to domain knowledge, application requirements, and environment constraints. For instance, in subsurface contaminant monitoring, high resolution, yet expensive chemical or biological sensors are only deployed at a few preselected sites, resulting in a sparse and disconnected network. To ensure coverage and also communication, lower-cost relay nodes may be inserted to forward readings from each individual sensor in multiple hops to the server or data sink. Given the required sensor locations in an area of interest, how to minimize the number of relay nodes and optimize node deployment locations are NP-complete problems (Hao et al., 2004). Node locations are considered optimal if the resulting network can satisfy sensing and communication coverage. In case of plume tracking, the predictions, provided by the models driven by the samples from existing sensors, may be used to dynamically deploy additional sensors or activate predeployed sensors.

Various relay node placement algorithms for wireless sensor networks have been proposed and reported in the literature that considers different network models, including single-tiered networks, two-tiered networks, and heterogeneous networks. In a single-tiered network, all the nodes have a similar transmission range and are equal partners. In a two-tiered network, nodes are grouped into clusters with one node in each cluster serving as the cluster head; the cluster heads are able to communicate among each other with some type of a backbone. A heterogeneous network has a more complex structure, e.g., with different sensor nodes possessing different transmission ranges and even different roles.

- In the single-tiered network model, the goal is to ensure connectivity for each pair of sensor nodes by adding a minimum number of relay nodes. This problem can be formulated as the Steiner minimum tree with minimum number of Steiner points and bounded edge length problem (SMT-MSP). To find a solution to this NP-hard problem, different approximation algorithms have been developed, including a 5-approximation algorithm (Lin and Xue, 1999), a 3-approximation algorithm (Chen et al., 2000), a faster 3-approximation algorithm and a randomized algorithm with an approximation ratio of 2.5 (Cheng et al., 2008). The number of relay nodes needed by an n-approximation solution is n times the optimal solution. Different from 1-connectivity scenario in these solutions, k-connectivity has been considered. Polynomial time $O(1)$-approximation algorithms for any fixed k (Bredin et al., 2005) and a 10-approximation algorithm ensuring 2-connectivity (Kashyap et al., 2006) were developed. While all of the above assume that relay nodes have the same transmission range as sensor nodes, a 7-approximation algorithm (Lloyd and Xue, 2007) was developed to work in the case where relay nodes have much greater transmission range than sensor nodes.
- In a two-tiered network, one formulation assumes that the communication range of a relay node R is at least four times that of a sensor node (r), i.e., $R \geq 4r$ (Hao et al., 2004). Two variants of the problem have been studied. One is connected relay node single cover, whose

objective is to deploy a minimum number of relay nodes so that (1) every sensor node is within distance r of a relay node, and that (2) between every pair of relay nodes, there is a connecting path consisting of relay nodes such that each hop of path is not longer than R. The other is 2-connected relay node double cover, whose objective is to place the minimum number of relay nodes such that (1) each sensor node can communicate with at least two relay nodes, and (2) the relay node network is 2-connected. A 4.5-approximation algorithm under the assumption that $R \geq 4r$ and that the sensor nodes are uniformly distributed was developed (Tang et al., 2006). Under the assumption that $R = r$, but with no restriction on the distribution of the sensor nodes, a $(6 + \varepsilon)$-approximation algorithm for connected relay node single cover problem and a $(24 + \varepsilon)$-approximation algorithm for the 2-connected relay node double cover problem, where $\varepsilon > 0$ is any given constant, were both developed (Liu et al., 2005). A $(5 + \varepsilon)$-approximation algorithm was proposed to solve the connected relay node single cover problem with the condition $R = r$ relaxed to $R \geq r$, (Lloyd and Xue, 2007). A better approximation algorithm under the assumption $R \geq 2r$ was then developed (Srinivas et al., 2006). Approximate algorithms were developed for the problems where each sensor node is covered by one or two relay nodes and the relay nodes form a connected network with the base station (Yang et al., 2012).

- The problem for connected relay node single cover problem in heterogeneous wireless sensor networks has been studied (Wang et al., 2007). Two problems have been considered: (1) full fault-tolerant relay node placement, which aims to deploy a minimum number of relay nodes to establish k ($k \geq 1$) vertex-disjoint paths between every pair of sensor and/or relay nodes; and (2) partial fault-tolerant relay node placement, which aims to deploy a minimum number of relay nodes to establish k ($k \geq 1$) vertex-disjoint paths only between every pair of sensor nodes (Han et al., 2007).

While minimizing the number of relay nodes is desirable, it may not always provide the most energy efficient solution. Replacing batteries or redeploying nodes can be impossible or labor intensive, therefore it is more desirable to reduce energy consumption in data collection even with a slight increase of the number of relay nodes. In this case, it is important to tradeoff between relay nodes placement (i.e., the number of relay nodes and their locations) and energy efficiency in collecting data from a set of sensors deployed at pre-determined locations. A similar problem was considered in

Ganesan et al. (2006), where data correlation was exploited to guide the placement of relay nodes. In subsurface monitoring, the sensor network is typically sparse and data correlation is rather weak, hence one can start with the minimal number of relay nodes using an existing network, and then observe the impact of gradually increasing the number of relay nodes on total energy consumption in raw sensor data collection. By optimizing the energy consumption and the number of relay nodes used, the most economical network deployment can then be identified (Zhu and Han, 2009).

Power harvesting technologies for sensor nodes using means such as solar, wind, electromagnetic radiation, and pressure have advanced significantly over the last decade (Beeby and White, 2014). Martinez et al. (2009) and Marcus and Bonnet (2010) present examples of environmental monitoring devices deployed in relatively harsh environments, but relying purely on harvested energy. Thus depending on cost-benefit tradeoffs, power harvested systems should be given serious consideration.

Although WSN networks were considered in the past to be separate from cellular networks, with inexpensive and low-power cellular modems becoming widely available, now it is even possible to consider sensor nodes communicating via cellphone network, or hybrid networks that combine the two. In case of remote locations or when the cost is not a main constraint, satellite communication links may be used as well. Careful cost–benefit analysis is needed as the tradeoffs involve deployment and operational cost, robustness, deployment ease, granularity of information collection, and long-term reliability.

4.4.2 Virtual Sensor Networks

Research on WSNs has primarily considered dedicated sensor networks each supporting one specific application. Independent sensor networks each dedicated to a specific task, however, may not be the best, most cost efficient, or the most practical deployment technique under a wide variety of conditions, e.g., for deployment of large-scale networks having thousands of nodes or covering large geographical areas or even crowded urban areas or difficult terrains. Instead, the applications can benefit by the use of a virtual sensor network (VSN) concept (Jayasumana et al., 2007).

We next use chemical plume monitoring as an example to present the concept of VSN and explain the VSN support functionality needed for its realization. A chemical plume can be considered as a 3D transient phenomenon that is spatially and temporally distributed, and which evolves in

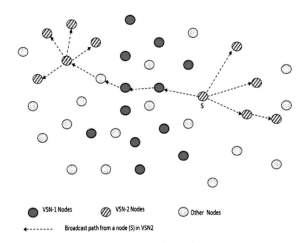

Figure 4.1 Multiple VSNs on a physical sensor network infrastructure

its intensity and extent. Hence, it is different from a phenomenon that is time varying in a fixed region (such as temperature/humidity changes in a room), or a phenomenon that varies in locations but not extents (such as a mobile object). For instance, plumes can change their configuration/shapes as a result of not only migration but also remedial treatment. In other words, two plumes can merge into one, and one plume can also be separated into two. As a result of this, a sensor node should adapt to plume dynamics and change its functionality to either active sensing (when they are embedded in plumes) or passive listening (when they are emerged out of plumes). This further implies that the sensor nodes should self-organize themselves to ensure that the right set of nodes collaborate at the right time for sensing and tracking a given plume, and the collected data can be delivered to the appropriate nodes, or perhaps even a remote server, for processing in an energy efficient manner.

Fig. 4.1 illustrates the VSN concept where two different networks VSN-1 and VSN-2 exist on a common physical sensor network. A VSN is formed by a subset of sensor nodes of a wireless sensor network (WSN), with the subset dedicated to a certain task or an application at a given time. In traditional dedicated sensor networks, all the nodes in the network collaborate more or less as equal partners to achieve the end result. In contrast, the subset of nodes belonging to VSN-1 collaborates to carry out a given application while those on VSN-2 carry out a different task. As the nodes in a VSN are distributed over the physical network, as is the case with VSN-2, they may not be able to communicate directly with each other.

Thus a VSN in general depends on the remaining nodes providing VSN support functionality to create, maintain and operate itself. Another advantage of VSN based deployment is that multimodal sensor nodes, i.e., each node equipped with multiple types of sensor, perhaps even for completely unrelated applications, can be present in the physical network, but VSNs may be formed to perform specific types of sensing functions (e.g., plume tracking) on demand by recruiting these nodes. Thus the membership node set of a VSN may change over time. Some nodes may serve one VSN at one time and another at a different time, or a node may be on multiple VSNs simultaneously, or in no VSN at all. Multiple VSNs may exist simultaneously on a physical wireless sensor network, and the membership of a VSN may change over time, and the number of VSNs itself may change.

Next we consider a few aspects of underground monitoring that use attributes of VSNs described above. Consider two plumes each being tracked by a VSN. If the two plumes merge, the corresponding VSNs will also merge to form one VSN tracking the newly formed plume. Similarly, breakup of a large plume into different plumes should result in the partition of VSN into multiple VSNs. Keeping the membership of a VSN limited is important as the complexity of tasks such as search, broadcast, and plume boundary detection grows with the number of nodes.

The sensor nodes monitoring a plume are not necessarily adjacent in terms of connectivity. This subgroup and any other sensors of interest for tracking the plume (for example, downstream nodes) are to be considered as a virtual sensor network. The shape and concentration profile of the plume as well as its migration path dictates the membership of the VSN. As the plume migrates, the membership of VSN changes. The sensor network, by itself or based on prediction of models, alerts additional sensor nodes in the predicted and possible paths. The plume monitoring task thus will be based on the concept of Virtual Sensor Networks (VSNs), where multiple VSNs exist on a physical sensor network. While some of the nodes in a VSN may be able to communicate directly with each other, the VSN may consist of multiple zones, where communication between zones has to rely on nodes that are not members of the VSN. Thus providing communication support for maintaining the VSN is a key support function that is necessary.

To support multiple applications on the same physical sensor network, there are several different approaches. One technique is to mimic TDMA (time-division multiple access) and enforce different parts of the network to be active for different applications at different times. While this is easy to implement, the scheduling of multiple applications makes this approach artificial, inefficient, and sometimes impractical. Another approach is to use different radio channels for different applications. The limitation of this approach is its reliance of the hardware of sensor nodes, and furthermore the number of concurrent applications is bounded by the number of existing radio channels. Most of the current platforms as well as the IEEE 802.15.4 standard (IEEE, 2006) for sensor networking can support both these approaches. However, the use of separate time slots or frequency channels imply that each network operating in a separate slot/channel has to work independent of other channels, i.e., they appear as a set of logically separate networks. This results in duplicated efforts for functions such as routing, thus resulting in inefficient use of resources. Furthermore, due to the fact that the members of a VSN may not be within the transmission range of others in the VSN, it may not always be possible to provide the node connectivity necessary with these approaches. The third approach is to rely on a shared communication channel (in time and frequency) for all the nodes as in the traditional dedicated sensor networks, yet provide VSN support using appropriate algorithms and protocols.

Other key issues involved in the implementation of VSN include:
- How to dynamically determine which node should join which VSN?
- How to maintain and support constant changes in the membership of VSN?
- How to ensure energy efficient communication between disjoint VSN segments?

The major functions of VSN can be divided into two categories: VSN maintenance and membership maintenance. The membership in VSN is dynamic, and the communications among VSN nodes frequently rely on other nodes. Functionality expected of a node will be dependent on whether or not it is currently a member of a VSN. Those that are not in the VSN need to support the maintenance of VSNs, while those within the VSN need to carry out tasks such as profile detection, pattern recognition, and tracking. The VSN maintenance functions include:
- Adding and deleting nodes (decision made by nodes other than that being added/deleted),
- Nodes entering and leaving VSN (decision made by node itself),
- Broadcasting within VSN,
- Joining two VSNs (e.g., when two plumes merge),
- Splitting VSNs (e.g., plume broken into parts), and
- Deriving contours of boundaries.

The supporting nodes (i.e., nodes that do not belong to the VSN at present) need to provide efficient message exchanges among the sensor nodes for implementation of those functions. These functions have to be implemented with minimal overhead, while taking other limiting factors in wireless sensor networks into account.

In addition, efficiently managing the membership changes in the VSN is critical for energy conservation. Sensors have different role assignments in the context of plume monitoring. For instance, sensors within the VSN actively participate in sensing for profile detection, pattern recognition and tracking; sensors outside the VSN may help relay data from VSN members to the server, or simply remain asleep, depending on whether or not they are on the path to the server. Different roles may impose different burdens on nodes. For example, nodes within the VSN consume more energy due to sensing and communicating its own readings to the server. These different sensor roles must be taken into account in supporting real-time plume monitoring, since nodes with certain critical roles may affect application level quality to a high degree while overburdened nodes might be more liable to energy starvation. Moreover, these roles need not be statically assigned to nodes in the system, since sensor roles will be changed when the VSN membership changes. A closed-loop implementation of plume tracking, where the sensor data is fed to numerical models, and the models in turn control the sensing process is described in Han et al. (2008) VSNs help manage heterogeneous networking technologies (Chowdhuri et al., 2010) and also can be extended to achieve other properties, such as sophisticated power management (Tynan et al., 2008).

4.4.3 Reliable Sensor Data Collection

Using WSNs to support environmental applications can provide application domain experts with unprecedented amount of data than before. However, in situ sensors can be prone to errors; links between nodes are often unreliable, and nodes may become unresponsive in harsh environments, leaving to researchers the onerous task of deciphering anomalous data. Thus significant attention has been paid to develop approaches to provide reliable sensor data collection.

To address packet losses, which have high probability of occurring in harsh environments, a reliable transport protocol is proposed that fragments and reassembles a data object reliably over the sensor network (Stann and Heidemann, 2003). Another solution is to use both retransmissions and erasure codes to reconstruct a message of M packets if any M from $M + R$ packets are received where M an R are integers (Kim et al., 2004).

Misrouting of packets and unreliable and inefficient routing of packets is possible in large sensor networks. Limiting the effects of such faults has been studied in the past using protocols such as TAG (Madden et al., 2002) and SKETCH (Considine et al., 2004) for aggregate queries. In TAG, a routing tree is formed during a query dissemination phase. Later, a sensor node selects a new parent if (1) the quality of the link with his parent is significantly worse than that of another potential parent, or (2) it has not heard from its parent for some period of time. SKETCH uses a DAG (directed acyclic graph) instead of a tree for data delivery. Given that most nodes have multiple parents in a DAG, an individual link or node failure has limited effects. A robust technique for computing duplicate sensitive aggregates was proposed by combining multipath routing and duplicate insensitive sketches (Considine et al., 2004).

Having identified the ill-defined semantics of current best-effort algorithms over dynamic networks (including P2P systems and sensor networks), researchers have formalized these with a correctness criterion called single-site validity and dealt with node faults (Bawa et al., 2004). Identifying the drawbacks of traditional best-effort algorithms, with the key observation that not all sensor values should be retrieved at all times to cut down on energy-expensive sensing and communication, others built a statistical model based on a subset of sensor values (Deshpande et al., 2004). The model had the added benefit of being able to predict "missing values." How to answer continuous selection queries over sensor data must consider the presence of faults. Reports produced by small sensors may not reach the querying node, resulting in an incomplete and ambiguous answer, as any of the nonreporting sensors may have produced a tuple which was lost. Fault tolerant evaluation of continuous selection queries (FATE-CSQ) is a protocol that guarantees a user-requested level of quality in an efficient manner and is designed to be resilient to different kinds of failures (Lazaridis et al., 2009).

Hop-by-hop error recovery in sensor networks was proposed by PSFQ (pump slowly, fetch quickly) (Wan et al., 2002). Driven by the purpose of controlling, managing or retasking sensors, PSFQ aims to provide in-sequence data delivery from the sink to the sensors. Along similar lines, GARUDA also provides sink-to-sensors reliability (Park et al., 2004). In addition, PSFQ assumes that message losses in sensor networks occur because of poor link quality rather than congestion. However, the urgent need for congestion control has been pointed out while discussing the infrastructure tradeoffs for wireless sensor networks (Tilak et al., 2002). ESRT (event-to-sink reliable transport) aims to provide congestion control in sensor networks by adjusting sensor reporting frequency based on current network congestion and application specific reliability requirements (Akan et al., 2003). With the same objective, CODA (congestion detection and avoidance) provides an energy efficient congestion control scheme which decouples application reliability from control mechanisms (Wan et al., 2003).

Providing reliable data delivery has also been addressed by routing protocols. Braided diffusion maintains multiple "braided" paths as backup (Ganesan et al., 2002). When a node on the primary path fails, data can go on an alternate path. GRAB (gradient broadcast) ensures robust data delivery through controlled mesh forwarding (Ye et al., 2003). It controls the "width" of the mesh, thus the degree of redundancy in forwarding data. Reliable routing does not differentiate data and enforces reliable delivery of each piece of data, which is neither efficient nor necessary.

Interesting work has been done to evaluate the impact of link quality estimation and neighborhood table management on reliable routing in sensor networks (Woo et al., 2003). Multichannel protocols have been developed in order to improve the reliable communication in wireless sensor networks, so the utility of multichannel operation for reliable routing has been systematically assessed (Khan et al., 2010).

SYMPATHY (Ramanathan et al., 2009) was developed at UCLA for debugging and detecting failures in sensor networks. It analyzes failures to uncover their causes. Information output by a tool like SYMPATHY could be used to react to faults more intelligently. A tele-diagnostic power-tracer was designed and developed as an in-situ troubleshooting tool that uses a low-cost power meter to determine the internal health status of an unresponsive node and the root cause (Khan et al., 2010). REDFLAG, a WSN fault-detection middleware service to manage network and data faults, is used to address anomalous data due to drifting sensor calibration, faulty electronics, and unforeseen transient environmental conditions (Urteaga et al., 2009). REDFLAG resolves two main causes of data uncertainty: abnormal and missing sensor readings. REDFLAG is capable of detecting unresponsive nodes with detection accuracy above 90% for different network topologies, network sizes, network densities, failure durations, and failure probabilities. It exposes faults as they occur by using distributed algorithms in order to conserve energy. REDFLAG was integrated with the numerical subsurface contaminant transport models to evaluate the impact of data uncertainty management on the application. It reduced the objective function (which measures the residuals between the data "received" from the WSN and the data computed by the predictive computational model) by almost three orders of magnitude when faulty data is removed by REDFLAG, bringing the values for the filtered data set quite close to the data set containing only theoretical noise levels.

Various tools have been developed to identify existing or potential bugs in software. For instance, T-Check (Li and Regehr, 2010) is a tool that finds safety and liveness errors in WSN applications running on TinyOS. KleeNet (Sasnauskas et al., 2010) is a debugging environment that discovers bugs before deployment that occur due to nondeterministic events. Sundaram et al. (2010) developed an intraprocedural and interprocedural control-flow tracing algorithm that generates the traces of all interleaving concurrent events. This way, faults can be reproduced at a later stage so that programmers can identify them effectively.

4.5 WIRELESS SENSOR NETWORKS FOR LONG-TERM MONITORING OF CONTAMINATED SITES

An area where the use of sensors and WSN has a great potential is in long-term management of sites contaminated with hazardous industrial waste. A recent National Research Council (NRC, 2013) report highlights that even though there have been success stories over the past 30 years of cleanup and closure of hazardous waste sites, a majority of these sites was relatively simple. More complex sites require much longer remediation time.

Volatile chlorinated solvents such as trichloroethylene (TCE) in the form of dense nonaqueous phase liquids (DNAPLs) have been widely used at DOD facilities and approximately 3000 individual sites require cleanup. The NRC study also reported that at least 126,000 sites across the country are documented to have residual contamination at levels preventing them from reaching closure and thus requiring long-term stewardship and monitoring.

A projection in 2004 from the US Environmental Protection Agency (EPA) states that as many as 350,000 contaminated sites will require cleanup over the next 30 years, costing $180–$280 billion (EPA, 2004). The same report estimates that out of these sites, 6400 are Department of Defense (DoD) sites, with an approximate cleanup cost about $33 billion. Of these sites, approximately 3000 require cleanup of chlorinated solvents. A 2005

National Research Council (NRC) report discussed the US Army's Installation Restoration Program, stating that 88% of the over 11,000 sites managed by the Army have reached "remedy-in-place/response completed," requiring long term management (LTM) options.

LTM of VOC-contaminated sites pose many technological and logistical challenges. Existing technologies that rely on grab water samples from monitoring wells require analysis using relatively expensive laboratory analytical methods that limit high-resolution spatial and temporal sampling for source characterization and to monitor generally spatially large (thousands of feet to miles) plumes for long time durations. Air Force Center for Engineering and the Environment (AFCEE) reported *"The network of monitoring wells at an installation can range from 250–3000 wells with a total of approximately 50,000 wells for the entire Air Force. At a cost of $2500 per sampling event per well the cost for the entire AF program can run into the hundreds of millions of dollars per year."* Another significant cost issue highlighted in the (NRC, 2013) report is that even with expected uncertainties, the estimated "cost to complete" of $110–127 billion is likely to be an underestimate of future liabilities. These estimates do not include the *"cost of long-term management of sites that will have contamination remaining in place at levels above those allowing for unlimited use and unrestricted exposure for the foreseeable future."* The geohydrology and bio-geochemistry conditions at some sites are conducive to natural attenuation. However, it was reported (Leeson and Stroo, 2001; Leeson et al., 2013) in a recent Strategic Environmental Research and Development Program (SERDP) workshop that *"In some cases, natural attenuation of sources and dissolved plumes alone may not be acceptable or sufficient for transition to closure within reasonable timeframes, but low-cost enhancements to natural source depletion and dissolved plume attenuation may allow passive management."* These observations and findings clearly suggest that the stewardship of these sites requires LTM to evaluate whether the site has the potential to reach conditions through natural attenuation or through low-cost attenuation enhancement without additional active treatment (passive management) to achieve restoration goals.

4.5.1 WSN for Underground Plume Monitoring

The authors proposed and tested a WSN scheme specific to LTM of aquifers contaminated with organic wastes. Fig. 4.2 schematically shows the basic setup of a WSN-based LTM technology. The basic features of the setup are: (1) wells where a string of sensors is placed, both in the unsaturated and saturated zones, (2) a mote (a microprocessor and a wireless

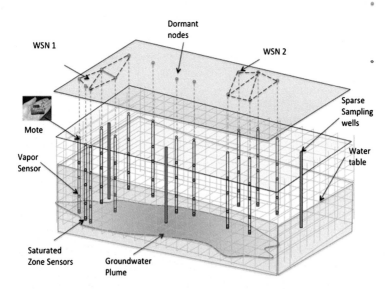

Figure 4.2 A configuration of a Wireless Sensor Network (WSN) showing strings of sensors placed in monitoring wells. As radio signals may not travel through soil adequately, surface nodes (i.e., motes) are placed at the wellhead to connect the sensors in the well to the above ground network

receiver/transmitter) and a power source that are placed on the ground at the well head, wired to the sensors in the well, and (3) an existing set of sparsely distributed monitoring wells for collecting water samples. The core feature of the LTM system is the integration of the WSN and well sample data to achieve the following: (1) continually improve the site conceptual model capitalizing on diverse data indicating both short-term variability and long-term plume behavior, (2) identify subzones where short-term variations are occurring, (3) identify subzones with persistent source conditions (e.g., pools or low permeability zone (LPZ) mass rebound) and estimate parameters that characterize net mass loading in these subzones, (4) estimate process parameters that contribute to plume attenuation and continually calibrate and update the plume simulation model with new data, and (5) identify evolving plume boundaries and use the model to estimate the natural contaminant assimilative capacity of the aquifer in the current plume area as wells as in the zones where the plume is expected to spread based on model predictions.

Proof-of-concept study for the development of such a technology was reported by Porta and coworkers (Porta et al., 2009). Subsequent work continued to pilot scale testing through the development of a three-dimensional intermediate scale test bed (Shulte, 2009), validation of real-time automatic calibration of groundwater transport models (Loden et al., 2009), plume tracking using WSN (Bandara et al., 2010) and the introduction of a cluster tree based self-organization of virtual sensor networks (Bandara et al., 2008).

a. *Proof of Concept in Intermediate Scale Tanks*

The goal of this study was to use a two-dimensional synthetic aquifer that is instrumented with sensors to collect time and space continuous data to conduct a proof of concept study of the WSN presented on Fig. 4.1 (Porta et al., 2009). The synthetic aquifer was built at a length scale that is intermediary between the traditional small scale laboratory systems and the field scale. This testing scale is referred to as the intermediate scale where the field-scale processes can be mimicked under controlled conditions (Lenhard et al., 1995). The two-dimensional intermediate scale test bed consisted of a Plexiglas tank with dimensions: 244 cm long × 44 cm deep × 8 cm wide. Sands for which the hydraulic characteristics are accurately known were packed in the tanks to create geology of the synthetic aquifer (Moreno-Barbero and Illangasekare, 2006). Sodium bromide (NaBr) solution was used as a tracer to simulate a conservative contaminant plume. An array of the $ECHO_2O$-TE sensors (manufactured by Decagon Devices, Inc.) that measure electrical conductivity indicating the salinity or resistivity was used to monitor the contaminant plume. A central component of the WSN is the mote: a wireless sensing board that transmits data via radio communication. The motes chosen for this study are the TPR2420CA TelosB motes that provide a low power IEEE 802.15.4 compliant wireless sensor module with a built-in antenna and with USB capable programming and data collection. The TelosB has compact design and is capable of integration of programming, computation, communication, and sensing onto a single device (Polastre et al., 2005). Fig. 4.3 shows the schematics of the tank setting and the sensor configuration and Fig. 4.4 shows the wireless communications system.

The results from one of the experiments are presented to demonstrate the working of the communication system and help identify bottlenecks in implementing the technology in the field. This experiment was setup to compare wireless sensing data with traditional sampling methods to validate this sampling approach. The synthetic aquifer configuration was created using three test sands (sieve size classification #30, #50, and #70). The heterogeneity of the synthetic aquifer was designed using a selected set of

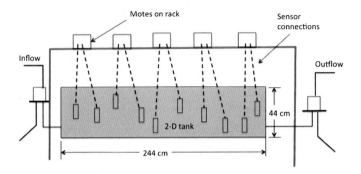

Figure 4.3 Test bed configuration (Porta et al., 2009)

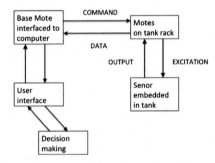

Figure 4.4 Wireless communication system (Porta et al., 2009)

geostatistical parameters. The parameter was used to generate a spatially correlated random field using an algorithm based on the turning bands (TBands) method (Mantoglou and Wilson, 1982). The geostatistical parameters are based on the hydraulic conductivity values K (m/d) of the sands. The two primary parameters that define the spatially correlated random field are the mean of $\ln K$ and the variance that were set at 3.5 and 1.2, respectively. The packing configuration of the synthetic aquifer and the sensor placement are given in Fig. 4.5.

Fig. 4.6 shows the plume breakthrough at sensors 1 and 10, respectively. The WSN transmitted concentration data is compared with the concentrations measured using the grab samples collected at the same locations as the sensors and analyzed in the laboratory using ion chromatography (IC). The following observations that were found to be useful in implementing WSN in the field for plume tracking could be made. In naturally heterogeneous aquifer systems such as the one physically modeled in this experiment, as the plume migrates away from the source, the spreading is controlled by

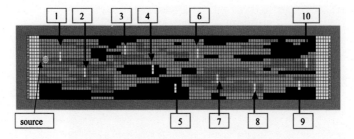

Figure 4.5 Heterogeneous packing of the synthetic aquifer and the sensor placement (Porta et al., 2009)

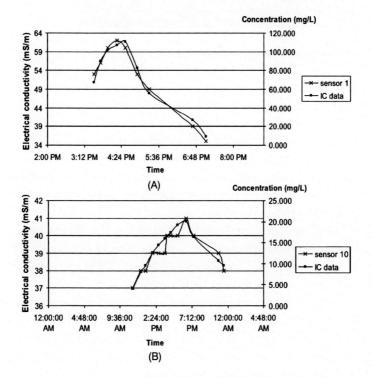

Figure 4.6 Comparison on WSN and IC concentration data at sensor locations 1 (A) and 10 (B)

hydrodynamic dispersion. In subsurface transport model calibration, the concentration breakthrough data as the ones shown in Fig. 4.5 are used to calibrate dispersivity, a parameter capturing hydrodynamic mixing. It is known that dispersivity exhibits scale dependence. That is, as the plume migrates, the dipersivity changes with the length scale of the plume until it reaches an asymptotic value. The WSN and IC data at sensor location 1

Figure 4.7 Three-dimensional synthetic aquifer test bed

that is closest to the source shows very close agreement. However, as the plume migrates further, at sensor location 10, the data does not match. This suggests that if WSN data is used to calibrate the dispersivity, the estimates will be more accurate when data from sensors close to the source is used, but the accuracy will diminish as the WSN data further away from the source are used. This finding suggests that to accurately capture the plume behavior using WSN, more distributed sensors have to be placed as the plume spreads to reduce the uncertainty of model parameter estimates.

With the goal of upscaling the knowledge gained in the 2D test bed to realistic field system, a three–dimensional synthetic aquifer was developed. The test bed shown in Fig. 4.7 has dimensions 8 feet (2.438 meters) wide by 16 feet (4.877 meters) long and can be filled with soil to a depth of 22.5 inches (0.5715 meters). Seven supply reservoirs are placed at the upstream and downstream boundaries of the aquifer. The hydraulic head in each of these reservoirs can be independently controlled through operable gates to create different flow patterns within the aquifer with fourteen degrees of freedom. Forty-four fully penetrating wells constructed of 1 1/4 inch in-side diameter PVC pipe which can accommodate insertion of a Decagon Devices, Inc. 5TE sensor were installed in the aquifer. The Decagon 5TE sensor was constructed for robust operation in harsh subsurface environ-ments to measure soil moisture, subsurface temperature, and bulk electrical conductivity. The measurement of temperature allows for the correction of the electrical conductivity measurements when the temperature changes occur within the aquifer during long-term testing. The internal sensor cir-cuitry processes the analog signal and provides a calibrated, digital response.

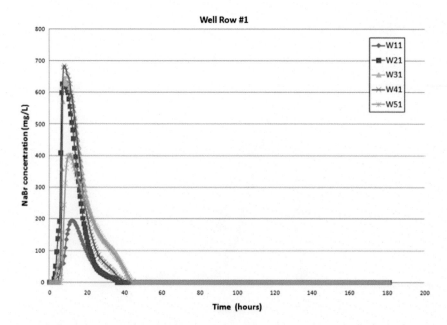

Figure 4.8 NaBr concentration BTC's for the wide centerline plume configuration, data from well row 1 (closest to the source)

The sensor excitation and data acquisition were accomplished by plugging the sensors into Decagon Devices model EM50 data loggers (Shulte, 2009). In addition to monitoring the plume, down-well pressures were monitored using transducers manufactured by Omega Engineering Inc. (model PX26-001DX). These were low cost, wet/wet sensors with a range of ±6895 Pa (±1 psi).

With the goal of identifying issues related to sampling three-dimensional plumes, sensor reliability and data interruption, a simple homogenous packing was used to create the 3D synthetic aquifer. The tank was filled with laboratory grade silica sand purchased from the Unimin Corporation in Idaho. This relatively uniform test sand has a porosity of 0.42 and a hydraulic conductivity of 0.0141 cm/s. The plume was created through point injection of NaBr solution of concentration 1000 mg/L under steady groundwater flow monitored using the pressure measurements system. As in the 2D test bed, the concentrations were measured using the sensors and the extracted aqueous samples analyzed using the IC. Several tracer experiments for the same flow but varying source conditions were conducted, generating a large data set to evaluate the sensor consistency, stability, and data interruption. Only a sample of the results are presented (Figs. 4.8

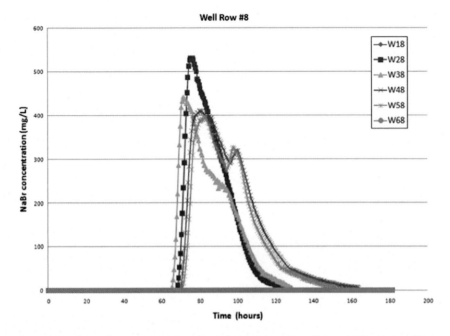

Figure 4.9 NaBr concentration BTC's for the wide centerline plume configuration, data from well row 8 (furthest from the source)

and 4.9) and the findings that are useful in designing WSN based monitoring systems are summarized in here.

A comparison of the spatial distribution of the breakthrough concentrations shows that the sensor system was able to capture the general behavior of the dispersing plume. Both the longitudinal and lateral spreading of the plume was captured. The movement of the contaminant as a slug could be captured using a self-organizing WSN system. The breakthrough curves in well row 1 were smooth as expected, but as the plume migrated downstream, a second peak developed in two adjacent wells (W48 and W58). This result suggests that the density sinking and rebound into the transporting plume is a possible cause for this behavior. A similar observation was made and results verified in another experiment where the concentration differences were seen when the sensor was moved vertically within the well. This observation suggests the need to have a string of sensors in the well as was shown in Fig. 4.2.

The proof of concept studies conducted in both 2D and 3D test beds showed that WSNs have a great potential and many advantages for field scale applications in the groundwater contamination field. Multiple issues that require further research and development are identified. These findings are summarized in the following sections that also address some of the new development efforts.

In the proof-of-concept studies, the plume was generated using sodium bromide as a tracer, thus allowing the use of generally inexpensive conductivity sensors. One of the constraints that need to be overcome in field deployment of WSN is the cost associated with implementing this technology in large aquifers where a large number of sensors may be needed to monitor plumes that could be large and persistent for long periods. Under those circumstances the hardware cost of the whole system could become significant. The currently available sensors to measure dissolved organic chemicals are very expensive (in thousand dollar range), particularly the chemical constituents of plumes at sites contaminated with industrial wastes such as volatile organics (VOC) and petroleum products. The cost of monitoring and sensing depends on the needed accuracy and resolution. For WSN to become viable to monitor VOC plumes either in remediation assessment or long term monitoring, the cost per sensor has to go down drastically. For instance, this requirement may be particularly acute, if the sensors are used to make regulatory decisions, the minimum concentration limits (MCLs) are very low (e.g., in toxic organics it is low as 5 ppb). Hence, until low-cost, miniature sensors become abundant, the opportunities to use WSNs for underground environmental applications may be limited to situations that may not necessarily require to deliver precision concentration measurements. One important observation that was made by Porta et al. (2009) is the potential to use the sensor networks to identify the "hot spots" in the plume where manual sampling may be needed for better characterization of the plume when accurate concentration measurements are needed for regulatory purposes. This hybrid approach could be used until the sensor technologies advances to the stage that precision measurements can be made in a sustainable manner.

The proof-of-concept studies pointed out to another issue that is of critical importance is sensor calibration. Calibration of sensors for laboratory applications are fairly well developed (e.g., Sakaki et al., 2011). However, maintaining the calibration of sensors in field settings is challenging mostly stemming from difficulties of physical access to installed sensors at locations of harsh environment and topography. Ramanathan et al. (2009) pointed out the importance of in-situ calibration of sensors in field applications. In situations when a sensor fails, algorithms should be able to detect data interruptions and loss of accuracy when the measurement accuracies fall outside the calibrated ranges. The development of such an algorithm was presented by reports of research in out team (Barnhart et al., 2010; Barnhart and Illangasekare, 2012).

Another issue that is of importance for economical adaptation of WSN for field applications is optimization of the number of sensors that are needed to capture the required behavior, such as that of a contaminant plume. The total implementation and the operational cost of a field deployed WSN depends on the number of sensors. Even though densely distributed sensors will provide efficient and high spatial resolution data, a large number of sensors will increase installations costs, power needs, calibration cost and maintenance, hence lower sustainability of the system and purpose it is targeted to serve. Bandara et al. (2010) conducted a numerical study to demonstrate how an environmental WSN system could be optimized to minimize the number of sensors while not sacrificing the monitoring accuracy significantly.

4.5.2 Integrating WSN to Transport Models

The transport of dissolved chemicals in the subsurface is described by the advection–dispersion equation

$$\frac{\partial C}{\partial t} = \nabla.(\boldsymbol{D}.\nabla C) - \nabla(\boldsymbol{v}C) + \frac{q_s}{n}C_s \tag{4.1}$$

where C is the dissolved chemical concentration, $\boldsymbol{v}$ is the Darcy velocity, $\boldsymbol{D}$ is the dispersion coefficient, q_s is the strength of a source term, and n is the porosity. This equation captures three fundamental processes that contribute to transport: (1) molecular diffusion, (2) hydrodynamic dispersion, and (3) advection. The first defines the Brownian motion of the molecules of the specific chemical species in water. This coefficient for a free-phase system gets modified for effective behavior in porous media systems due to the tortuosity of the connected pathways of the pores. The hydrodynamic dispersion combines molecular diffusion and mixing of the molecules due to the microscopic variations of the velocity within the pores. In a three-dimensional system, this transport process is parameterized through coefficients of hydrodynamic dispersion:

$$D_L = D_d + \alpha_L v, \tag{4.2a}$$

$$D_{TH} = D_d + \alpha_{TH} v, \tag{4.2b}$$

$$D_{TV} = D_d + \alpha_{TV} v, \tag{4.2c}$$

where D_d is the coefficient of molecular diffusion, D_L, D_{TH}, and D_{TV} are the dispersion coefficients in the longitudinal, and the two lateral directions defined as horizontal and vertical, respectively. The parameters a_L, a_{TH} and a_{TV} [L] are characteristic lengths called longitudinal dispersivity, horizontal transverse dispersivity, and vertical transverse dispersivity, respectively. Since dispersivities quantify mixing through pore-scale velocity variations, they are characteristic properties of the porous medium. Advection is the result of the dissolved molecules transported through the porous medium due to the bulk movement of the water controlled by its seepage velocity. Dissolved contaminant plumes are monitored through observing the temporal and spatial variability of the dissolved concentrations.

Underground monitoring systems can be used for two primary tasks in soil and groundwater contamination studies. The first is to track the migration of the plume so that chemicals do not threat an existing water source (e.g., drinking water wells or surface water bodies that are hydraulically connected to the aquifer). The second is to use the monitoring data to calibrate numerical models that can be used for the prediction of the future behavior of the plume to assess risks or help make other management decisions. The model calibration is a process that involves the use of observations to estimate the parameters that captures the flow and transport processes. If the goal of monitoring is estimating both the flow and transport parameters, then in addition to the concentrations the spatial and time distribution of water pressure has to be observed. The water pressure data can then be used to determine the hydraulic conductivity and the concentration data to estimate the dispersivities. This approach of determining is referred to as inversion (Poeter and Hill, 1997). A number of algorithms are available to conduct such inversion analysis (e.g., PEST by Doherty and Hunt, 2010). It should be noted that the hydraulic conductivity K determined through inverting the water pressure data is not the value that is defined at the macroscopic scale, but the effective value influenced by the of spatial distribution of K, ubiquitous in natural aquifers. Similarly, when the concentration data are used to estimate the dispersivity values they often result in estimations three to four orders of magnitude larger than the macroscopic values obtained in the laboratory, exhibiting strong scale dependence due to the natural heterogeneity of a formation (Gelhar et al., 1992). It was also observed that these dispersivity values are scale dependent (Sauty, 1980). In other words the estimates of dispersivities increase with distance and with time as the plume expands.

Use of WSNs provides some advantages in estimating both the effective K values and scale dependent dispersivities. As the sensors in a WSN are expected to be more densely distributed compared to traditional monitoring wells where head values are recorded, the inversion methods will provide much higher resolution effective K and dispersivity values.

4.5.3 Network Optimization

Recent advances in areas such as compressive sensing and pattern detection can be exploited for network optimization. Bandara et al. (2010) presented an approach to significantly reduce the number of sensors used for a required spatial and temporal accuracy and/or improving the resolution for a given number of sensors in case of contaminant plume tracking. The energy needed for determining the status of the network and then redistributing the gathered information back to the network was shown to be in magnitude less than when using conventional WSN approaches without any optimization. In their approach it was shown that data fusion had no effect to the message length, thus it required no additional bandwidth.

The theoretical foundation for the approach is the discrete wavelet transform based data compression (DWT). The concentration data collected by the sensors are treated as pixels of an image. The underlying image of the chemical concentration is viewed as pixels on a fine grid. The sensors in the network are randomly deployed and only sparsely populate this grid. The goal is to reconstruct the image within the fine grid from the sensor measurements. The data collected by each sensor was compressed before processing to minimize the communication volume. The details of the development of the two-dimensional DWT are provided in Bandara et al. (2010).

The network optimization method is illustrated for a plume tracking problem using synthetic concentration data created using a groundwater flow MODFLOW (McDonald and Harbaugh, 1988) and a transport model (MT3DMS) (Zheng and Wang, 1999). The data set emulated the propagation of a plume for a three-year period. The sensors were assumed to collect concentration data daily. The program developed for synthesizing the data from the plume for any desired locations of the sensors forming a network. To develop the base case, sensors were placed on all nodes of a grid with high spatial resolution (small node spacing) the whole plume, allowing the sensor network to recognize the actual plume. This sensor field was represented as a 64×64 pixel image. The measurements were compressed into a 16×16 pixel image using a two-level Daubeschie-4 wavelet.

Table 4.2 Effect of partial availability on the error

Error compared against	Dead nodes 25% Mean over entire sensing period		Max over entire sensing period	
	Mean over a snapshot	Max over a snapshot	Mean over a snapshot	Max over a snapshot
Actual	3.2	65.9	10.5	96.8
Approximation	2.7	24.9	4.9	47.7
Dead nodes 50%				
Actual	4.9	76.7	13.6	98.8
Approximation	5.3	41.1	9.0	82.6
Dead nodes 75%				
Actual	7.0	87.9	18.4	103.5
Approximation	7.9	56.1	11.8	84.9

To compare the base case with the performance of any other sensor configuration, a percent error measure was defined as the deference between the calculated value and the actual value normalized to the largest measured concentrations. For a given daily data snapshot, the mean of the errors and the maximum of the errors were calculated. These error values for each day were used to calculate the mean and the maximum for the three-year period. Data transmission cost associated with each sensor configuration was evaluated in terms of the number of transmissions. The tests used double precision floating point values for both measurements and coefficient matrices. Thus the actual transmission cost became a factor of the number of transmissions made.

To test the effect of reduction of the number of sensors deployed on the monitoring accuracy, three scenarios of 25%, 50% and 75% of the node unavailability were simulated. Table 4.2 summarizes the error performance of these simulations. Table 4.3 shows the mean and the standard deviation of the mean error over 100 random network settings of each of the sensor reduction scenarios. It can be seen that mean error in general is small and it varies very little from one sensor configuration to another.

During the reconstruction of the plume image, the coefficients automatically approximate values for the locations where the sensor are missing. Fig. 4.9 demonstrates for the scenarios with 25% of available sensors how the plume gets reconstructed. In this case only 25% of randomly distributed sensors measured the plume concentrations. The distribution of the sensors for this case is shown in Fig. 4.10B. The measured concentrations are shown

Table 4.3 Mean and standard deviation of the accuracy

Dead node %	Error compared against	Mean	Standard deviation
25	Actual	3.2	0.09
	Approximation	2.7	0.16
50	Actual	4.9	0.14
	Approximation	5.3	0.19
75	Actual	7.0	0.15
	Approximation	7.9	0.16

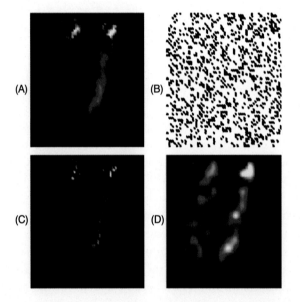

Figure 4.10 (A) The actual plume, (B) a sample deployment of sensors, (C) nonzero reading provided by 25% of sensors, (D) approximate plume reconstructed

in Fig. 4.9C. Using the coefficients for these nonzero measurements, the plume is approximated as shown in Fig. 4.9D. It should be noted that the mean error between the approximated reconstruction using only 25% of the sensors covering the plume compared to the base case with a full sensor deployment is only 7% as shown in Table 4.3.

As concluded by Bandara et al. (2010) the developed scheme estimated the state of the plume within 7% error using only 25% of the total number of potential sensor locations that would allow for the full definition of the plume. This reduction in number of sensors reduced the communication cost of the network by a factor of 10. This model demonstrated that using the optimization approach, acceptable measurement resolution was obtained within a given error bound while reducing the number of sensors to be used for a sustainable application for underground environmental monitoring.

4.6 WIRELESS SENSOR NETWORKS FOR REMEDIATION OF SITES CONTAMINATED WITH ORGANIC WASTES

Chlorinated solvents such as trichloroethylene (TCE) and tetrachloroethylene (PCE) are contaminants of concern in groundwater at many hazardous-waste sites. Each are present at about 60% of National Priority List sites and are ranking 16th and 30th, respectively, on the 2003 CERCLA Priority List of Hazardous Substances (ATSDR, 2003). These chemicals have been used extensively since the early 20th century as cleaning solvents in the dry-cleaning, metal cleaning, vapor degreasing, and textile processing industries (ATSDR, 2004). Widespread use of chlorinated solvents has resulted in accidental release and improper disposal of these compounds in the subsurface. The aqueous solubility of these organic liquid contaminants is low enough for them to exist in the subsurface as nonaqueous phase liquids but large enough to exceed safe drinking water standards. Chlorinated solvents such as tetrachloroethene (PCE) and trichloroethene (TCE) have densities greater than water and are classified as dense nonaqueous phase liquids (DNAPL).

Chlorinated solvents form primary source zones when present in the form of dense nonaqueous phase liquid (DNAPL) and secondary source zones when present as dissolved and/or absorbed contaminants in low permeability media (LPM) (Anderson et al., 1992; Pankow and Cherry, 1996). Contaminants are termed nonaqueous because they are present as a separate immiscible liquid phase having low solubilities in water, typically on the order of $10^2–10^3$ mg/L or parts-per-million (ppm). On the other hand, the drinking water maximum contaminant levels (MCLs) of these contaminants are typically on the order of $10^0–10^2$ μg/L or parts-per-billion (ppb). The combined result is that aqueous phase concentrations exceed MCLs by several orders of magnitude, and the DNAPL source can persist for decades producing risk to human health at downgradient receptor points.

A variety of remediation technologies have been developed and implemented during the last two decades with various levels of reported success. These technologies rely on physical, chemical, thermal and biological transformations that either directly removes the source mass (in the

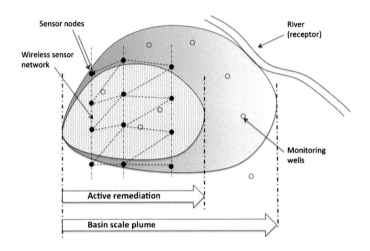

Figure 4.11 Schematic of active remediation zone and contaminant plume

form of DNAPL) or the dissolved constituents in the plume. The goal of removing the source mass is to reduce down gradient plume concentrations. However, the benefits and feasibility of removing the source mass are often debated, e.g., Sale and McWhorter (2001), Rao and Jawitz (2003), Illangasekare et al. (2007). At some sites, even after complete source mass removal, the plume persists due to rebound of diffused mass from low permeability zones (Ball et al., 1997; Liu and Ball, 2002; Parker et al., 2008; Sale et al., 2006; Wilkin et al., 2013). In some cases, mass removal success is hampered by challenges of treatment agent delivery in entrapment zones or low permeability zones (LPZs). The basic issue still remains on how to deal with sites where remedial action has not been successful in reaching the goals in a reasonable time frame at acceptable costs. Some sites with complex sources and hydrogeologic conditions require reassessment of ongoing remedial action. This reassessment requires monitoring of different constituents in the contaminant plume to determine remediation progress, limitations of the technology being implemented, and the long-term prospects of reaching cleanup goals. WSNs have the potential to be used for long-term monitoring of the performance of the remediation scheme, particularly for the long-term persistent contaminants such as TCE and PCE.

In designing monitoring systems for contaminated sites with DNAPLs, two zones of the dissolved plume: (1) the basin scale dissolved plume zone (BSZ) and (2) the source influenced active remediation zone (ARZ)

(Fig. 4.11). The BSZ at complex sites is expected to be very large (100s of meters to kilometers) and the plume concentrations are expected to be above the very low regularly levels of few ppb. On-going remediation is required at these locations as the potential exists for the plume to reach receptors and exposure pathways. The ARZ is the subzone of the plume that is closer to the source where the concentrations are expected to be higher and where most of the mass flux is generated (untreated primary products, remediation by products and rebounding mass from the LPZ). The reasons for demarcating these two zones in the application of WSN based monitoring technology are as follows: (1) The plume concentrations in the BSZ are not expected to change rapidly, hence not requiring high space and time frequency sampling. As regulatory decisions are made based on plume concentrations in this zone, more accurate data (ppb range) are required. Monitoring in the BSZ for regulatory purposes and assessment of long-term response to remediation will continue to be done using existing monitoring wells, grab samples and laboratory analysis; (2) It can be argued that the critical information for optimization of the remediation scheme comes from the ARZ where much more immediate response to remedial action is expected to occur. Contaminant concentrations in the ARZ are expected to be relatively high. The goal of remediation optimization is to reduce mass flux generation within the ARZ, which then loads the plume in the BSZ. Relatively low cost sensors will have adequate resolution to detect the concentration changes during remediation. Currently available pressure transducers that are generally inexpensive can be incorporated into WSNs to calculate groundwater velocities that can be used to estimate mass flux loading from the ARZ to BSZ.

4.7 WIRELESS SENSOR NETWORKS FOR CARBON LEAKAGE

Primary drivers that have contributed to global warming are atmospheric loading of greenhouse gases such as carbon dioxide (CO_2). The primary factor that contributed to anthropogenic emission of CO_2 since the beginning of the industrial revolution is the combustion of fossil fuels. The atmospheric concentrations of CO_2 that are linearly related to radiative forcing (IPCC, 2007) have increased from a preindustrial value of about 280 ppm to the 400 ppm reached in May 2013. Recent estimates of CO_2 loading are about 30 Gt/year (10^{12} kg) worldwide (Boden et al., 2010). One of the strategies proposed by IPCC (2007) is capture and storage of CO_2 in geologic repositories. Carbon capture and storage (CCS) has the potential to achieve the desired goals of stabilizing loading in a relatively short time scales is undergoing extensive research and is at the stage of field implementation. However, leakage of CO_2 from underground formations poses risk to the storage permanence goal of 99% of injected CO_2 remaining sequestered from the atmosphere.

Leaked CO_2 that invades overlying shallow aquifers may cause deleterious changes to groundwater quality (Apps et al., 2010; Kharaka et al., 2010; Little and Jackson, 2010) that pose risks to environmental and human health. For these reasons, technologies for monitoring, verification and accounting (collectively known as MVA) of injected CO_2 are necessary for regulation and permitting of EPA class VI CO_2 injection wells. While the probability of leakage related to CO_2 injection is thought to be small (Friedmann, 2007), it is still very important to ensure storage permanence and protect our groundwater resources.

Leakage from underground formations where CO_2 is stored by capillary and dissolution trapping may occur along abandoned wells or through faults and fractures (known and unknown). While the location of many of these features may be known there may be leakage pathways with unknown locations or the exact point of leakage from a well bore that spans the depth of an aquifer or spatially extensive fault system may not be known. Monitoring plans maybe required for large areas of review, e.g., Birkholzer et al. (2009), Saripalli et al. (2006) and up to 1000 m or more from the surface to the subsurface. Methods proposed for leakage detection include remote sensing such as hyperspectral imaging used to detect vegetation stress (Rouse et al., 2010), soil gas monitoring (Lewicki et al., 2010), and eddy covariance measurements (Lewicki, 2009) at the surface, geophysical techniques (Bohnhoff et al., 2010) and pressure monitoring (Chabora and Benson, 2009) for deep subsurface movement of CO_2, and geochemical monitoring of groundwater in shallow subsurface (Kharaka et al., 2010; Trautz et al., 2012). Each of these methods is optimal at different length and temporal scales. Cost-effective, intelligent monitoring systems with real-time data integration have the potential to be used in MVA. Such monitoring systems will rely on quantitative data from technologies at all points along a potential leakage pathway and long-term deployment of instrumentation.

Deep subsurface monitoring is the first stage of detection but technological advances are required before large-scale systems can be deployed. Geochemical analysis of shallow groundwater has been identified as a primary technology for near surface monitoring (Srivastava et al., 2009), which can be done with adaptation of current instrumentation, and will be required for any MVA scheme to track leaked CO_2 even if first detection occurs deeper. However, active sampling campaigns that require someone to pull a sample from wells and perform subsequent laboratory analysis will be costly for the long-term monitoring needed to evaluate CO_2 storage permanence. Additionally, samples will degas as they are brought to the surface changing the chemistry, thus requiring sophisticated and expensive sampling techniques (Trautz et al., 2012). In order for geochemical monitoring to be a reliable method for leak detection, statistical methods for identifying individual measurements or time series data of geochemical parameters outside natural variation are needed (Romanak et al., 2012). Placement of sensors in wells for in situ measurement overcomes some of limitations, but methods are needed to automate analysis of time series data collected from these samples to reduce costs. While detecting leakage is desirable, placement of sensors in deep subsurface formations with harsh environments is expected to pose many technological and operational (e.g., maintenance) challenges.

4.8 CONCLUSIONS

This chapter presented the past and ongoing research on the monitoring of underground environment in soil, rock, and groundwater using sensors and wireless sensor networks. Issues related to network deployment, virtual sensor network, reliable sensor data collection, and network optimization in supporting environmental applications have been discussed. Application of WSNs to monitor common subsurface contaminant plumes involving dissolved volatile organic chemical was presented. If developed, this technology will be an attractive and cost efficient alternative for long term monitoring of contaminated sites. Potential also exists for the technology to be applied to problems in monitoring gas leakage resulting from deep geologic storage of carbon dioxide and methane leakage from unconventional energy development.

ACKNOWLEDGMENTS

The material used in the development of this chapter came from the joint research conducted by the coauthors with their students and collaborators. They include Lisa Porta, Paul Schulte, Kevin Barnhart, Philip Loden, Inigo Urteaga, Ying Zhu, Dilum Bandara, Vidarshana Bandara, Dulanjalie Dhanapala, and Alexis Sitchler. The funding for this research came from by the National Science Foundation and the Army Research Office.

The contributions by all the coworkers and support from the funding agencies are gratefully acknowledged.

REFERENCES

Agency for Toxic Substance and Disease Registry (ATSDR), 2003. https://www.atsdr.cdc.gov/spl/previous/03list.html.

Agency for Toxic Substance and Disease Registry (ATSDR), 2004. https://www.atsdr.cdc.gov/toxprofiles/TP.asp?id=206&tid=37.

Akan, O.B., Sankarasubramaniam, Y., Akyildiz, I.F., 2003. Esrt: event-to-sink reliable transport in wireless sensor networks. In: Proc. of ACM MobiHoc.

Akyildiz, I.F., Su, W., Sankarasubramaniam, Y., Cayirci, E., 2002. Wireless sensor networks: a survey. Comput. Netw. 38 (4), 393–422.

Anderson, M.R., Johnson, R.L., Pankow, J.F., 1992. Dissolution of dense chlorinated solvents into ground water: 1. Dissolution from a well-defined residual source. Ground Water 30 (2), 250–256.

Apps, J.A., Zheng, L., Zhang, Y., Xu, T.F., Birkholzer, J.T., 2010. Evaluation of potential changes in groundwater quality in response to CO_2 leakage from deep geological storage. Transp. Porous Media 82, 215–246.

Ball, W., Liu, C., Xia, G., Young, D.F., 1997. A diffusion-based interpretation of tetrachloroethene and trichloroethene concentration profiles in a groundwater aquitard. Water Resour. Res. 33 (12), 2741–2757.

Bandara, H.M.N.D., Jayasumana, A.P., Illangasekare, T.H., 2008. Cluster Tree Based Self Organization of Virtual Sensor Networks, paper presented at IEEE Globecom Workshop on Wireless Mesh and Sensor Networks, Nov. 2008. New Orleans, LA.

Bandara, V., Jayasumana, A.P., Pezeshki, A., Illangasekare, T.H., Barnhardt, K., 2010. Subsurface plume tracking using sparse wireless sensor networks. Int. Electron. J. Struct. Eng. (EJSE). Special Issue.

Barnhart, K., Illangasekare, T., 2012. Automatic transport model data assimilation in Laplace space. Water Resour. Res. 48 (1).

Barnhart, K., Urteaga, I., Han, Q., Jayasumana, A., Illangasekare, T., 2010. On integrating groundwater transport models with wireless sensor networks. Groundwater 48 (5), 771–780.

Bawa, M., Gionis, Aristides, Garcia-Molina, Hector, Motwani, Rajeev, 2004. The price of validity in dynamic networks. In: SIGMOD Conference.

Beeby, S., White, N., 2014. Energy Harvesting for Autonomous Systems. Artech House.

Birkholzer, J.T., Zhou, Q., Tsang, C.-F., 2009. Large-scale impact of CO_2 storage in deep saline aquifers: a sensitivity study on pressure response in stratified systems. Int. J. Greenh. Gas Control 3 (2), 181–194.

Boden, T.A., Marland, G., Andres, R.J., 2010. Global, regional, and national fossil-fuel CO_2 emissions. In: Trends: A Compendium of Data on Global Change. Carbon Dioxide Information Analysis. Center, Environmental Sciences Division, Oak Ridge National Laboratory, Oak Ridge, Tennessee 37831–6290, USA.

Bohnhoff, M., Zoback, M.D., Chiaramonte, L., Gerst, J.L., Gupta, N., 2010. Seismic detection of CO_2 leakage along monitoring wellbores. Int. J. Greenh. Gas Control 4 (4), 687–697.

Bredin, J.L., Demaine, E.D., Hajiaghayi, M., Rus, D., 2005. Deploying sensor networks with guaranteed capacity and fault tolerance. In: Proceedings of the 6th ACM International Symposium on Mobile Ad Hoc Networking and Computing. MobiHoc '05, pp. 309–319.

Carrera, J., Alcolea, A., Medina, A., Hidalgo, J., Slooten, L.J., 2005. Inverse problem in hydrogeology. Hydrogeol. J. 13 (1), 206–222.

Chabora, Ethan R., Benson, S.M., 2009. Brine displacement and leakage detection using pressure measurements in aquifers overlying CO_2 storage reservoirs. Energy Proc. 1 (1), 2405–2412.

Chen, D., Du, D.-Z., Hu, X.-D., Lin, G.-H., Wang, L., Xue, G., 2000. Approximations for Steiner trees with minimum number of Steiner points. J. Glob. Optim. 18 (1), 17–33.

Cheng, X., Du, D.-Z., Wang, L., Xu, B., 2008. Relay sensor placement in wireless sensor networks. Wirel. Netw. 14 (3), 347–355.

Chinni, R.C., Gold, D.M., Brown, S.B., Chang, J.T., Angel, S.M., Colston, B.W. Jr., 2004. A non-lensed fiber-optic resonance-enhanced multiphoton ionization probe. Appl. Spectrosc. 58 (9), 1038–1043.

Chowdhury, N.M., Kabir, M., Boutaba, R., 2010. A survey of network virtualization. Comput. Netw. 54 (5), 862–876.

Considine, J., Li, F., Kollios, G., Brers, J., 2004. Approximate aggregation techniques for sensor databases. In: Proc. of IEEE ICDE.

Culler, D., Estrin, D., Sirivastava, M., 2004. Overview of sensor networks. IEEE Comput., 41–49.

Davis, W.M., Myers, K.F., Wise, M.B., Thompson, C.V., 2001. Environmental Laboratory Tri-Service Site Characterization and Analysis Penetrometer System (SCAPS) Validation of the Hydrosparge Volatile Organic Compound Sensor. Rep., US Army Engineer Research and Development Center.

Deshpande, A., Guestrin, Carlos, Madden, Samuel, Hellerstein, Joseph, Hong, Wei, 2004. Model driven data acquisition in sensor networks. In: VLDB Conference.

Dhanapala, D.C., Jayasumana, A.P., 2011. Geo-logical routing in wireless sensor networks. In: Proc. 8th IEEE Communications Society Conference on Sensor, Mesh and Ad Hoc Communications and Networks (SECON). Salt Lake City, Utah.

Dhanapala, D.C., Jayasumana, A.P., 2014. Topology preserving maps – extracting layout maps of wireless sensor networks from virtual coordinates. IEEE/ACM Trans. Netw. 22 (3), 784–797.

Doherty, J., Hunt, R., 2010. Approaches to Highly Parameterized Inversion – A Guide to Using PEST for Groundwater-Model Calibration, vol. 59. U.S. Geological Survey Scientific Investigations Report No. 2010-5169, US Geological Survey, p. 59.

EPA, 2004. A Review of Emerging Sensor Technologies for Facilitating Long-Term Ground Water Monitoring of Volatile Organic Compounds. Rep., 61 pp.

Frenzel, L., 2016. Twelve Wireless Options for IOT/M2M: Diversity of Dilemma? Electron. Des.

Friedmann, S.J., 2007. Geological carbon dioxide sequestration. Elements 3 (3), 179–184.

Ganesan, D., Cristescu, R., Beferull-Lozano, B., 2006. Power-efficient sensor placement and transmission structure for data gathering under distortion constraints. ACM Trans. Sensor Netw. (TOSN) 2 (2), 155–181.

Ganesan, D., Govindan, R., Shenker, S., Estrin, D., 2002. Highly-resilient, energy-efficient multipath routing in wireless sensor networks. ACM MCCR 1 (2).

Gelhar, L.W., Welty, C., Rehfeldt, K.R., 1992. A critical review of data on field-scale dispersion in aquifers. Water Resour. Res. 28 (7), 1955–1974.

Haenggi, M., 2005. Opportunities and challenges in wireless sensor networks. In: Ilyas, M., Mahgoub, I. (Eds.), Handbook of Sensor Networks: Compact Wireless and Wired Sensing Systems. CRC Press.

Han, X., Cao, X., Lloyd, E., Shen, C.-C., 2007. Fault-tolerant relay node placement in heterogeneous wireless sensor networks. In: 26th IEEE International Conference on Computer Communications. INFOCOM 2007. IEEE, pp. 1667–1675.

Han, Q., Jayasumana, A.P., Illangaskare, T., Sakaki, T., 2008. A wireless sensor network based closed-loop system for subsurface contaminant plume monitoring. In: IEEE International Symposium on Parallel and Distributed Processing. IPDPS 2008. IEEE, pp. 1–5.

Hao, B., Tang, H., Xue, G., 2004. Fault-tolerant relay node placement in wireless sensor networks: formulation and approximation. In: Workshop on High Performance Switching and Routing. HPSR 2004, pp. 246–250.

Ho, C.K., 2003. Chemiresistor Microsensors for in-Situ Monitoring of Volatile Organic Compounds. Rep., Sandia National Laboratories.

IEEE Standard for Information Technology-Telecommunications and Information Exchange Between Systems–Local and Metropolitan Area Networks-Specific Requirements Part 15.4: Wireless Medium Access Control (MAC) and Physical Layer (PHY) Specifications for Low-Rate Wireless Personal Area Networks (WPANs) IEEE Std 802.15.4-2006 (Revision of IEEE Std 802.15.4-2003) 2006. In: IEEE Std 802.15.4-2006 (Revision of IEEE Std 802.15.4-2003), pp. 301–305.

Illangasekare, T.H., et al., 2007. Mass Transfer from Entrapped DNAPL Sources Undergoing Remediation: Characterization Methods and Prediction Tools. Rep., ISERDP Project CU-1294.

IPCC, 2007. Contribution of Working Groups I, II and III to the Fourth Assessment Report of the Intergovernmental Panel on Climate Change. Rep., 104 pp.

Jayasumana, A., Han, Q., Illangasekare, T.H., 2007. Virtual sensor networks – a resource efficient approach for concurrent applications. In: The 4th IEEE International Conference on Information Technology – Next Generations (ITNG). Las Vegas, NV, April 2007.

Jayasumana, A.P., Paffenroth, R., Ramasamy, S., 2016. Topology maps and distance free localization from partial virtual coordinates for IoT. In: Proc. IEEE Communications Conference. ICC2016, Kuala Lumpur, Malaysia, May 2016.

Kanu, A.B., Gribb, M.M., Hill, H.H., 2008. Predicting optimal resolving power for ambient pressure ion mobility spectrometry. Anal. Chem. 80 (17), 6610–6619. http://dx.doi.org/10.1021/ac8008143.

Karp, B., Kung, H.T., 2000. GPSR: greedy perimeter stateless routing for wireless networks. In: Proc. 6th ACM/IEEE Int. Conf. Mobile Computing and Networking (MobiCom), pp. 243–254.

Kashyap, A., Khuller, S., Shayman, M., 2006. Relay placement for higher order connectivity in wireless sensor networks. In: Proceedings of the 25th IEEE International Conference on Computer Communications. INFOCOM 2006, pp. 1–12.

Khan, M.M.H., Le, Hieu K., LeMay, Michael, Moinzadeh, Parya, Wang, Lili, Yang, Yong, Noh, Dong K., Abdelzaher, Tarek, Gunter, Carl A., Han, Jiawei, Jin, Xin, 2010. Diagnostic powertracing for sensor node failure analysis. In: IPSN 2010.

Kharaka, Y.K., et al., 2010. Changes in the chemistry of shallow groundwater related to the 2008 injection of CO_2 at the ZERT field site Bozeman, Montana. Environ. Earth Sci. 60 (2), 273–284.

Kim, S., Fonseca, Rodrigo, Culler, David, 2004. Reliable transfer in wireless sensor networks. In: The First IEEE International Conference on Sensor and Ad Hoc Communications and Networks.

Lazaridis, I., Han, Qi, Mehrotra, Sharad, Venkatasubramanian, Nalini, 2009. Fault-tolerant evaluation of continuous queries over sensor data. Int. J. Dist. Sensor Netw. (IJDSN) 5 (4).

Leeson, A., Stroo, H., 2001. SERDP and ESTCP Workshop on Investment Strategies to Optimize Research and Demonstration Impacts in Support of DoD Restoration Goals. Rep., SERDP/ESTCP.

Leeson, A., Stroo, H., et al., 2013. SERDP and ESTCP Workshop on Long Term Management of Contaminated Groundwater Sites. Rep., SERDP/ESTCP, 38 pp.

Lenhard, R.J., Oostrom, M., Simmons, C.S., White, M.D., 1995. Investigation of density-dependent gas advection of trichloroethylene: experiment and a model validation exercise. J. Contam. Hydrol. 19, 47–67.

Lewicki, J.L., 2009. Eddy Covariance Observations of Surface Leakage During Shallow Subsurface CO_2 Releases. Rep., Lawrence Berkeley National Laboratory.

Lewicki, J.L., Hilley, G.E., Dobeck, L., Spangler, L., 2010. Dynamics of CO_2 fluxes and concentrations during a shallow subsurface CO_2 release. Environ. Earth Sci. 60 (2), 285–297.

Li, M., Liu, Yunhao, 2012. Underground structure monitoring in wireless sensor networks. In: IPSN 2012.

Li, P., Regehr, John, 2010. T-check: bug finding for sensor networks. In: IPSN 2010.

Lieberman, S.H., 1998. Direct-push, fluorescence-based sensor systems for in situ measurement of petroleum hydrocarbons in soils. Field Anal. Chem. Technol. 2 (2), 63–73.

Lieberman, S.H., 2007. Direct Push Chemical Sensors for DNAPL. Final Report ESTCP Project: ER-0109. 155 pp.

Lin, G.-H., Xue, G., 1999. Steiner tree problem with minimum number of Steiner points and bounded edge-length. Inf. Process. Lett. 69 (2), 53–57.

Little, M.G., Jackson, R.B., 2010. Potential impacts of leakage from deep CO_2 geosequestration on overlying freshwater aquifers. Environ. Sci. Technol. 44 (9225–9232).

Liu, C., Ball, W.P., 2002. Back diffusion of chlorinated solvent contaminants from a natural aquitard to a remediated aquifer under well-controlled field conditions: predictions and measurements. Groundwater 40 (2), 175–184.

Liu, H., Wan, P.J., Jia, X., 2005. Fault-Tolerant Relay Node Placement in Wireless Sensor Networks, vol. 3595. Springer, Berlin/Heidelberg.

Lloyd, E.L., Xue, G., 2007. Relay node placement in wireless sensor networks. IEEE Trans. Comput. 56 (1), 134–138.

Loden, P., Han, Q., Porta, L., Illangasekare, T., Jayasumana, A.P., 2009. A wireless sensor system for validation of real-time automatic calibration of groundwater transport models. J. Syst. Softw. 82 (11), 1859–1868.

Madden, S., Franklin, M.J., Hellerstein, J.M., Hong, W., 2002. Tag: a tiny aggregation service for Ad-Hoc sensor networks. In: USENIX OSDI.

Mantoglou, A., Wilson, J.L., 1982. The turning bands method for simulation of random fields using line generation by a spectral method. Water Resour. Res. 18 (5), 1379–1394.

Marcus, C., Bonnet, P., 2010. IEEE Pervasive Computing.

Markham, A., Trigoni, Niki, 2012. Magneto-Inductive Networked Rescue System (MIN-ERS): taking sensor networks underground. In: IPSN 2012.

Martinez, K., Hart, J.K., Ong, R., 2009. Deploying a wireless sensor network in Iceland. In: Proc. Geosensor Networks. In: Lecture Notes in Computer Science, vol. 5659, pp. 131–137.

McDonald, M.G., Harbaugh, A.W., 1988. A Modular Three-Dimensional Finite-Difference Ground-Water Flow Model. Rep., US Geological Survey.

Montenegro, G., Kushalnagar, N., Hui, J., Culler, D., 2007. Transmission of IPv6 packets over IEEE 802.15.4 networks. RFC4944.

Moreno-Barbero, E., Illangasekare, T.H., 2006. Influence of dense nonaqueous phase liquid pool morphology on the performance of partitioning tracer tests: evaluation of the equilibrium assumption. Water Resour. Res. 42.

Musaloiu, E.R., Terzis, A., Szlavecz, K., Szalay, A., Cogan, J., Gray, J., 2006. Life under your feet: wireless sensors in soil ecology. In: Proceedings of the Third Workshop on Embedded Networked Sensors.

NRC, 2013. Alternatives for Managing the Nation's Complex Contaminated Groundwater Sites. Rep. National Academies. 422 pp.

Pankow, J.F., Cherry, J.A., 1996. Dense Chlorinated Solvents and Other DNAPL's in Groundwater: History, Behavior, and Remediation. Rep. Portland, Oreg., Waterloo Press. 522 pp.

Park, S.J., Vedantham, Ra., Sivakumar, R., Akyildiz, I.F., 2004. A scalable approach for reliable downstream data delivery in wireless sensor networks. In: ACM MobiHoc Conference.

Parker, B.L., Chapman, S.W., Guilbeault, M.A., 2008. Plume persistence caused by back diffusion from thin clay layers in a sand aquifer following TCE source-zone hydraulic isolation. J. Contam. Hydrol. 102 (1–2), 86–104.

Plampin, M., Illangasekare, T., Sakaki, T., Pawar, R., 2014. Experimental study of gas evolution in heterogeneous shallow subsurface formations during leakage of stored CO_2. Int. J. Greenh. Gas Control 22, 47–62.

Poeter, E.P., Hill, M.C., 1997. Inverse models: a necessary next step in groundwater modeling. Ground Water 35 (2), 50–260.

Polastre, J., Szewczyk, R., Culler, D., 2005. Telos: Enabling ultra-low power wireless research., paper presented at The Fourth International Conference on Information Processing in Sensor Networks: Special Track on Platform Tools and Design Methods for Network Embedded Sensors, ISPN.

Porta, L., Illangasekare, T., Loden, P., Han, Q., Jayasumana, A., 2009. Continuous plume monitoring using wireless sensors: proof of concept in intermediate scale tank. J. Environ. Eng. 135 (9), 831–838.

Priya, S., Inman, D.J., 2009. Energy Harvesting Technologies, vol. 21. Springer.

Ramanathan, N., Schoelhammer, T., Kohler, E., Whitehouse, K., Harmon, T.C., Suelo, T.C., 2009. Human-assisted sensing for exploratory soil monitoring studies. In: Proceedings of the ACM SenSys.

Rao, P.S.C., Jawitz, J.W., 2003. Steady state mass transfer from single-component dense nonaqueous phase liquids in uniform flow fields. Water Resour. Res. 39 (3).

Romanak, K.D., Smyth, R.C., Yang, C., Hovorka, S.D., Rearick, M., Lu, J., 2012. Sensitivity of groundwater systems to CO_2: application of a site-specific analysis of carbonate monitoring parameters at the SACROC CO_2−enhanced oil field. Int. J. Greenh. Gas Control 6, 142–152.

Rouse, J.H., Shaw, J.A., Lawrence, R.L., Lewicki, J.L., Dobeck, L.M., Repasky, K.S., Spangler, L.H., 2010. Multi-spectral imaging of vegetation for detecting CO_2 leaking from underground. Environ. Earth Sc. 60 (2), 313–323.

Sakaki, T., Limsuwat, A., Illangasekare, T.H., 2011. A simple method for calibrating dielectric soil moisture sensors: laboratory validation in Sands. Vadose Zone J. 10, 526–531.

Sale, T., Illangasekare, T., Zimbron, J., Rodriguez, D., Wilking, B., Marinelli, F., 2006. AFCEE Source Zone Initiative. Rep.

Sale, T.C., McWhorter, D.B., 2001. Steady state mass transfer from single-component dense nonaqueous phase liquids in uniform flow fields. Water Resour. Res. 37 (2), 393–404.

Saripalli, P., Amonette, J., Rutz, F., Gupta, N., 2006. Design of sensor networks for long term monitoring of geological sequestration. Energy Convers. Manag. 47 (13–14), 1968–1974.

Sasnauskas, R., Landsiedel, Olaf, Alizai, Muhammad Hamad, Weise, Carsten, Kowalewski, Stefan, Wehrle, Klaus, 2010. KleeNet: discovering insidious interaction bugs in wireless sensor networks before deployment. In: IPSN 2010.

Sauty, J.-P., 1980. An analysis of hydrodispersive transfer in aquifers. Water Resour. Res. 16 (1), 145–158.

Shulte, P., 2009. Design of Experiments in a Three Dimensional Synthetic Aquifer for Evaluation of Wireless Sensor Network Technologies Applied to Real-Time Model Calibration and Plume Management. Colorado School of Mines, Golden, Colorado.

Srinivas, A., Zussman, G., Modiano, E., 2006. Mobile backbone networks – construction and maintenance. In: MobiHoc '06: Proceedings of the 7th ACM International Symposium on Mobile Ad Hoc Networking and Computing. ACM, New York, NY, USA, pp. 166–177.

Srivastava, R., Brown, B., Carr, T., Vikara, D., McIlvried, H., 2009. Monitoring, Verification, and Accounting of CO_2 Stored in Deep Geologic Formations. Rep. DOE National Energy Technology Laboratory.

Stann, F., Heidemann, John, 2003. Rmst: reliable data transport in sensor networks. In: Proceedings of the First International Workshop on Sensor Net Protocols and Applications. Anchorage, Alaska, USA, April 2003, pp. 102–112.

Stoianov, I., Nachman, L., Madden, S., 2007. Pipenet: A wireless sensor network for pipeline monitoring. In: Proceedings of IEEE Conference on Information Processing in Sensor Networks.

Stuntebeck, E., Pompili, D., Melodia, T., 2009. Underground wireless sensor networks using commodity terrestrial motes, paper presented at IEEE Conference on Sensor and Ad Hoc Communications and Networks (SECON).

Sundaram, V., Eugster, Patrick, Zhang, Xiangyu, 2010. Efficient diagnostic tracing support for wireless sensor networks. In: SenSys.

Tang, J., Hao, B., Sen, A., 2006. Relay node placement in large scale wireless sensor networks. Comput. Commun. 29, 490–501. Elsevier Science, Amsterdam, PAYS-BAS 1978, Revue.

Tilak, S., Abu-Ghazaleh, N.N., Heinzelman, W., 2002. Infrastructure tradeoffs for sensor networks. In: Proc. of WSNA.

Trautz, R.C., et al., 2012. Effect of dissolved CO_2 on a shallow groundwater system: a controlled release field experiment. Environ. Sci. Technol. 47 (1), 298–305.

Trubilowicz, J., Cai, K., Weile, M., 2009. Viability of motes for hydrological measurement. Water Resour. Res. 45, W00D22.

Tynan, R., O'Hare, G.M., O'Grady, M.J., Muldoon, C., 2008. Virtual sensor networks: an embedded agent approach. In: 2008 IEEE International Symposium on Parallel and Distributed Processing with Applications. IEEE.

Urteaga, I., Barnhart, K., Han, Q., 2009. REDFLAG: a real-time, distributed, flexible, lightweight, and generous fault detection service for data-insensitive sensor applications. J. Pervasive Mob. Comput.

Vuran, M.C., Silva, A.R., 2009. Communication through soil in wireless underground sensor networks – theory and practice. In: Sensor Networks: Where Theory Meets Practice. Springer.

Wan, C.Y., Campbell, A.T., Krishnamurthy, L., 2002. Psfq: A reliable transport protocol for wireless sensor networks. In: Proc. of ACM WSNA.

Wan, C.Y., Eisenman, S.B., Campbell, A.T., 2003. Coda: congestion detection and avoidance in sensor networks. In: SenSys Conference.

Wang, Q., Xu, K., Takahara, G., Hassanein, H., 2007. Transactions papers – device placement for heterogeneous wireless sensor networks: minimum cost with lifetime constraints. Wirel. Commun. 6 (7), 2444–2453.

Werner-Allen, G., Lorincz, K., Welsh, M., Marcillo, O., Johnson, J., Ruiz, M., Lees, J., 2006. Deploying a wireless sensor network on an active volcano. IEEE Internet Comput., 1069–7801.

Wilking, B.T., Rodriguez, D.R., Illangasekare, T.H., 2013. Experimental study of the effects of DNAPL distribution on mass rebound. Groundwater 51 (2), 229–236.

Wolfbeis, O.S., 2004. Fiber-optic chemical sensors and biosensors. Anal. Chem. 76, 3269–3284.

Woo, A., Tong, T., Culler, D., 2003. Taming the underlying challenges of reliable multihop routing in sensor networks. In: SenSys Conference.

Yang, D., Misra, Satyajayant, Fang, Xi, Xue, Guoliang, Zhang, Junshan, 2012. Two-tiered constrained relay node placement in wireless sensor networks: computational complexity and efficient approximations. IEEE Trans. Mob. Comput. (TMC) 11 (8), 1399–1411.

Ye, F., Zhong, G., Lu, S., Zhang, L., 2003. Gradient broadcast: a robust data delivery protocol for large scale sensor networks. In: ACM WINET, vol. 11.

Yoon, S.-U., Ghazanfari, E., Cheng, L., Pamukcu, S., Suleiman, Muhannad T., 2012. Subsurface event detection and classification using wireless signal networks. Sensors (Basel) 12 (11), 14862–14886.

Zheng, C., Wang, P.P., 1999. A Modular Three-Dimensional Multispecies Transport Model. Rep. US Army Corps of Engineers, Washington, DC 20314-1000.

Zhu, Ying, Han, Q., 2009. The more Relay Nodes, the More Energy Efficient? In: Proceedings of Embedded and Ubiquitous Computing.

CHAPTER 5

EM-Based Wireless Underground Sensor Networks

Abdul Salam, Mehmet C. Vuran
University of Nebraska-Lincoln, Lincoln, NE 68588, USA

5.1 INTRODUCTION

Wireless underground sensor networks (WUSNs) consist of buried nodes, which are connected to subsurface underground sensors, and use electromagnetic (EM) waves for underground communications. WUSNs are capable of operating in environments where no other computer network has functioned before, and has the potential to provide real-time, robust, and energy efficient sensing, and communication in these environments.

The characteristics of the wireless underground (UG) channel are much different as compared to the conventional over-the-air (OTA) wireless communication channel. These differences are caused by the wave propagation mechanism in the UG channel. Electromagnetic (EM) waves interacting with the soil medium exhibit distinct characteristics, and experience higher attenuation. Physical properties of the soil texture, soil moisture, soil temperature, and bulk density impact UG wave propagation. These interactions introduce channel impairments, which varies with space and time.

These spatiotemporal variations present challenges to maintain connectivity in WUSNs, and necessitate, to maintain connectivity, adaptive solutions tailored to the variations in soil moisture. To model the characteristics of the underground channel, and to validate interactions among different component of the WUSN, experimental studies are essential. Conducting underground channel characterization experiments is a very challenging and time consuming task (Salam et al., 2016; Silva and Vuran, 2010a). First, it is difficult to get a wide range of soil moisture levels, which are essential to ascertain its impact on UG channel, in a very short span of time because of climatic conditions. Digging process becomes difficult in extreme weather conditions due to soil becoming

Underground Sensing.
DOI: http://dx.doi.org/10.1016/B978-0-12-803139-1.00005-9

247

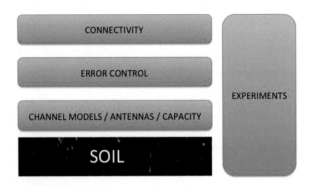

Figure 5.1 Contents of the chapter

hardened to excavating. Also controlling and getting the desired soil moisture for the experiments is not a trivial task. Finding different soil types,
installation, and replacement of equipment further adds to this complexity. Moreover, experiments cannot be conducted all year round because
severe weather and temperatures make it infeasible for the wellbeing of the
personnel and the functionality of equipment.

Electromagnetic waves in soil experience losses by absorption in soil
and by diffusion attenuation due to soil permittivity, which is higher than
that of air. Moreover, permittivity of soil changes with soil moisture, which
causes changes in the wavelength. These changes in the wavelength impact the resonance of an underground antenna. Resonant frequency of
the antenna varies with soil moisture variations, which can be tackled by
cognitive radio adjustments of operating frequency based on soil moisture
changes.

WUSN have many applications in precision agriculture, border patrol,
and environment monitoring. In precision agriculture WUSNs are being
used for sensing and monitoring of the soil moisture and other physical properties of soil (Dong et al., 2003; Gutiérrez et al., 2014; Kim et
al., 2008; Tiusanen, 2013; Hopkins, 2017; Bogena et al., 2010; Akyildiz
and Stuntebeck, 2006; Guo and Sun, 2014; Markham and Trigoni, 2012;
Sun and Akyildiz, 2010; Salam and Vuran, 2016, 2017a, 2017b, 2017c;
Salam et al., 2017; Vuran et al., 2016; Tiusanen, 2006). Border monitoring
is another important application area of WUSNs where these networks are
being used to enforce border and stop infiltration (Akyildiz et al., 2009;
Sun et al., 2011a). Monitoring applications of WUSNs include land slide
monitoring and pipeline monitoring (Sun et al., 2011b; Sun and Aky-

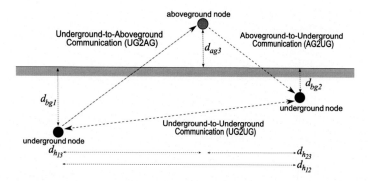

Figure 5.2 Three links of underground and aboveground communication (Silva and Vuran, 2010b)

ildiz, 2010; Guo and Sun, 2014). The structure of this chapter is shown in Fig. 5.1. In Section 5.2, soil as communication media in wireless underground sensor networks is discussed. Detailed analysis of the underground channel propagation in WUSNs, along with impulse response, and path loss analysis are presented in Section 5.3. Impacts of soil on the underground antenna are provided in Section 5.4 and capacity of wireless underground channel is analyzed. Connectivity of wireless sensor nodes in WUSN is discussed in Section 5.6. In Section 5.7, design of indoor field testbeds and outdoor testbeds is described for conducting WUSN experiments.

5.2 SOIL AS A COMMUNICATION MEDIA

EM wave-based communication in the underground channel consists of three types of links (Silva and Vuran, 2010a), namely underground to aboveground (UG2AG), aboveground to underground (AG2UG), and underground to underground (UG2UG) (see Fig. 5.2).

Soil medium is involved in communication through these three links. Wavelength of an EM wave incident into soil is affected by dielectric properties of the soil. Soil texture and its water holding capacity, bulk density, and salinity affect the propagation of waves. To understand the propagation of waves in soil, it is important to understand the physical processes occurring in soil. Soil medium consists of soil particles, pore space, and water content. Soil particles are divided into silt, sand, and clay based on their size. Soils are classified based on these particles sizes. Complex dielectric constant of soil ϵ_s consists of real part ϵ_s' and imaginary part ϵ_s''. Dielectric constant of soil that is fully dried is not dependent on frequency and can be determined from Ulaby and Long (2014):

$$\epsilon'_s = [1 + 0.44\rho_b]^2, \tag{5.1}$$

where ρ_b is the bulk density of soil. Bulk density is related to the ratio of dry soil mass and bulk volume of the soil. This ratio also contains pore spaces inside the bulk volume. Dialectic spectra of the soil becomes more complex with the increase in moisture content. Water content inside the soil is classified into bound water and unrestrained water. Bound water refers to water held by soil particles in the top layers of soil, and depends on particle surface area which is defined by the soil composition. Water content in the soil can be ascertained by either volumetric or gravimetric bases (Ulaby and Long, 2014; Foth, 1990; Hillel, 2003).

Electromagnetic waves traveling in the soil interacts with soil particles, air, restrained and unrestrained water. Free and restrained water molecules when in interaction with electromagnetic wave exhibits different dielectric dispersion, and the dielectric constant depends on the frequency of EM waves. In addition to the water content and frequency, other factors such as bulk density and soil texture also affects the permittivity of soil (Ulaby and Long, 2014; Foth, 1990; Hillel, 2003).

In Dobson et al. (1985), a model of dielectric properties of soil has been developed for frequencies higher than 1.4 MHz. In Peplinski et al. (1995), Peplinski modified the model through extensive measurements to characterize the dielectric behavior of the soil in the frequency range of 300 MHz to 1.3 GHz. Peplinski model for complex dielectric permittivity of the soil is given as (Peplinski et al., 1995):

$$\epsilon_s = \epsilon'_s - j\epsilon''_s , \tag{5.2}$$
$$\epsilon'_s = 1.15[1 + \frac{\rho_b}{\rho_s}(\epsilon_s^{\alpha'}) + m_v^{\beta'} \eta'^{\alpha'}_{f_w} - m_v]^{1/\alpha'} - 0.68,$$
$$\epsilon''_s = [m_v^{\beta''} \eta''^{\alpha'}_{f_w}]^{1/\alpha'} ,$$

where ϵ_s is the dielectric constant of mixture of soil and water, and it depends on the soil texture, volumetric water content, bulk density, frequency and particle density. In (5.2), m_v is the volumetric water content, ρ_b is the bulk density of soil given in g/cm^3, $\rho_s = 2.66$ g/cm^3 is the solid soil particle density, $\alpha' = 0.65$ is a constant obtained through empirical evaluations, and values β' and β'' depend on the soil texture (Peplinski et al., 1995):

$$\beta' = 1.2748 - 0.519S - 0.152C, \tag{5.3}$$

$$\beta'' = 1.33797 - 0.603S - 0.166C, \tag{5.4}$$

where S and C represent the percentage of sand and clay particles, respectively. Real and imaginary parts of the relative dielectric constant of free water are represented by η'_{fw} and η''_{fw}.

The complex propagation constant of the EM waves in soil is given as (Dong and Vuran, 2011)

$$C = X + jY, \tag{5.5}$$

where

$$X = \omega \sqrt{ \frac{\mu\,\epsilon'_s}{2} \left[\sqrt{1 + (\frac{\epsilon''_s}{\epsilon'_s})^2} - 1 \right] }, \tag{5.6}$$

$$Y = \omega \sqrt{ \frac{\mu\,\epsilon'_s}{2} \left[\sqrt{1 + (\frac{\epsilon''_s}{\epsilon'_s})^2} + 1 \right] }, \tag{5.7}$$

where angular frequency is represented $\omega = 2\pi f$, μ is the magnetic permeability, and ϵ'_s and ϵ''_s are the real and imaginary parts of the dielectric constant, respectively, as given in (5.2). Due to complex dielectric constant of soil, propagation constant in soil is affected by both ϵ'_s and ϵ''_s. Hence two types of losses happen in soil. Diffusion attenuation which comes from real part ϵ'_s and water absorption attenuation which comes from ϵ''_s. Therefore, EM waves propagation in soil exhibits higher losses as compared to over-the-air cases in soil.

It can be observed that propagation constant of EM waves in soil is a function of soil moisture. Soil moisture can be represented by volumetric water content (VWC) or soil matric potential. The ratio of soil's water to a combined mixture of soil–water is called volumetric water content (VWC). VWC is percentage of water content in soil and a higher value represents a higher soil moisture. VWC is a major indicator of soil moisture. Variations in VWC strongly impact the resonant frequency. Soil moisture is also expressed as soil matric potential with units of centibar (CB). Movement of water in soil is affected by matric forces called capillarity and adsorption. These forces are measured in soil matric potential. Greater matric potential values indicate lower soil moisture and zero matric potential represents near saturation conditions.

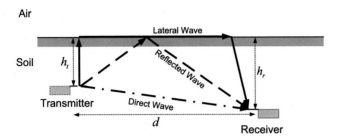

Figure 5.3 Layout of EM waves propagation in the underground channel (Dong and Vuran, 2011)

5.3 PROPAGATION IN THE UNDERGROUND CHANNEL

Propagation in the underground channel is carried out through three different paths (Fig. 5.3). Direct wave and reflected wave paths are through the soil, whereas lateral wave travels through both air and soil. Lateral waves travel along the soil–air interface. A detailed characterization of propagation of these three waves in the underground channel is provided in this section, as we review different underground communication channel models.

5.3.1 Two-Wave UG Channel Model

The two–wave UG channel model considers communication is carried out mainly through two paths, direct and reflected (Fig. 5.3) (Vuran and Akyildiz, 2010). Direct path represents the direct component between transmitter and receiver, whereas the reflected path represents the path where waves reaching at the receiver are reflected by the ground surface. At higher burial depths, due to longer wave propagation length, reflected waves are attenuated and communication is carried out by the direct path only. Path loss in the soil is represented by the Friis free space path loss equation adjusted to account for the losses in soil

$$P_r = G_t + G_r + P_t - L_p ,\qquad(5.8)$$

where P_r is the received power, G_t and G_r are transmitter and receiver antenna gains, L_p is sum of L_0 and L_s, where L_0 is the free space path loss and path loss in soil is represented by L_s. L_s is given by the sum of L_β and L_α, where L_α is transmission loss and L_β is the attenuation loss due to difference of wave length in soil and free space. L_p is given as (in dB)

$$L_p = 20\log_{10}(d) + 20\log_{10}(B) + 8.69 \cdot A \cdot d + 6.4,\qquad(5.9)$$

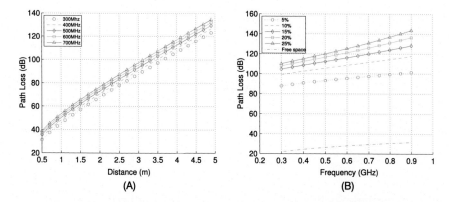

Figure 5.4 Two-wave path loss with: (A) distance and frequency, (B) VWC and frequency (Vuran and Akyildiz, 2010)

where d is distance in meters, A is the attenuation constant, B is the phase shift constant. Path loss results are shown in Fig. 5.4.

The effects of soil moisture can be observed in Fig. 5.4B, where path loss for soil moisture values has been compared with free space. At 0.3 GHz, when VWC is 5% path loss in soil is 64 dB higher as compared to free space and it increases further to 104 dB with the increase in soil moisture level from 5% to 25%.

5.3.2 Three-Wave UG Channel Model

In addition to the direct and reflected paths, the three-wave (3W) UG channel model takes into account the lateral waves (Dong and Vuran, 2011). Lateral wave travels along the soil–air interface and its propagation path consists of both soil and air (Fig. 5.3). These waves are modeled as follows.

Direct Wave

The direct wave is the "line-of-sight" wave between transmitter and receiver and its complete travel path is in the soil. It is expressed as

$$\mathscr{P}_{av}^d = \left(\frac{D_d}{4\pi r_1^2} \right) e^{i2k_s r_1} G_t , \tag{5.10}$$

where $\mathscr{P}_{av}^d$ is time averaged Poynting vector, transmitter receiver distance is r_1; D_d is a constant which depends on the permittivity of the soil, the wave number in soil is $k_s = \beta_s + i\alpha_s = \omega\sqrt{\mu_0 \hat{\epsilon}_s}$, and β_s, α_s represent phase shifting and attenuation, respectively; $\omega = 2\pi f$, where f is the wave frequency; μ_0 is permeability, and $\hat{\epsilon}_s$ is the soil permittivity.

Reflected Wave

The wave reflected from the soil–air interface due to different permittivity of the soil and air is called the reflected wave. The total travel path of the reflected wave is also through the soil. The reflected wave is expressed as

$$\mathscr{P}_{av}^r = D_r \left(\frac{1}{4\pi r_2^2} \right) e^{i2k_s r_2} G_t \Gamma^2 , \tag{5.11}$$

where D_r is permittivity dependent constant. The length of reflection path r_2 is given as

$$r_2 = \sqrt{(h_t + h_r)^2 + d^2} \tag{5.12}$$

and Γ is the reflection coefficient given by Johnk (1988)

$$\Gamma = \frac{\frac{1}{n}\cos\theta_{ri} - \cos\theta_{rt}}{\frac{1}{n}\cos\theta_{ri} + \cos\theta_{rt}} , \tag{5.13}$$

where θ_{ri} is the incident angel, θ_{rt} is the refracted angle, and n is the refractive index of soil. Based on Snell's law,

$$\sin\theta_{ri} = \frac{d}{r_2} , \quad \cos\theta_{ri} = \frac{h_t + h_r}{r_2} ,$$

$$\sin\theta_{rt} = n\sin\theta_{ri} , \quad \cos\theta_{rt} = \sqrt{1 - \sin^2\theta_{rt}}.$$

The soil's refractive index of is determined from the real and imaginary parts of the soil's relative permittivity and is given as

$$n = \sqrt{\frac{\sqrt{\epsilon'^2 + \epsilon''^2} + \epsilon'}{2}} . \tag{5.14}$$

Lateral Wave

The lateral wave propagates horizontally from the transmitter, travels in air along the soil–air interface, and reaches the receiver through the soil. As such, the propagation media of the lateral wave is a combination of soil and air. The lateral wave is given as

$$\mathscr{P}_{av}^L = \left(\frac{D_l}{4\pi d^4} \right) e^{i2k_s(h_t + h_r)} e^{i2k_0 d} T^2 G_t , \tag{5.15}$$

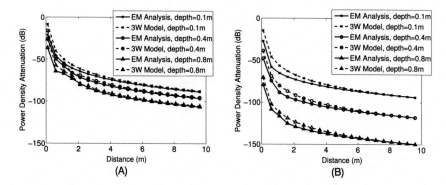

Figure 5.5 Model comparison at different burial depths: (A) VWC = 10%; (B) VWC = 35% (Dong and Vuran, 2011)

where D_l is a constant depending on permittivity of the soil, T is the refraction coefficient given by

$$T = \frac{2\cos\theta_{li}}{n\cos\theta_{li} + \cos\theta_{lt}} .$$

(5.16)

The power received by direct, lateral, and reflected components is given as:

$$P_r^d = P_t + 20\log_{10}\lambda_s - 20\log_{10}r_1 - 8.69\alpha_s r_1 - 45,$$

$$P_r^r = P_t + 20\log_{10}\lambda_s - 20\log_{10}r_2 - 8.69\alpha_s r_2 + 20\log_{10}\Gamma - 45, \text{ and}$$

$$P_r^L = P_t + 20\log_{10}\lambda_s - 40\log_{10}d - 8.69\alpha_s(h_t + h_r) + 20\log_{10}T - 30,$$

and the overall received power is expressed as

$$P_r = 10\log_{10}(10^{\frac{P_r^d}{10}} + 10^{\frac{P_r^r}{10}} + 10^{\frac{P_r^L}{10}}) .$$

(5.17)

In Fig. 5.5, a comparison of attenuation of model and EM wave analysis with distance has been shown for two different soil moisture levels (VWC 10% and VWC 30%) for 10, 40, and 80 cm depth and $D_d = D_r = 0.005$ and $D_l = 0.15$. EM wave analysis is based on numerical analysis of Maxwell equations and electromagnetic fields from King et al. (1992). It can seen that as depth increases from 0.1 to 0.8 m at 2 m distance at 10% VWC, 21 dB higher attenuation is observed due to the longer path traveled by EM waves. Attenuation for the same depth and distance increases 130 dB with increase in soil moisture from 10% to 35%.

5.3.3 Impulse Response Analysis of the UG Channel

To design efficient communications schemes, it is important to gain physical insights into the underground channel model. UG channel impulse response model provides all the statistics which can fully characteristic the UG channel and also provides the information necessary to design physical layer and link layer protocols. Moreover, effective communication bandwidths can also be ascertained from this analysis. In this section we present the impulse response model of the wireless UG channel (Salam et al., 2016).

UG channel impulse response of a wireless underground channel is given as (Salam et al., 2016)

$$h(t) = \sum_{l=0}^{N-1} \alpha_l \delta(t - \tau_l) , \tag{5.18}$$

where N is the number of multipaths, α_l are the gains associated with the multipaths, τ_l are the arrival delay of these multipath components.

Impulse response of the UG model, when reflected, lateral and direct components of the three-wave model are considered, is given as:

$$h_{ug}(t) = \sum_{l=0}^{L-1} \alpha_l \delta(t - \tau_l) + \sum_{d=0}^{D-1} \alpha_d \delta(t - \tau_d) + \sum_{r=0}^{R-1} \alpha_r \delta(t - \tau_r) , \tag{5.19}$$

where L, D, and R are the numbers of lateral, direct and refracted wave multipaths, respectively. α_l, α_d, and α_r are the complex gains, and arrival delay of these three components are represented by τ_l, τ_d, and τ_r, respectively.

The received power for these three components is determined by using the magnitude squared method. Powers of direct, reflect and lateral components are summed together over the entire delays associated with these components:

$$P_r = \sum_{l=0}^{L-1} |\alpha_l|^2 + \sum_{d=0}^{D-1} |\alpha_d|^2 + \sum_{r=0}^{R-1} |\alpha_r|^2 . \tag{5.20}$$

Path loss can be determined once received powers has been calculated and is given as

$$PL(dBm) = G_r(dBi) + G_t(dBi) + P_t(dBm) - P_r(dBm) , \tag{5.21}$$

where transmitter and receiver antenna gains are represented by G_t and G_r respectively, transmitted power is represented by P_t, and received power is P_r.

Metrics for Impulse Response Characterization

Different metrics are used to characterize the impulse response analysis of the wireless underground channel which includes mean access delay and RMS delay spread.

Excess delay of the channel is defined as time delay between the first component and last component arriving at the receiver. Mean excess delay is the first moment of the power delay profile (PDP) and is defined as (τ) (Rappaport et al., 1991)

$$\tau = \sum_k P_k \tau_k \Big/ \sum_k P_k \,, \tag{5.22}$$

where kth bin delay is τ_k and P_k is the absolute instantaneous power. A multipath component is considered when its power is within 30 dB component of the strongest component in the power delay profile.

Root mean square (RMS) delay spread is used to get insight into the inter-symbol interference (ISI) of the wireless underground channel. The RMS delay spread is defined as (Rappaport et al., 1991)

$$\tau_{rms} = \sqrt{(\tau^2) - (\tau)^2} \,, \tag{5.23}$$

where $(\tau^2) = \sum_k P_k \tau_k^2 / \sum_k P_k$, with τ_k being the delay of the kth bin and P_k the absolute instantaneous power at the kth bin.

5.3.4 Testbed Design for Impulse Response Parameters Analysis

An indoor testbed facility has been developed in University of Nebraska–Lincoln in a greenhouse setting (Salam et al., 2016). The purpose of this indoor testbed is to conduct WUSN channel modeling experiments. In WUSNs, deployment of underground communication devices is limited to 50 cm depths (Bogena et al., 2010). These burial depths are close to the surface of the earth. Four dipole antennas are buried 10, 20, 30, and 40 cm depths and these antenna settings are extended at 50 cm and 1 m distances (Salam et al., 2016). Due to higher attenuation in soil, only distance up to 1 m is considered. This structure helps in verifying the accuracy

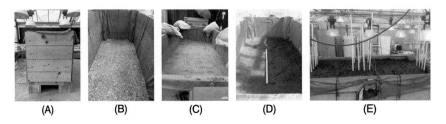

(A)   (B)   (C)   (D)   (E)

Figure 5.6 Testbed Development: (A) Testbed box, (B) Packed soil, (C) Layer of gravel at the bottom of the testbed, (D) Antenna placement, (E) Final outlook (Salam et al., 2016)

Table 5.1 Physical Properties of Testbed Soils (Salam et al., 2016)

Textural Class	%Sand	%Silt	%Clay
Sandy Soil	86	11	3
Silt Loam	33	51	16

and repeatability of experiments such that role of receiver and transmitter antennas can be switched and channel symmetry can be investigated. Moreover, multiple burial depths help capture the depth effects on the channel. In Fig. 5.6, testbed development and final outlook is shown. Pipes are installed at the bottom of the testbed to support water drainage. Gravel in the bottom of box further improves drainage. Soil is packed to achieve the bulk density of the field.

Variations in soil moisture affects the underground communications. Therefore, it is important to monitor the soil moisture with each experiment to accurately characterize channel behavior. In the testbed, Watermark sensors are used to log the soil moisture with time. It is an efficient method with less chance of error. An outdoor testbed has been prepared in silty clay soil with dipole antennas buried at 20 cm depth covering distances up to 12 m. Properties of soil used in testbed are shown in Table 5.1. Change in soil moisture with time at four burial depths, and experimental setup has been shown in Fig. 5.7. In Section 5.3.5, we show metrics derived using these testbeds in silt loam, and silty clay loam soils.

5.3.5 UG Channel Impulse Response Parameters

UG channel is completely characterized by using impulse response metrics (excess delay, mean access delay (5.22), RMS delay spread Saleh and Valenzuela, 1987; Rappaport et al., 1991; Cassioli et al., 2002, given in (5.23) and coherence bandwidth in relation to RMS delay spread Howard and Pahlavan, 1990).

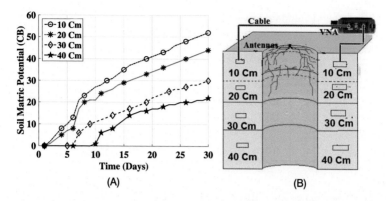

Figure 5.7 (A) Change in soil moisture vs. time, (B) layout of antennas in the experiment (Salam et al., 2016)

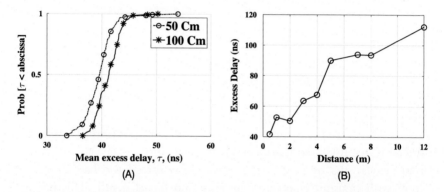

Figure 5.8 (A) Distribution of mean excess delay τ in indoor testbed (silt loam) experiment, (B) excess delay with distance at 20 cm depth in field (silty clay loam) experiment (Salam et al., 2016)

Flat fading channels differ from frequency selective channels in degree of distortion to which different frequency components undergoes. In flat fading channels all components have the same distortion, whereas in frequency selective fading channels different components undergoes through different distortion. Moreover, for a channels to be considered flat, delay spread must be less than the symbol period. Soil texture, moisture, roots and organic matter (OM), soil–air interface, are the major disturbances in the wireless UG channel, in contrast to the OTA (over-the-air channels) where mobility causes the channels to become frequency selectiveness.

In the UG channel, excess delay has strong correlation with the transmitter receiver distance as it can be observed in Fig. 5.8A, where at dif-

Table 5.2 Mean (μ) and Standard Deviation (σ) in nanoseconds for the mean excess delay and RMS delay spread in indoor testbed (silt loam) experiment (Salam et al., 2016)

Depth	Mean Excess Delay τ				RMS Delay Spread τ_{rms}			
	50 cm		1 m		50 cm		1 m	
	μ	σ	μ	σ	μ	σ	μ	σ
10 cm	33.53	1.24	36.09	0.80	20.05	2.24	21.94	2.32
20 cm	34.66	1.07	37.12	1.00	24.93	1.64	25.10	1.77
30 cm	35.87	0.72	37.55	0.65	24.84	2.17	25.34	3.41
40 cm	36.43	0.74	40.18	0.94	23.91	2.84	25.62	1.87

ferent depths in silt loam soil, distribution of mean excess delay is given at transmitter-receiver distance of 50 cm and 1 m. When transmitter–receiver (T–R) distance is increased from 50 cm to 1 m, a 2–3 ns increase in mean excess delay is observed. In Table 5.2, more mean excess delay statistics (mean (μ) and standard deviation (σ) values) are shown at 10, 20, 30, and 40 cm depths. In addition to increase in mean excess delay by increasing distance, deep burial depths also lead to the higher excess delays. Excess delays as function distance for silty clay loam soil where antennas are buried at depth of 20 cm is shown in Fig. 5.8B. It can be observed that for up to 12 m distance a mean excess delay of 116 ns is exhibited.

In Fig. 5.9A, in silt loam soil at different depths, distribution of RMS delay spread is given at transmitter–receiver distance of 50 cm and 1 m. Similar to the statistics of excess delay, RMS delay spread also has strong correlation with transmitter receiver distance and burial depth. This increase in delay spread is caused by the higher delays associated with reflected and lateral waves. These delays increase with increasing distance. More detailed RMS delay spread statistic at different distances and depths are given in Table 5.2.

Coherence bandwidth is another important parameter of the impulse response of the wireless underground channel. Coherence bandwidth is defined as the inverse relationship of the RMS delay spread of the UG channel. For 50 cm and 1 m distances, coherence bandwidth is shown in Fig. 5.9B, change in coherence bandwidth with distance is shown. In UG channel, for distances up to 1 m, coherence bandwidth changes from 1 MHz to 600 kHz and then to 400 kHz for distances up to 12 m range. Increased RMS delay spread causes the coherence bandwidth of the UG channel to reduce. These coherence bandwidth of the underground channel gives very useful insight into the UG channel and can be used as guidelines for selection of different system parameters of a practical UG

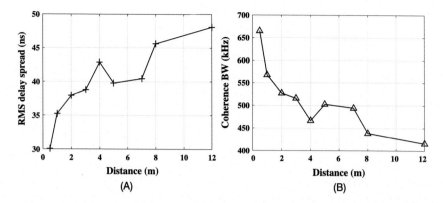

Figure 5.9 (A) RMS delay spread, τ_{rms}, with distance in field (silty clay loam) experiment, (B) coherence bandwidth with distance in field (silty clay loam) experiment (Salam et al., 2016).

communication system design. In Section 5.3.5.1, we discuss the impact of soil moisture variations on the these statistics of the wireless underground channel impulse response. In Section 5.3.5.2, impacts of soil type on underground communications are shown. Impacts of change of operation frequency are discussed in Section 5.3.5.3.

5.3.5.1 Impact of Soil Moisture Changes on Impulse Response

Water in soil is arranged into bound water and free water. Water contained in the initial few molecule layers of the soil is called bound water, which is strongly held by soil particles because of the impact of osmotic and matric forces (Foth, 1990). Beneath these layers, impacts of osmotic and matric powers is lessened, which brings about unlimited water development. EM waves experience scattering when interfaced with bound water. Since dielectric constant of soil changes with time because of the variations in soil moisture, wavelength in soil changes, which impacts the attenuation of EM waves in the soil. Effects of soil moisture variations on amplitudes of the delay profiles are shown for 50 cm distance in silt loam soil is shown in Fig. 5.10A. With higher soil moisture, lower gains are exhibited and this decrease is consistent over all delay ranges and varies between 5–8 dB for the whole power delay profile.

The path loss variations with change in soil moisture at 50 cm and 1 m distances at 10 cm depth in silt loam soil is shown in Fig. 5.10B. A 3–4 dB (7%) decrease in path loss is observed with 0 to 50 cb (centibars) changes in the soil moisture.

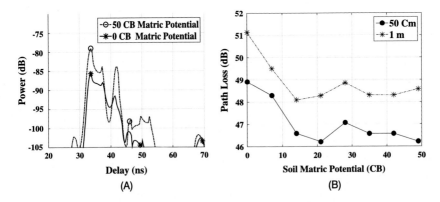

Figure 5.10 Indoor testbed (silt loam) experiment: (A) power delay profile, (B) path loss with vs. soil moisture at 10 cm depth.

To design an efficient wireless underground communication system, these changes in amplitudes and RMS delay spread of wireless underground channel play an important role, as system parameters which work best at low soil moisture levels may not work effectively for high soil moisture levels. Therefore, an underground wireless communication system should have the ability to optimize system parameters in adaptation to these parameters (Akyildiz et al., 2006).

5.3.5.2 Impact of Soil Texture

Textural classes of soil are determined from its particle size distribution (Hillel, 2003). Soil texture impacts communication in the wireless underground channel. Path loss, mean excess delay, and RMS delay spread statistics for three different soils (silt loam, silty clay loam, and sandy) are shown in Table 5.3. Differences in soil texture, pore size, and space result in different capacity of each soil to which water is retained in soil (Foth, 1990). Attenuation in soil depends on this capability. Higher water retention capable soil results in higher attenuation of EM waves. Similarly, attenuation is reduced by lower water content present in the soil. Moreover, there is an important trade-off of RMS delay spread and attenuation, as some coarser soils have lower attenuation with higher RMS delay spreads, as compared to finer soils, which have higher attenuation and low delay spreads.

Table 5.3 Mean (μ) and Standard Deviation (σ) for the Mean Excess Delay, RMS delay spread and Path Loss for 50 cm and 1 m distances, and 20 cm depth for three soils. Values are in nanoseconds

Soil Type	Mean Excess Delay				RMS Delay Spread				Path Loss	
	Distance				Distance				Distance	
	50 cm		1 m		50 cm		1 m		50 cm	1 m
	μ	σ	μ	σ	μ	σ	μ	σ		
Silty Clay Loam	34.77	2.44	38.05	0.74	25.67	3.49	26.89	2.98	49 dB	52 dB
Silt Loam	34.66	1.07	37.12	1.00	24.93	1.64	25.10	1.77	48 dB	51 dB
Sandy Soil	34.13	1.90	37.87	0.80	27.89	2.76	29.54	1.66	40 dB	44 dB

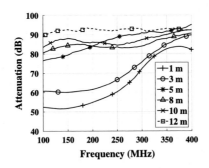

Figure 5.11 Attenuation with frequency (Salam et al., 2016)

5.3.5.3 Impact of Operation Frequency

Attenuation with frequency at different distances of up to 12 m are presented in Fig. 5.11 for transmitter and receiver depths of 20 cm. At 2 m distance, attenuation increases by 24 dB when frequency increases from 200 to 400 MHz. Similarly, for 200 MHz, attenuation is increased from 51 to 92 dB (80%) when distance increases from 50 cm to 12 m.

When EM waves propagate in the soil, higher frequencies suffer more attenuation because their wavelength shortens due to higher permittivity of soil than the air. Therefore, due to less effects of permittivity of soil on lower frequency spectrum, it is more desirable for underground to underground communication as larger communication distances can be achieved. To minimize attenuation, a suitable operation frequency should be selected for each distance and depth. Deployment needs to customized to the soil type and frequency range of sensors being used for deployment from WUSN topology design perspective.

5.3.6 Impulse Response Model Validation Through Experiments

In this section, we present an analysis to calculate the wave speed in soil and then use this formulation to compare this model with measurement results.

Speed of electromagnetic waves traveling in soil is calculated by using the following equation (Handbook, 2014):

$$S = c/\eta, \tag{5.24}$$

where
- c is the speed of light in m/s and is given as $c = 3 \times 10^8$ m/s,

Table 5.4 Speed of the wave in all three soils, calculated by refractive indices n based on particle size distribution (Salam et al., 2016)

Soil Type	Speed in Soil m/s	% of c	Refractive Index n
Silt Loam	5.66×10^7	18.89	5.28
Sandy Soil	5.01×10^7	16.71	5.98
Silty Clay Loam	5.67×10^7	18.91	5.29

- η represents the refractive index of soil $\eta = \sqrt{\sqrt{\epsilon''^2 + \epsilon'^2} + \epsilon'/2}$,
- ϵ' is real part and ϵ'' is imaginary parts of soil's permittivity.

With the formulation of EM waves speed in soil, the time of arrival of each of three components, i.e., direct (τ_d), lateral (τ_l), and reflected components (τ_r), at the receiver, in underground communications is given as:

$$\tau_d = (\delta_s/S) + 2 \times (L/S_c) , \tag{5.25}$$

$$\tau_r = 2 \times (\delta_s/S) + 2 \times (L/S_c) , \tag{5.26}$$

$$\tau_l = 2 \times (\delta_s/S) + (\delta_a/c) + 2 \times (L/S_c) . \tag{5.27}$$

As some part of these components passes through the soil therefore, in (5.25), (5.26) and (5.27), δ_s, S account for the distance and speed of the wave traveling in the soil, respectively. Similarly, S_c and L represent the speed of EM waves and length of waves traveling in coaxial cables.

Refractive indices (Table 5.4) of the soils used in testbeds (Table 5.1) are calculated using (5.25), (5.26) and (5.27), and then the exact time of arrival of these components is determined, which are then used to identify and compare the same three components obtained through experimental measurements.

These comparison results are shown in Fig. 5.12 for different sandy, silt loam, and silty clay loam soils at burial depth of 40 cm. Results of these analysis shows a very good agreement of three-wave model of underground wave propagation in the UG channel with the measurement data. Model results are shown as dotted lines overlay to the measurements data. Due to higher speed of the lateral wave in the air, it arrives first at the receiver for all distances except for the 10 cm depth at 50 cm transmitter–receiver distance.

From the UG channel impulse response model parameters values given in Table 5.5, we observe that the lateral wave is the most dominant wave

Table 5.5 Model parameters: peak amplitude, delays, and number of multipaths statistics for direct, lateral, and reflected components for three soils (Salam et al., 2016)

	Silty Clay Loam			Silt Loam			Sandy Soil		
	Distance 1 m			Distance 1 m			Distance 1 m		
	Peak α dB	τ ns	N	Peak α dB	τ ns	N	Peak α dB	τ ns	N
Direct Component	−90	18–28	3	−103	15–23	2	−87	11–19	4
Lateral Component	−80	30–40	2	−82	26–43	3	−63	22–45	5
Reflected Component	−91	41–47	2	−94	47–59	4	−70	47–61	6

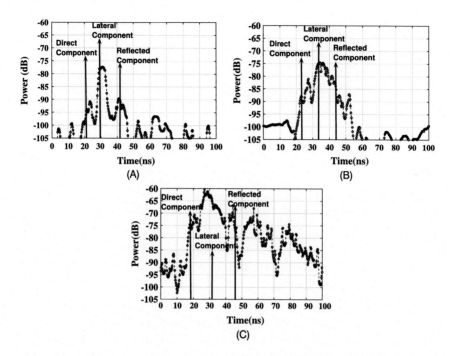

Figure 5.12 Measured impulse response (dotted lines) and impulse response model (solid lines) in: (A) silt loam, (B) silty clay loam soil, and (C) sandy soil (Salam et al., 2016)

among the three components as it undergoes lower attenuation due to its travel path in the air as compared to the direct and reflected components which completely go through the soil.

5.4 EFFECTS OF SOIL ON ANTENNA AND CHANNEL CAPACITY

Channel capacity is employed as the criterion to analyze the effects of the soil on data rates in the underground channel (Dong and Vuran, 2013a). In this section, we provide the analysis of the impact of soil moisture on the channel capacity in underground communications. Capacity is given as follows (Dong and Vuran, 2013a):

$$C = B \log_2 \left(1 + \frac{S}{N_0 B} \right) , \tag{5.28}$$

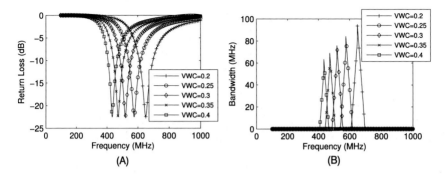

Figure 5.13 (A) The return loss of the dipole antenna for different soil moisture values, (B) change in the bandwidth of the dipole with change in soil moisture (Dong and Vuran, 2013a)

where system bandwidth is represented by B, S is signal strength received, and N_0 is the noise power density.

Soil moisture affects wireless underground communications and channel capacity depends upon the variation in soil moisture. In addition to the change in soil moisture, the return loss, and bandwidth of underground antenna also change with the soil moisture, and causes variations in the underground channel capacity (Dong and Vuran, 2013a). Bandwidth of underground antenna is determined by the return loss which is taken at 10 dB threshold value. For the antenna operating at resonant frequency, the bandwidth is higher. Resonant frequency is defined as the frequency range where the antenna has the lowest return loss (less than the threshold). When the soil moisture increases, the resonant frequency decreases. Moreover, the bandwidth of the underground antenna depends upon the operation frequency. When the UG antenna is operating at the resonant frequency, the higher bandwidth is available, and once the antenna is operating at frequencies other than the resonant frequency, bandwidth decreases. These factors are discussed in detail below (Dong and Vuran, 2013a).

Resonant Frequency of the UG Antenna

In Fig. 5.13A, return loss of a dipole underground antenna has been shown for 100 to 1000 MHz frequency range, for volumetric water content range (VWC) of 20% to 40%. With change in soil moisture (VWC increase from 20% to 40%), resonant frequency decreases from 650 to 430 MHz. It can be observed that soil moisture variations causes significant changes in the resonant frequency of the underground antenna (Dong and Vuran, 2013a).

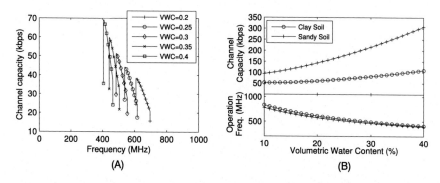

Figure 5.14 (A) UG2AG channel capacity, (B) the change in the optimal operation frequency and the corresponding capacity with change in water content

Bandwidth of the UG Antenna

In Fig. 5.13B, bandwidth of the underground-to-aboveground (UG-to-AG) link has been shown in the UG channel for frequency range of 400 to 700 MHz for volumetric water content (VWC) values range of 20% to 40% taken at 10 dB threshold. At 20% soil moisture level, the bandwidth is 94 MHz and when soil moisture increases to 30%, bandwidth has decreased to 74 MHz. With further increase in soil moisture to 40%, bandwidth decreases to 62 MHz. This happens because as the soil moisture increases, the operational frequency becomes different from resonant frequency which results in low available bandwidth.

Channel Capacity

Channel capacity of the wireless underground channel depends on the soil moisture, operation frequency, and bandwidth of the antenna. Impact of different factors on the channel capacity are shown in the following.

In Fig. 5.14A, channel capacity of underground communication system operating at different frequencies is shown for different soil moisture values. The channel capacity is calculated as a function of the operation frequency according to (5.28). It can be observed that capacity range is from 38 to 70 kbps when system operates at optimum frequency. It can also be observed that for a given soil moisture level, there is an optimal operational frequency at which high capacity can be achieved.

The channel capacity is also a decreasing function of the soil moisture. Therefore, UG communication devices should be able to work at a wide range of spectrum in to adapt to the soil moisture changes. When the an-

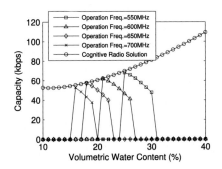

Figure 5.15 The cognitive radio system and fixed frequency channel capacity comparison (Dong and Vuran, 2013a)

tenna is operating at a low operation frequency, the bandwidth reduces, however, at lower operation frequency, lower path losses are observed. In Fig. 5.14B, channel capacity in sandy, and clay soils is shown as a function of change in soil moisture. In clay soil, with decrease in soil moisture from 40% to 10%, operation frequency increases from 408 to 832 MHz. Similarly, in sandy soil, with the same change in soil moisture, operation frequency changes from 394 to 779 MHz. Sandy soils have higher capacity as compared to the clay soils, in fact, capacity in sandy soil is approximately three times higher (308 kbps) as compared to clay soil (110 kbps).

In Fig. 5.15, the comparison of capacity of a UG system operating at a fixed frequency, and a UG communication system with cognitive capabilities is shown in clay soil. It can be observed that cognitive solution where devices can adjust their operation frequency based on the soil moisture changes, achieves the highest capacity. Therefore, cognitive solution can work on a wide range of frequencies by keeping the capacity at the optimal level, to adapt to wider range of variations in soil moisture.

5.5 ERROR CONTROL

Error control schemes are used in wireless communication systems to reduce errors, and to increase reliability of data transfer between devices. In WUSNs, due to variations in soil moisture, packet error rate changes with change in soil moisture (Dong and Vuran, 2013b). Therefore, error control schemes which can adapt to soil moisture changes are used in underground wireless communications. High bit error rate in wireless underground com-

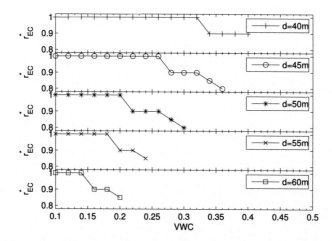

Figure 5.16 The optimal LDPC code rate and the energy efficiency at different soil moisture levels for the 10 dBm transmit power (Dong and Vuran, 2013c)

munications is challenging due to the environment effects, particularly the soil moisture.

Error control schemes in underground communications (Dong and Vuran, 2013b) rely on channel estimation without the use of the feedback. This can be achieved by sensing the soil moisture information to estimate the quality of the channel. This sensing approach uses signal to noise ratio to estimate channel quality. Once the channel estimate is available, adaptive forward error correction (FEC) and adaptive transmit power schemes can be used. Adaptability in FEC is achieved by dynamically puncturing an FEC code based on the channel sensing information. If the FEC does not adjust to the channel estimate, then it results in bit errors which will require retransmission of the packets and hence the energy efficiency is decreased. If the transmit power control is available in underground nodes, then it can be combined with transmit power control. These factors are explained below.

Energy Efficiency of FEC Codes

In Fig. 5.16, energy efficiency of LDPC rate is analyzed with a 10 dBm transmit power and BER threshold of 10^{-4}. Rates at which LDPC codes have optimum performance at different transmitter–receiver distances for different soil moisture levels have been shown. The rate changes from 0.9 to 0.85 with an increase of soil moisture from 32% to 34% at UG2AG

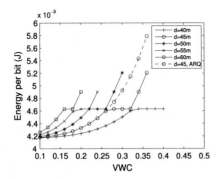

Figure 5.17 The optimal LDPC code rate and the energy efficiency at different soil moisture levels for the 10 dBm transmit power (Dong and Vuran, 2013c)

T–R distance of 45 m. With further increase, bit error rate surpasses the threshold hence leading to no code being used. It can be observed that lower rates of LDPC codes are desirable to compensate the high bit error rates.

Energy efficiency for different codes is shown in Fig. 5.17. It can be observed that for a particular soil moisture level, use of FEC codes results in reduction of energy per bit since by avoiding retransmissions. At T–R distance of 45 m, for 30% soil moisture, FEC requires 4.6 mJ energy which is less than the energy required by ARQ (4.9 mJ). Moreover, fixing the energy required gives higher probability of successful transmission for FEC case as compared to ARQ case.

Transmit Power Control

Transmit power control is used to keep the SNR at a level at which low error rate communications are possible. For this, the underground devices should be able to sport a wider range of transmit power levels. While improving reliability, transmit power adaptation leads to increasing collision because of the hidden node problem. On the other hand, transmit power control is also required to compensate the soil moisture variations.

With support of power control, optimization problem involves power and code rates and it becomes

$$[P_{ut}^*, r_{EC}^*] = \operatorname*{argmin}_{P_{ut},\, r_{EC}} \frac{P_{ut}}{R \cdot r_{EC}}. \tag{5.29}$$

These approaches can be reconfigured in the underground devices for adaptive error control. In Fig. 5.18, results of power control are shown,

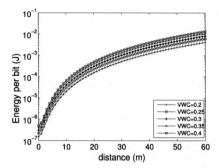

Figure 5.18 The energy efficiency and optimal transmit power with communication distance (Dong and Vuran, 2013c)

and it can be observed that to have optimum energy optimality, transmitter should be capable of multiple power levels. If this multiple power levels support is not available at the receiver, then optimum code rate of FEC is determined by

$$r_{EC}^* = \underset{r_{EC}}{\operatorname{argmin}} \frac{P_{ut}}{R \cdot r_{EC}},$$
(5.30)

such that

$$\text{BER} < \Delta,$$

where r_{EC} is the code rate and BER is corresponding the bit error rate when the FEC code rate is r_{EC}.

5.6 NETWORK CONNECTIVITY

Connectivity in WUSNs is an important issue (Dong and Vuran, 2013c). Dynamic processes undergoing in soil effects the connectivity in WUSNs which consists of underground and above ground devices. The connectivity architecture (Dong and Vuran, 2013c) of WUSN is shown in Fig. 5.19. Two types of approaches are used to maintain connectivity in wireless underground communications: transmit power control and routing. In wireless underground communications, networks cannot be deployed in a worst case scenario because it will result in loss of hardware resources (Dong and Vuran, 2013c) due to excess deployment. Moreover, if these networks are deployed at best case scenario, then changes in the soil moisture will result in a disconnected. Therefore, environment aware routing and connectivity

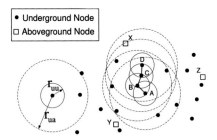

Figure 5.19 WUSN connectivity architecture (Dong and Vuran, 2013c)

approaches (Dong and Vuran, 2013c) are required to maintain connectivity efficiently based on the changes in the soil moisture.

Modeling Cluster Size Distribution in WUSN

In this approach, first, cluster size is modeled and then the connectivity is captured which accounts for the dynamic changes in the soil moisture. By using the cluster size distribution and above-ground node coverage range, a map of a network connectivity is obtained which depends upon the transmit power control and soil moisture. For the cluster size distribution, the density of the underground node is selected such that all nodes remain connected for the worst case scenario. These distributions are determined empirically and then compared with the model that we have developed for underground communications connectivity. Cluster size distribution is approximated as (Dong and Vuran, 2013c)

$$p_{cs}(n) \approx \alpha n^{-\tau} e^{-n/n_\xi} , \qquad (5.31)$$

where α and τ are determined empirically, n_ξ is the average size of cluster which depends on range of transmission r_{uu} and density, λ_u. A comparison of this model with simulated results is shown in Fig. 5.20, which results in close agreement between model and simulated results.

Communication Coverage Model

The coverage model is defined as

$$\tilde{\mathbf{E}}[A_C] = L_w \times L_h , \qquad (5.32)$$

where A_C is coverage of the communication in the above ground node and L_w is given as

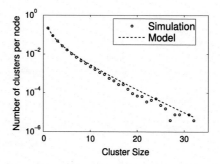

Figure 5.20 Results of cluster size vs. number of nodes per cluster (Dong and Vuran, 2013c)

$$L_w = \frac{d}{2} + r_{ua} - \frac{d}{N+1}$$
$$= \frac{\lfloor\sqrt{C/\lambda_u}\rfloor}{2} + r_{ua} - \frac{\lfloor\sqrt{C/\lambda_u}\rfloor}{\lceil\sqrt{C}\rceil + 1}, \qquad (5.33)$$

where C is cluster area, λ_u is underground node density, r_{ua} is underground to aboveground communication range, N is the column width of the grid. l_h is obtained by replacing N with M (row width of the grid). By using this model of communication coverage, results are compared with simulations for soil moisture values of 10% to 30%. These results are shown in Figs. 5.21A and 5.21B. A good match is shown between the estimated and simulated results. A 10% difference is observed between the model and simulations at soil moisture level of 10%, and this difference increases to 12% when soil moisture is increased to 30%. Underground node density and communication and link range, soil moisture level, and transmit power have strong impact on the communication coverage area.

WUSN Connectivity

Based on the cluster size and coverage parameters, network connectivity $\varphi(G_h)$ in wireless underground sensor networks is modeled as

$$\varphi(G_h) = \frac{\sum_n \{|\mathcal{U}|p_{cs}(n)\Upsilon\}}{|\mathcal{U}|} = \sum_n p_{cs}(n)\Upsilon, \qquad (5.34)$$

$$\Upsilon = \begin{cases} \lambda_a|\mathcal{S}| & \text{if } \lambda_a|\mathcal{S}| < 1, \\ 1 & \text{otherwise,} \end{cases}$$

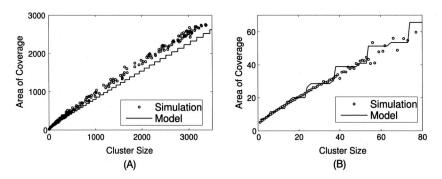

Figure 5.21 Communication coverage area: (A) for VWC = 10%, (B) for VWC = 30%, (Dong and Vuran, 2013c)

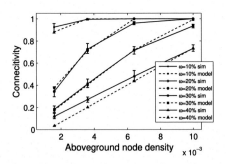

Figure 5.22 Network connectivity vs. aboveground node density for different soil moisture levels (Dong and Vuran, 2013c)

where λ_a is above ground node density, and $\varphi(G_h)$ is cluster size distribution. Simulation results of network connectivity are shown in Fig. 5.22, where network connectivity comparison between model and simulations is shown for underground node density of 0.15 for 4 soil moisture levels for four above-ground densities. The model shows good agreement with simulations, with one exception at 40% soil moisture level and low density. However, for practical density values, it shows a very good agreement.

Energy Consumption Analysis

Connectivity for different soil moisture values by using different transmit power levels is shown in Fig. 5.22. It can be observed that desired connectivity values can be determined by employing an appropriate power level. Energy consumption is another major concern in underground communications due to unavailability of external power sources in underground

Table 5.6 Optimal transmit power for different soil moisture levels and node densities (Dong and Vuran, 2013c)

ω	P_u^* (dBm)		
	$\lambda_a = 0.0036$	$\lambda_a = 0.0064$	$\lambda_a = 0.01$
10%	8	1	1
20%	11	9	7
30%	13	11	10
40%	13	12	11

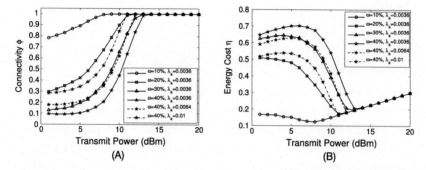

Figure 5.23 (A) Connectivity vs. transmit power, (B) energy cost function vs. transmit power

devices. Therefore, connectivity schemes uses transmit power control such that devices which are not connected to the network can stop their communications.

Energy consumption of the network depends on the power level used. On the one hand, while increasing the transmit power can lead to higher connectivity (Fig. 5.23B), it also results in increased energy consumption of the underground nodes which is proportional to the transmit power. Therefore, for each soil moisture level, best transmit power control should be used to balance the connectivity and energy consumption trade-off. In Table 5.6, optimal transmit power for different soil moisture levels and nodes densities is shown. It can be observed that density of above-ground nodes also affect the transmit power levels as shown in Table 5.6. However, soil moisture variations can be accounted for if higher density of above ground nodes is used.

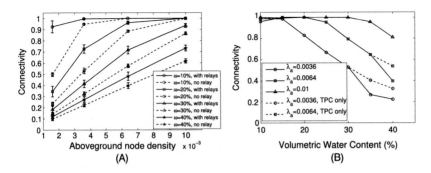

Figure 5.24 (A) Connectivity by using relays, (B) connectivity in the composite scheme (Dong and Vuran, 2013c)

Routing Using Neighbor Node

In environment aware routing, routing decisions are based on the information obtained from the environment in which UG devices are deployed. Nodes can exploit the information from their neighbors to form a multihop network which results in higher connectivity. The multihop network is used in low soil moisture scenario. Through these multihop networks, an underground node relay information to the above ground node. But relaying in these multihop networks also depends on the soil moisture. When soil moisture increases, multihop routing is not the optimum solution (Dong and Vuran, 2013b). In Fig. 5.24A, use of relays have shown to lead to significant improvement in network connectivity for low soil moisture values. Moreover, it can be observed that when the soil moisture increases, relays do not make a major impact in connectivity improvement.

A New Connectivity Approach

A composite scheme is required where energy can be saved by minimizing the transmit power and indirect nodes can stop their communications based on channel state. This information can be obtained by using the communication, and soil moisture parameters. Moreover, soil moisture information can be obtained locally by using the soil moisture sensors. In Fig. 5.24B, connectivity is shown as a function of soil moisture. It can be observed that for 10% to 15% low soil moisture levels, a connectivity of greater than 95% can be maintained with low node density. However, for 35% to 45% soil moisture levels, maintaining connectivity results in higher energy consumption. Therefore, routing is halted to conserve energy.

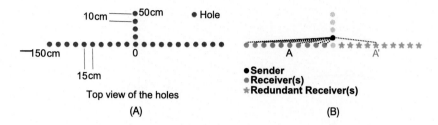

Figure 5.25 Overview of the field testbed (Silva and Vuran, 2010a)

5.7 WUSN TESTBEDS AND EXPERIMENTAL RESULTS

Experimentation in the UG channel is important to gain insights into UG channel propagation mechanisms, and to develop, and verify theoretical models, and results. Therefore, testbeds with flexible control over different experiment parameters are required. In this section, testbeds which can be used to perform underground channel experiments, in outdoor field settings, are discussed. Results of experiments done in these testbeds are also presented.

5.7.1 Field Testbed

A field testbed has been developed in Silva and Vuran (2010a) to conduct wireless underground sensor network (WUSN) experiments. Both underground-to-underground, and underground-to-aboveground experiments can be performed using this testbed. Design of this testbed is shown in Fig. 5.25. Top view of the testbed is shown in Fig. 5.25A, with holes of diameter of 10 cm to bury underground devices. Sender nodes use middle five vertical holes, whereas the rest of the holes in the horizontal line are used by receiving devices. Redundant nodes are buried on the right side of 0 sign. A′ and A are two nodes with equal distance from the center. In Fig. 5.25B, layout is shown where distance between nodes is varied from 18 to 150 cm. Results of the field experiments are presented in Section 5.7.2.

5.7.2 Results of WUSN Experiments

In this section, we present results of underground, and aboveground experiments.

Figure 5.26 Testbed used in software defined radio experiments

Aboveground Experiments

Both UG2AG and AG2UG experiments in WUSNs have been done in Silva and Vuran (2010). Effects of change in soil moisture, and burial depth are investigated in detail. It has been shown that with increase in soil moisture, a 70% decrease in communication range is observed. A wide-band antenna is employed to some dynamic changes happening in soil, and a 350% improvement in communication range is reported. Further details about these experiments results can be found in Silva and Vuran (2010, 2009).

Software-Defined Radio Experiments

A testbed in SCAL (South Central Agricultural Laboratory) is used for software–defined radio (SDR) experiments. It consists of 4 sets of buried dipole antennas. Each set has four antennas buried at 50, 200, and 400 cm distance from each other, respectively. Burial depths are 10, 20, 30, and 40 cm, respectively. Fig. 5.26 shows the testbed used for software defined radio experiments.

A wideband Gaussian signal (2 MHz) RF waveform is transmitted from an underground dipole antenna buried at 40 cm depth by using one USRP (Universal Serial Radio Peripheral). Signal is received on other buried dipole antennas by using different USRPs connected to antennas buried at four different depths (10, 20, 30, and 40 cm) with a fixed transmitter–receiver distance of 50 cm. Experiment are repeated for all these depth by varying the distance to 200 cm and 400 cm. For each frequency, the sender

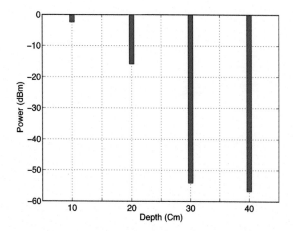

Figure 5.27 Received signal strength: $d = 50$ cm, and $f = 225$ MHz

sends for one second. Receivers collect IQ data of 4 mega samples. Receivers acknowledge the sender after finishing, and then transmitter moves to the next frequency. For each depth, and distance, three measurements are taken. Post processing, and analysis is done in Matlab (MATLAB, 2017). We used the GNU Radio (GNU Radio Website, 2017) and USRPs (Ettus Research Website, 2017) to conduct these software–defined radio experiments.

For spectral estimation and path loss analysis, Welch's method (Welch, 1967) is followed. This method is enhanced form of periodogram analysis. By using the computationally efficient discrete Fourier transform, data is divided into fixed blocks to calculate periodograms, and modified periodograms. These modified periodograms are averaged to calculate the power spectrum. The analysis follows details from Welch (1967).

Power spectral density (PSD) is the average of these periodograms and is given by Welch (1967)

$$\mathbf{P}(f_n) = \frac{1}{K} \sum_{k=0}^{K} I_k(f_n) \tag{5.35}$$

In Fig. 5.27, received signal strength for 10, 20, 30, and 40 cm depths is shown. Received signal strength decreases significantly as burial depth is increased. In Fig. 5.28, experimental results are compared with theoretical model of (5.17). Experimental results for 30 and 40 cm depths are in good agreement with theoretical model (Dong and Vuran, 2011).

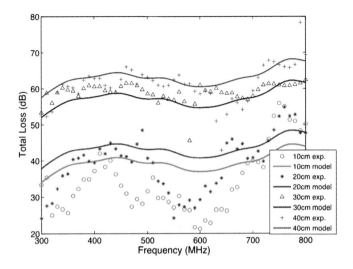

Figure 5.28 Comparison of experimental results with theoretical model

5.8 CONCLUSIONS

In this chapter, an in-depth coverage of EM-based wireless underground sensor networks is presented. Effects of the permittivity of soil on the propagation are analyzed in detail, and it has been shown that water absorption of EM waves in soil causes higher attenuation in underground communications as compared to free space communications. Characteristics of soil as a communication medium are discussed, and 2- and 3-wave channel models have been shown to accurately estimate the path loss of the UG channel. Impulse response model of wireless underground channel is characterized with RMS delay spread and coherence bandwidth statistics. It has been shown that soil moisture variations affect the RMS delay spread, and coherence BW. UG channel has coherence bandwidth range of less than 1 MHz. Effects of soil type, and soil moisture variations on underground antenna, and channel capacity are analyzed. Channel frequency, and UG antenna bandwidth are shown to change with soil moisture variations. Advantages of adaptive error control, and connectivity schemes in WUSNs are described in detail. Chapter concludes with discussion of design of WUSN testbeds, and results of field experiments.

REFERENCES

Akyildiz, I.F., et al., 2006. Next generation/dynamic spectrum access/cognitive radio wireless networks: a survey. Comput. Netw. J. (Elsevier) 50, 2127–2159.

Akyildiz, I.F., Stuntebeck, E.P., 2006. Wireless underground sensor networks: research challenges. Ad Hoc Netw. J. (Elsevier) 4, 669–686.

Akyildiz, I.F., Sun, Z., Vuran, M.C., 2009. Signal propagation techniques for wireless underground communication networks. Phys. Commun. J. (Elsevier) 2 (3), 167–183.

Bogena, H.R., et al., 2010. Potential of wireless sensor networks for measuring soil water content variability. Vadose Zone J. 9 (4), 1002–1013.

Cassioli, D., Win, M., Molisch, A., 2002. The ultra-wide bandwidth indoor channel: from statistical model to simulations. IEEE J. Sel. Areas Commun. 20 (6), 1247–1257. http:// dx.doi.org/10.1109/JSAC.2002.801228.

CRC Handbook, 2014. CRC Handbook of Chemistry and Physics, 95th edition. CRC Press.

Dobson, M., et al., 1985. Microwave dielectric behavior of wet soil—Part II: dielectric mixing models. IEEE Trans. Geosci. Remote Sens. GE-23 (1), 35–46. http://dx.doi.org/ 10.1109/TGRS.1985.289498.

Dong, X., Vuran, M.C., 2011. A channel model for wireless underground sensor networks using lateral waves. In: Proc. of IEEE Globecom '11. Houston, TX.

Dong, X., Vuran, M.C., 2013a. Impacts of soil moisture on cognitive radio underground networks. In: Proc. IEEE BlackSeaCom. Batumi, Georgia.

Dong, X., Vuran, M.C., 2013b. Exploiting soil moisture information for adaptive error control in wireless underground sensor networks. In: 2013 IEEE Global Communications Conference (GLOBECOM), pp. 97–102.

Dong, X., Vuran, M.C., 2013c. Environment aware connectivity for wireless underground sensor networks. In: Proc. IEEE INFOCOM '13. Turin, Italy.

Dong, X., Vuran, M.C., Irmak, S., 2003. Autonomous precision agricultrue through integration of wireless underground sensor networks with center pivot irrigation systems. Ad Hoc Netw. (Elsevier) 11 (7), 1975–1987. http://dx.doi.org/10.1016/j.adhoc. 2012.06.012.

Ettus Research Website, 2017 (accessed March 2017), http://www.ettus.com.

Foth, H.D., 1990. Fundamentals of Soil Science, 8th edition. John Wiley and Sons.

GNU Radio Website, 2017 (accessed February 2017), http://www.gnuradio.org.

Guo, H., Sun, Z., 2014. Channel and energy modeling for self-contained wireless sensor networks in oil reservoirs. IEEE Trans. Wirel. Commun. 13 (4), 2258–2269. http:// dx.doi.org/10.1109/TWC.2013.031314.130835.

Gutiérrez, J., Villa-Medina, J.F., Nieto-Garibay, A., Porta-Gándara, M.Á., 2014. Automated irrigation system using a wireless sensor network and GPRS module. IEEE Trans. Instrum. Meas. 63 (1), 166–176. http://dx.doi.org/10.1109/TIM.2013.2276487.

Hillel, D., 2003. Introduction to Environmental Soil Physics. Elsevier Science.

Hopkins, J., 2017. USDA ERS – ARMS farm financial and crop production practices: tailored reports: crop production practices. Available: http://www.ers.usda.gov/ data-products/arms-farm-financial-and-crop-production-practices/tailored-reports-crop-production-practices.aspx.

Howard, S., Pahlavan, K., 1990. Measurement and analysis of the indoor radio channel in the frequency domain. IEEE Trans. Instrum. Meas. 39 (5), 751–755. http://dx.doi.org/ 10.1109/19.58620.

Johnk, C.T., 1988. Engineering Electromagnetic Fields and Waves, 2nd edition. John Wiley & Sons.

Kim, Y., Evans, R.G., Iversen, W.M., 2008. Remote sensing and control of an irrigation system using a distributed wireless sensor network. IEEE Trans. Instrum. Meas. 57 (7), 1379–1387. http://dx.doi.org/10.1109/TIM.2008.917198.

King, R.W.P., Owens, M., Wu, T.T., 1992. Lateral Electromagnetic Waves. Springer-Verlag.

Markham, A., Trigoni, N., 2012. Magneto-inductive networked rescue system (miners): taking sensor networks underground. In: Proc. 11th ICPS. IPSN '12. ACM, New York, NY, USA, pp. 317–328. http://doi.acm.org/10.1145/2185677.2185746.

MATLAB, 2017 (accessed March 2017), http://www.matlab.com.

Peplinski, N., Ulaby, F., Dobson, M., 1995. Dielectric properties of soil in the 0.3–1.3 GHz range. IEEE Trans. Geosci. Remote Sens. 33 (3), 803–807.

Rappaport, T., Seidel, S., Takamizawa, K., 1991. Statistical channel impulse response models for factory and open plan building radio communicate system design. IEEE Trans. Commun. 39 (5), 794–807. http://dx.doi.org/10.1109/26.87142.

Salam, A., Vuran, M.C., 2016. Impacts of soil type and moisture on the capacity of multicarrier modulation in internet of underground things. In: Proc. of the 25th International Conference on Computer Communication and Networks (ICCCN 2016). Waikoloa, Hawaii, USA, 2016 (Best Student Paper Award).

Salam, A., Vuran, M.C., 2017a. Wireless underground channel diversity reception with multiple antennas for internet of underground things. In: Proc. IEEE ICC 2017. Paris, France.

Salam, A., Vuran, M.C., 2017b. Smart underground antenna arrays: a soil moisture adaptive beamforming approach. In: Proc. 36th IEEE INFOCOM 2017. Atlanta, USA.

Salam, A., Vuran, M.C., 2017c. Smart Underground Antenna Arrays: A Soil Moisture Adaptive Beamforming Approach. Tech. Rep. TR-UNL-CSE-2017-0001. Department of Computer Science and Engineering, University of Nebraska-Lincoln.

Salam, A., Vuran, M.C., Irmak, S., 2017. Towards internet of underground things in smart lighting: a statistical model of wireless underground channel. In: Proc. 14th IEEE International Conference on Networking, Sensing and Control (IEEE ICNSC). Calabria, Italy.

Salam, A., Vuran, M.C., Irmak, S., 2016. Pulses in the sand: impulse response analysis of wireless underground channel. In: Proc. 35th Annual IEEE International Conference on Computer Communications (INFOCOM 2016). San Francisco, USA.

Saleh, A., Valenzuela, R., 1987. A statistical model for indoor multipath propagation. IEEE J. Sel. Areas Commun. 5 (2), 128–137. http://dx.doi.org/10.1109/JSAC.1987.1146527.

Silva, A.R., Vuran, M.C., 2009. Empirical evaluation of wireless underground-to-underground communication in wireless underground sensor networks. In: Proc. of IEEE DCOSS '09. Marina del Rey, CA, pp. 231–244.

Silva, A.R., Vuran, M.C., 2010a. Development of a testbed for wireless underground sensor networks. EURASIP J. Wirel. Commun. Netw.. http://dx.doi.org/10.1155/2010/620307.

Silva, A.R., Vuran, M.C., 2010b. (CPS)^2: integration of center pivot systems with wireless underground sensor networks for autonomous precision agriculture. In: Proc. of ACM/IEEE International Conf. on Cyber-Physical Systems. Stockholm, Sweden, pp. 79–88.

Silva, A.R., Vuran, M.C., 2010. Communication with aboveground devices in wireless underground sensor networks: an empirical study. In: Proc. of IEEE ICC'10. Cape Town, South Africa, pp. 1–6.

Sun, Z., et al., 2011a. Border patrol through advanced wireless sensor networks. Ad Hoc Netw. 9 (3), 468–477.

Sun, Z., et al., 2011b. MISE–PIPE: magnetic induction-based wireless sensor networks for underground pipeline monitoring. Ad Hoc Netw. 9 (3), 218–227.

Sun, Z., Akyildiz, I., 2010. Channel modeling and analysis for wireless networks in underground mines and road tunnels. IEEE Trans. Commun. 58 (6), 1758–1768. http://dx.doi.org/10.1109/TCOMM.2010.06.080353.

Tiusanen, M.J., 2006. Wideband antenna for underground Soil Scout transmission. IEEE Antennas Wirel. Propag. Lett. 5 (1), 517–519.

Tiusanen, M.J., 2013. Soil scouts: description and performance of single hop wireless underground sensor nodes. Ad Hoc Netw. 11 (5), 1610–1618. http://dx.doi.org/10.1016/j.adhoc.2013.02.002.   http://www.sciencedirect.com/science/article/pii/S157087051300022X.

Ulaby, F.T., Long, D.G., 2014. Microwave Radar and Radiometric Remote Sensing. University of Michigan Press.

Vuran, M.C., Akyildiz, I.F., 2010. Channel model and analysis for wireless underground sensor networks in soil medium. Phys. Commun. 3 (4), 245–254. http://dx.doi.org/10.1016/j.phycom.2010.07.001.

Vuran, M., Dong, X., Anthony, D., 2016. Antenna for wireless underground communication. US Patent 9,532,118 (Dec. 27 2016). https://www.google.com/patents/US9532118.

Welch, P.D., 1967. The use of fast Fourier transform for the estimation of power spectra: a method based on time averaging over short, modified periodograms. IEEE Trans. Audio Electroacoust. 15 (2), 70–73. http://dx.doi.org/10.1109/TAU.1967.1161901.

CHAPTER 6

Fiber-Optic Underground Sensor Networks

Kenichi Soga*, Cedric Kechavarzi†, Loizos Pelecanos‡,
Nicholas de Battista†, Michael Williamson†, Chang Ye Gue†,
Vanessa Di Murro†, Mohammed Elshafie†,
David Monzón-Hernández Sr.§, Erika Bustos¶, J.A. García‖

*University of California, Berkeley, USA
†University of Cambridge, Cambridge, United Kingdom
‡University of Bath, Bath, United Kingdom
§Centro de Investigaciones en Óptica A.C., León Guanajuato, Mexico
¶Centro de Investigación y Desarrollo Tecnológico en Electroquímica, S.C., Querétaro, Mexico
‖Instituto Tecnológico de Atitalaquia, Atitalaquia, Mexico

SUBCHAPTER 6.1

Distributed Fiber-Optic Strain Sensing for Monitoring Underground Structures – Tunnels Case Studies

K. Soga*, C. Kechavarzi†, L. Pelecanos‡, N. de Batista†, M. Williamson†, C.Y. Gue†,
V. Di Murro†, M. Elshafie†

*University of California, Berkeley, USA
†University of Cambridge, Cambridge, United Kingdom
‡University of Bath, Bath, United Kingdom

6.1.1 INTRODUCTION

The main motives driving the trend toward increased implementation of structural monitoring systems are the need for structural health monitoring of existing and ageing structures and the desire for a better understanding of increasingly complex designs through performance monitoring of new structures. This drive is sustained by rapid progress in research and technology development on sensors and communications.

In the UK, much of the existing infrastructure is old and no longer fit for purpose; indeed, the Institution of Civil Engineers in its *State of*

Underground Sensing.
DOI: http://dx.doi.org/10.1016/B978-0-12-803139-1.00006-0

the Nation Infrastructure 2014 reported that none of the sectors analyzed were "fit for the future" and only one sector was "adequate for now." The problem is, however, that global and ageing infrastructure worldwide requires monitoring and remedial interventions in order to extend life and prevent catastrophic failures. There is also a progressive need to be able to increase the load and usage of existing structures, be that number of passengers carried, numbers of vehicles or volume of water used. The high level of maintenance required to preserve these assets and run them at full or increased capacity has created a demand for long term structural health monitoring or performance based maintenance of legacy infrastructure. In addition, increasingly congested urban centers and the lack of usable space to develop have meant a trend towards the construction of new structures closer to existing structures. In some cases the construction of the new structure is only possible by the temporary monitoring of movement and the associated structural integrity of the existing structure.

For new structures, the monitoring of construction and operational processes allows assessments of performance against engineering design parameters or predictive models. In situ monitoring is crucial to ensure that the design is sound and satisfies the specifications set by the clients but also that the construction is executed safely. Innovation in construction technology and the need to meet societal demand mean that underground structures are getting larger, being built at deeper locations and closer to each other or to existing structures. In addition, ongoing advances in numerical methods allow designing structures with increasingly complicated geometries and boundary conditions. However, design methodologies have not always changed as much and in many cases they were originally developed for smaller, shallower constructions. Uncertainty is addressed by the use of safety factors, design standards and laboratory and field tests. Nevertheless, extrapolation of design methods to more complex structures and predicted behavior or performance requires validation against in situ data in order to develop confidence or improve the design (Soga, 2014). This performance based design approach can also help delivering efficiency and reducing over specification.

There is long history of robust monitoring methods for geotechnical engineering problems but recent advances in field instrumentation and monitoring systems provide great opportunities to advance the understanding of the performance of geotechnical structures. Distributed fiber-optic strain (DFOS) sensing, notably, has recently proved useful in characterizing the performance of a variety of geotechnical structures. Engineering design standards are often based on limiting the strain and/or stress developing in a structure. It is important that strain is limited because it is strain rather than movement that is a measure of how close the material of a structure is to its strength capacity. It is strain, and the strain history, that indicates the potential degree of damage or health of a structure. In addition, for structures interacting with soil such as underground infrastructure (foundations, tunnels or pipelines), the ground loads are spatially distributed and not point loads. This is why distributed strain sensing using fiber optic systems is proving to be a very valuable tool for better understanding the state and behavior of geotechnical structures and pinpoint localized problem areas such as cavity collapse, nonuniformly distributed soil–structure interaction loads, and joint movements.

DFOS sensing has, for example, been used in rail tunnels to monitor joint movements in concrete tunnel lining (Cheung et al., 2010), in an old masonry tunnel to monitor the effect of proximity tunneling (Mohamad et al., 2010) and for monitoring twin tunnels interaction (Mohamad et al., 2012). It has been used on existing pile foundations to assess pile reusability (Bell et al., 2013), during pile tests (Pelecanos et al., 2017) and on thermal piles to study the effect of thermal load (Bourne-Webb et al., 2009 and 2013; Ouyang et al., 2011; Amatya et al., 2012; Mohamad et al., 2014). Movement during basement excavation has been assessed in secant piled walls (Mohamad et al., 2011; Schwamb et al., 2011) and in diaphragm wall panels to monitor the effect of excavation on a deep circular shaft (Schwamb et al., 2014). DFOS has also been used in embankments to detect ground movements (Janmonta et al., 2008) and in soil nails on steep highway cut slopes (Amatya et al., 2008).

This chapter discusses the basic principles of distributed sensing using Brillouin-based techniques and presents three case studies focusing on the monitoring of tunnels using DFOS sensing. The first case study presents the results of strain development in the sprayed concrete lining (SCL) of a new Crossrail station tunnel during excavation of a cross-passage. The second case study focuses on the short term monitoring of a century old cast iron tunnel during the proximity construction of a large platform tunnel for the Crossrail project. The final case study describes the early stage implementation of a long-term structural health monitoring program to assess the integrity of tunnels at CERN. For all these projects Brillouin-based time domain techniques were used. A brief description of the basic principles of the method and the fiber-optic systems used is presented before the case studies.

6.1.2 DISTRIBUTED FIBER-OPTIC SENSING (DFOS) BASED ON BRILLOUIN SCATTERING

Distributed fiber-optic sensing (DFOS) based on Brillouin scattering is a technology that allows spatially continuous strain and/or temperature measurements along optical fibers attached onto or embedded in structures, potentially replacing a very large number of discrete sensors. Strain or temperature measurements are simply made along a single cable connected to a spectrum analyzer. This single cable approach is very attractive because it greatly simplifies the installation of the monitoring system and measurements can be made over kilometers of cable through a single measurement port. There are other application specific advantages. Fiber optics are intrinsically passive and immune to electromagnetic interference (such as interference due to lightning, for example) and can be used in explosive or high voltage environments. They are chemically inert, water and corrosion resistant, and do not pose contamination risks.

Although there are various fiber-optic based distributed sensing technologies, reviewed by Galindez-Jamioy and López-Higuera (2012), the case studies presented here make use of time domain techniques called Brillouin optical time-domain reflectometry or analysis (BOTDR and BOTDA) as described by Kechavarzi et al. (2016).

Basic Principle

The BOTDR and BOTDA techniques rely on the fact that when light travels in an optical fiber a small amount is backscattered due to small refractive-index or density fluctuations. The backscattered light spectrum has various components including the Brillouin frequency peaks. Brillouin scattering is due to the interaction of light with propagating density waves or acoustic phonons. The frequency of the Brillouin peak v_b is proportional to the material's local thermodynamic properties such as the acoustic velocity, v_a, and phase refractive index, n, two quantities which in turn depend on local temperature and strain (the acoustic velocity being highly sensitive to minute changes in material density):

$$v_b = \frac{2n v_a}{\lambda} \qquad (6.1.1)$$

where λ is the wavelength of the incident light. The Brillouin frequency is shifted by about 10–11 GHz from the incident light wave frequency at a wavelength of 1550 nm. This is called the reference Brillouin frequency v_{b0}.

This Brillouin frequency shifts linearly with changes in longitudinal strain and temperature in the fiber core/cladding so that

$$\Delta v_b = \frac{\partial v_b}{\partial \varepsilon}\bigg|_{T=const} \Delta \varepsilon + \frac{\partial v_b}{\partial T}\bigg|_{\varepsilon=const} \Delta T$$

or

$$\Delta v_b = C_\varepsilon \Delta \varepsilon + C_T \Delta T \qquad (6.1.2)$$

where Δv_b is the change in Brillouin frequency due to a simultaneous change in strain, $\Delta \varepsilon$, and in temperature, ΔT. The coefficients C_ε and C_T are referred to as the strain and the temperature coefficients of the Brillouin frequency shift, respectively (Horiguchi et al., 1989). These co-efficients depend on the properties of the optical fiber only and not on the type of cable. Hence, C_ε and C_T will vary slightly at around values of 500 MHz/% and 1 MHz/°C, respectively, for standard telecommunication single mode fibers and at the operating wavelength of 1550 nm as used with BOTDR and BOTDA. For a given fiber, C_ε can be determined by calibrating strain cables in strain rigs equipped with accurate linear actuators. The coefficient C_T varies little from a standard single mode fiber to another and a value of 1.0 MHz/°C is commonly used.

The shift in frequency due to strain applied to a section of a fiber is illustrated Fig. 6.1.1.

The frequency, v_b, at the peak power of the Brillouin spectrum, is determined by curve fitting the spectrum. This allows the calculation of strain from that frequency at every point along the optical fiber. Therefore the principle used in this type of sensing system is not based on intensity measurements, but on the value of the frequency corresponding to the peak power of the Brillouin scattered light.

The distance z between the location where the laser pulse is launched and the position where scattered light is generated can be calculated using the following equation (Ohno et al., 2001):

$$z = \frac{ct}{2n} \qquad (6.1.3)$$

where c is the light velocity in a vacuum and t is the time interval between launching the pulsed light and receiving the light scattered at a given position along the optical fiber. Hence the strain distribution along the optical fiber can be obtained through a time-domain analysis by measuring the propagation times of the light pulse traveling in the fiber. This gives a

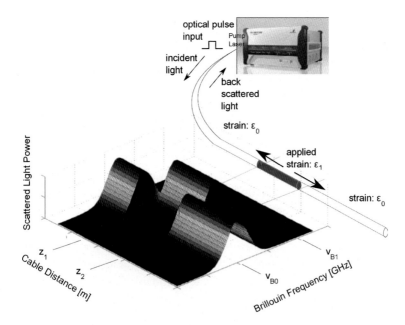

Figure 6.1.1 Brillouin frequency shift caused by strain

continuous strain distribution along the fiber without the need for a null indicator as the shift change, Δv_b, is based on the characteristic Brillouin frequency shift, v_b, of the fiber used.

BOTDR and BOTDA

With BOTDR a short light pulse is launched into one end of the optical fiber and the backscattered signal due to spontaneous Brillouin scattering is detected from the same end. A BOTDR spectrum analyzer detects the frequency component of the backscattered signal with a high spectral resolution analysis and the location of this frequency component from a time domain analysis as described above.

With BOTDA stimulated Brillouin scattering can be generated by launching a short pump pulse into one end of the optical fiber as well as a continuous Stokes wave probe beam into the other end. Interference between the incident pump wave and Stokes waves traveling in the opposite direction form a beat signal at the Brillouin frequency. The resulting electromagnetic field of this beat pattern applies an electrostrictive force to the molecular structure of the medium, resulting in periodic fluctuations in its density. If the induced density fluctuations add up constructively, they will form an acoustic wave, consequently increasing the number of phonons. This results in an increased efficiency of the scattering process which is said to be operating in the stimulated regime (Nöther, 2010).

This results in a better strain resolution of ±4 με with BOTDA whereas it is in the order of ±30 με with BOTDR. However, the BOTDA is a double ended system which requires the cable to be installed as a continuous loop which may be a major drawback in harsh environments where cable breakages are possible. With BOTDR, if breakage occurs and both ends are accessible, it is still possible to obtain the whole strain profile.

Strain data at any point is calculated based on the frequency shift of the Brillouin backscattering spectrum within the spatial resolution L ahead of that point. The spatial resolution is determined by the pulse width of the incident light τ and it is expressed as

$$L = \frac{v\tau}{2}. \qquad (6.1.4)$$

Higher spatial resolution can be obtained by narrowing the pulse width, however, this leads to a broadened Brillouin gain spectrum and a weaker Brillouin signal and a drop in strain resolution. With current BOTDR and BOTDA measurement systems, the spatial resolution is limited to between 0.5 and 1 m. However, note that measurements can be made every 0.05 m, providing a moving average over the spatial resolution at points 0.05 m apart along the fiber.

As mentioned before, with both techniques, strain measurement can be carried out over several kilometers along a single optical fiber. However, over such distances, a larger pulse width and therefore larger spatial resolution is needed. This also increases the acquisition time. For typical geotechnical installations of a few hundred meters, the measurement time is of a few minutes with conventional analyzers. Hence, at present, most commercial Brillouin-based DFOS systems are inadequate for dynamic strain and temperature measurements.

Temperature Compensated Strain

Since the Brillouin frequency is affected by both temperature and strain, it is essential to decouple these two effects when only measuring changes in Brillouin frequency. To resolve this issue, several methods have been proposed to measure temperature and strain simultaneously (Mohamad, 2012). One common solution to this problem is to use a separate temperature compensation cable placed adjacent to the strain cable. In this temperature cable the fiber is isolated from any mechanical strain effects. This is achieved by using loose tube cables. In the loose tube cable the frequency shift can be assumed to be caused by temperature variations only so that Eq. (6.1.2) reads:

$$(\Delta v_b)_T = C_\varepsilon \Delta \varepsilon_T + C_T \Delta T. \qquad (6.1.5)$$

The residual strain, $\Delta \varepsilon_T$, on the fiber in the loose tube is caused by the thermal expansion of glass and that of the 250 μm primary coating due to the change ΔT. If α_T is the equivalent net thermal expansion coefficient of the coated fiber, the effect of temperature can be a lumped into a single coefficient C_{T_T}:

$$(\Delta v_b)_T = C_\varepsilon \alpha_T \Delta T + C_T \Delta T = C_{T_T} \Delta T \qquad (6.1.6)$$

so that the change in temperature can be calculated by measuring the frequency shift in the fiber in the loose tube cable as

$$\Delta T = \frac{(\Delta v_b)_T}{C_{T_T}}. \qquad (6.1.7)$$

Substituting Eq. (6.1.7) into Eq. (6.1.2) gives the temperature compensated total strain

$$\Delta \varepsilon = \frac{1}{C_\varepsilon} \left[(\Delta v_b)_S - C_T \frac{(\Delta v_b)_T}{C_{T_T}} \right]. \qquad (6.1.8)$$

C_{T_T} can be obtained through calibration of the loose tube cables by using a water bath where the water is heated in incremental stages usually up to 80–90°C and the change in frequency along the loosely arranged submerged section of cable is measured.

Thermal Expansion of Concrete

Under nonisothermal conditions, it may be necessary to differentiate strain due to parameters of interests such stress, creep, and shrinkage of the structure, from strain due the thermal expansion of the host material such as concrete. The total mechanical strain ε can be divided into two components: the strain due to the structure's deformation ε_s and that due to thermal expansion, ε_T, so that $\varepsilon = \varepsilon_s + \varepsilon_T$. Eq. (6.1.2) reads

$$\Delta v_b = C_\varepsilon (\Delta \varepsilon_s + \Delta \varepsilon_T) + C_T \Delta T. \qquad (6.1.9)$$

The strain ε_T is caused by the net thermal expansion of all the materials making up the various optical fiber cable components (glass, all coatings, and Jacket) and that of the host material. The equivalent net thermal expansion coefficient α can be written as

$$\alpha = \frac{\sum A_i E_i \alpha_i}{\sum A_i E_i} \tag{6.1.10}$$

so that $\Delta \varepsilon_T = \alpha \Delta T$, where E_i is the Young's modulus, A_i is the cross-section area, and α_i is the thermal expansion coefficient of all the material i making up the cable and host materials surrounding the cable. Eq. (6.1.9) now reads:

$$\Delta v_b = C_\varepsilon \Delta \varepsilon_s + C_\varepsilon \frac{\sum A_i E_i \alpha_i}{\sum A_i E_i} \Delta T + C_T \Delta T,$$

$$\Delta v_b = C_\varepsilon \Delta \varepsilon_s + (C_\varepsilon \alpha + C_T) \Delta T. \tag{6.1.11}$$

α is unknown; however, because the cross-section area of a structure is often infinitely larger than that of an optical fiber cable, the thermal expansion coefficient of the cable is negligible compared to that of the structure. It can therefore be assumed that the overall thermal expansion is due to the host structure such that $\alpha_{structure} \approx \frac{\sum A_i E_i \alpha_i}{\sum A_i E_i}$ for $A_{structure} \gg A_{cable}$ where $\alpha_{structure}$ is the coefficient of thermal expansion of the host structure which is around 10 $\mu\varepsilon/°C$ for concrete. In this case both ε_s and ε_T can be calculated providing that the effect of temperature change on the Brillouin frequency shift are compensated for as shown above. Also note that if the cables are bonded onto the structure or used as gauge length rather fully embedded, they may expand or contract at a different rate and the method above should be used with caution if at all.

Cables

As opposed to telecommunication cables, strain sensing cables require that any strain applied to the jacket is transferred to the fiber core without any loss. However, strain transfer can be affected by slippage between the different protective layers. Therefore, although strain is, at least, partially transferred to the core in tight buffer telecom cables, they do not qualify as strain sensing cables. Several specialist strain sensing cables have been developed in recent years (Iten, 2011), with varying degree of protection depending on the intended application. For temperature compensation measurements, standard telecom loose tubed cables filled with gel can be used.

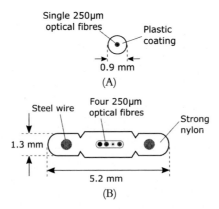

Figure 6.1.2 Specialist strain cables used in the case studies: (A) single fiber cable and (B) steel reinforced cable

The strain cables used in the case studies presented in this chapter are shown in Fig. 6.1.2. These vary from a single fiber cable (0.9 mm diameter) surface mounted in Case Studies 2 and 3 to a reinforced cable used in Case Study 1 which has steel wires on both sides and strong nylon sheathing to protect the fragile optical fibers when embedding in concrete.

6.1.3 CASE STUDY 1: MONITORING OF A SPRAYED CONCRETE TUNNEL LINING AT THE CROSSRAIL LIVERPOOL STREET STATION

Project Background

Crossrail is a 136 km-long railway route across London, running from Reading and Heathrow in the west, to Shenfield and Abbey Wood in the east (Fig. 6.1.3). The £16bn ($25bn) Crossrail construction project included 21 km of new twin-bore tunnels, excavated using eight tunnel boring machines, at depths of up to 40 m. Tunneling began in 2012 and was completed in 2015, resulting in around six million metric tons of earth being excavated.

The Crossrail route passes through 39 rail stations, including ten newly built stations, of which six are underground in central London and the Docklands area. Liverpool Street Station is one of the underground stations, linking a new ticket hall in Broadgate with the existing Moorgate and Liverpool Street Stations. The new station provides an interchange between Crossrail, London Underground's Northern, Central, Metropolitan,

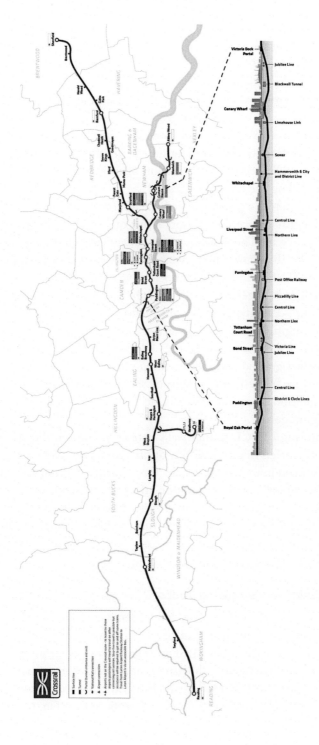

Figure 6.1.3 The Crossrail route map (top) and cross-section of the part running under central London (bottom) (© Crossrail Ltd.)

Circle and Hammersmith & City lines, and train connections to Stansted Airport and National Rail services.

The Crossrail Liverpool Street Station consists of three main tunnels: two 250 m–long platform tunnels and a concourse tunnel between them, inter-connected by 12 cross-passages. The tunnels were constructed using a sprayed concrete lining (SCL) technique based on a sequential cycle of meter by meter mechanical excavation and spraying of steel fiber reinforced concrete onto the open face to support the ground. The three main tunnels were constructed first, followed by the formation of around 40 service and access junctions, including the cross-passages.

Prior to the excavation of a junction, the SCL tunnel walls in the vicinity were thickened and reinforced with more layers of steel fiber and rebar reinforced sprayed concrete. The amount of reinforcement, and the thickness and extent of the thickening SCL, were derived from design charts based on a series of 3D finite element models of generic junctions having varying aspect ratios between the parent and child tunnels. The models take into account the sequential staged construction to estimate the expected changes in longitudinal and hoop stress within the parent tunnel lining, during and after the excavation of the child tunnel. Due to the uncertainties in the actual ground and tunnel lining conditions, conservative assumptions have to be made in the design of the SCL around the tunnel junctions. In addition, there are few reliable field data with which to calibrate the design models.

Standard monitoring of SCL tunnel walls consists of optical displacement measurements carried out periodically at discrete points. Whilst discrete displacement measurements are useful for ensuring the safety and convergence of the tunnel during excavation, they are sensitive to measurement errors and cannot be used to provide an accurate estimate of the strain within the tunnel lining. In order to truly understand the behavior of a tunnel SCL in the vicinity of a junction, a direct and distributed measurement of strain within the tunnel wall is required.

Distributed Fiber-Optic Strain Sensor Installation

A BOTDR monitoring system was embedded within the SCL of the CH5 enlargement chamber at the western end of the Liverpool Street Station concourse tunnel (de Battista et al., 2015). The objective was to measure the changes in strain that occurred within the parent tunnel walls during the excavation of cross-passages CP1 and CP2, which link the concourse tunnel to the westbound and eastbound platform tunnels, respectively (Fig. 6.1.4).

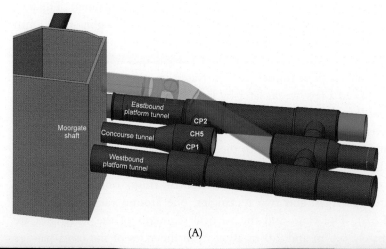

(A)

(B)

Figure 6.1.4 (A) Model of the western end of the Crossrail Liverpool Street Station; (B) the CH5 enlargement chamber of the concourse tunnel, where the DFOS instrumentation was installed

The tunnel at CH5 is entirely within London Clay, and it has an excavated profile diameter of 10.8 m (vertical) to 11.1 m (horizontal). The primary and thickening SCL are both 400 mm thick, resulting in an 800 mm thick tunnel lining which extends for a distance of approximately 20 m around the cross-passage junctions. The cross-passages were constructed in two phases: first a pilot tunnel was excavated to approximately 5.0 m diameter, followed a few weeks later by its enlargement to the full excavated profile of 7.0 m (vertical) to 7.2 m (horizontal) diameter. The excavation in both phases was carried out in meter by meter advances, and a 200 mm thick primary SCL was applied to the final cross-passage tunnels.

The layout of the distributed fiber-optic monitoring system in CH5 consists of four circuits: two circumferential rings, one on each side of the cross-passages, and two rectangles, one around each cross-passage (Fig. 6.1.5A). The rings and the vertical segments of the rectangles are approximately 0.5 m and 2.0 m away from the sides of the cross-passage junctions, respectively, and the horizontal segments of the rectangles are approximately 0.5 m above/below the crown/invert of the cross-passage junctions.

Each fiber-optic sensor circuit was made up of a loose-tube fiber optic cable for estimating temperature (single mode Excel 8-core 9/125 OS1) and a fully bonded fiber-optic cable for measuring total strain (single mode Fujikura 4-core 9.5/125 JBT 03813). The cable pairs were installed next to each other on the surface of the primary SCL, by passing them through steel eyebolts fixed at approximately 1 m intervals to the concrete surface (Fig. 6.1.5B). The strain (bonded) cables were pretensioned to approximately 1000 $\mu\varepsilon$ using Fujikura cable clamps at 8 to 10 m intervals (Fig. 6.1.5B). The cables were routed from the end of each circuit, along the tunnel lining surface, to a junction box at the eastern edge of the CH5 chamber, where they were fusion spliced together to form one single circuit. One end of this circuit was then fusion spliced to a 140 m-long reinforced fiber-optic cable which was used as an extension cable. This was routed to a nearby area within the concourse tunnel, where the monitoring could be carried out at a safe distance during the cross-passage excavation.

Once the installation of the distributed fiber-optic system was completed, the steel rebar reinforcement of the thickening lining was installed and the thickening SCL was sprayed over the fiber-optic cables. Thus the fiber-optic circuits were embedded at approximately the center of the 80 cm-thick SCL. During the fixing of the steel reinforcement, the rectangular circuit around CP2 was accidentally damaged and it could not be

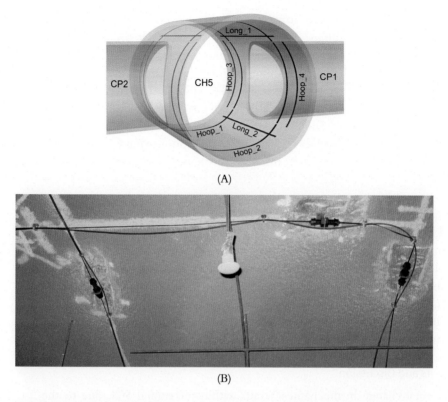

(A)

(B)

Figure 6.1.5 Details of the DFOS sensor installation: (A) layout of the embedded fiber optic cable circuits; (B) fiber optic cables fixed to the tunnel lining with eyebolts and pretensioned with cable clamps

repaired due to the time constraints imposed by the tight construction schedule. Therefore this circuit was excluded from the distributed fiber-optic system and only the three remaining circuits were used for monitoring the tunnel lining. These are labeled in Fig. 6.1.5A as Hoop_1,2,3,4 which measure hoop strain, and Long_1,2 which measure longitudinal strain.

Monitoring Regime and Data Analysis

The cross-passages CP1 and CP2 were excavated during June and August 2014. The concourse tunnel lining was monitored with the distributed fiber optic system throughout this excavation, starting from a few days before and ending a few days after each of the excavation phases. The monitoring was carried out using a Yokogawa AQ8603 BOTDR analyzer, connected to the end of the fiber optic extension cable. The analyzer was set to scan a frequency range from 10.40 to 11.30 GHz, in 0.01 GHz steps, averaged over 2^{15} readings. The data were discretized at 0.1 m intervals, with each data point representing a moving average of the Brillouin frequency peak over a 1 m length along the fiber optic circuits.

An initial baseline measurement was taken immediately after the breaking out of the tunnel SCL for the CP1 pilot tunnel. Subsequently the measurement sessions, which took about 15 minutes to complete, were set to initiate automatically every 15 minutes. The monitoring frequency was later changed to once every 30 minutes for the final enlargement phase of CP2.

Once the data were downloaded manually from the analyzer, they were postprocessed to extract the data points corresponding to each of the embedded fiber optic circuits. The temperature of the tunnel lining was estimated from the loose-tube fiber optic cables and used to separate the total strain from the bonded fiber optic cable into thermal and mechanical strains, as explained in Section 6.1.2. By matching the timestamp on each data set with the construction shift report, it was possible to identify the changes in mechanical strain that occurred along the fiber-optic circuits during each individual advance of the cross-passages excavation.

Results and Discussion

The total change in mechanical strain within the tunnel SCL due to the excavation of the two cross-passages (relative to the baseline after the CP1 tunnel break-out) is shown as a strain map in Fig. 6.1.6. The results indicate that, as the parent tunnel lining was being broken out and the cross-passages were being excavated, the superimposed load was redistributed around the circumference of the junctions. The strain profile is indicative of arching action, with an increase in compressive hoop strain to the sides of the junctions and a localized increase in longitudinal tensile strain under the invert of the CP1 junction.

During each excavation phase, the strain was observed to increase rapidly as the parent tunnel lining was being broken out. Once the breaking out was completed, the strain stabilized and only increased at a smaller rate as the cross-passage excavation advanced. The maximum strains observed throughout the monitoring were of 820 $\mu\varepsilon$ compressive in the hoop direction and 567 $\mu\varepsilon$ tensile in the longitudinal direction, both estimated from the readings taken after the two cross-passages were completed.

The strain maps obtained during the monitoring, such as the one in Fig. 6.1.6, indicate that the effect of the cross-passage excavation on the

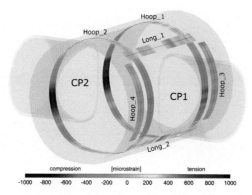

Figure 6.1.6 Total strain increase observed in the tunnel lining after the two cross-passages were completed

parent tunnel lining is localized around the perimeter of the junction. For example, the compressive strain increases in the hoop direction are concentrated in the parts of the Hoop_1 and Hoop_2 circuits that are directly adjacent to the sides of the cross-passage junctions. In comparison, a much smaller strain was observed in the rest of the hoop circuits that are farther away from the junctions.

These findings suggest that the thickening SCL layer is being utilized in a relatively small area around the junctions and that beyond approximately the first 3 m horizontally on either side of the junctions, the thickening SCL is no longer required. In this case, about one third of the total volume of concrete used in the thickening SCL of CH5, amounting to some 75 m^3, would be superfluous. If these results can be corroborated by similar investigations with distributed fiber-optic sensors embedded in other SCL tunnel junctions, leaner and more efficient junction designs could be developed. This would lead to significant savings in SCL tunnel construction, not only in material, labor and time, but also in terms of reducing the amount of dangerous steel fixing and concrete spraying operations carried out at heights.

6.1.4 CASE STUDY 2: LIVERPOOL STREET STATION – ROYAL MAIL TUNNEL

Project Background

The Royal Mail Tunnel, with 8 stations and 10.5 km long, was completed in 1917 to transport mail from eastern to western sorting offices in London, UK. It was constructed entirely in London Clay at an average depth of

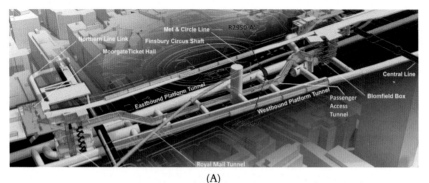

(A)

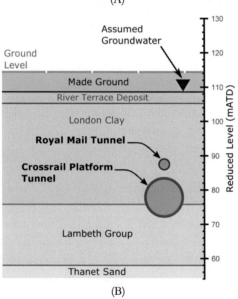

(B)

Figure 6.1.7 (A) Underground 3D view of Royal Mail and Crossrail tunnels (© Crossrail Ltd.) with highlighted zone of instrumentation. (B) Typical cross-sectional view of soil with location of tunnels. (C) Longitudinal soil profile and view of tunnels (Gue et al., 2015)

27 m below ground level. Each ring along the running tunnel consists of seven grey cast iron segments, approximately 0.5 m wide, bolted together to produce an external diameter of 2.97 m. Operation of the Royal Mail tunnel was stopped in 2003 due to high running costs and the tunnel has remained closed ever since.

A new 11 m diameter platform tunnel at Liverpool Street Station was constructed directly beneath the Royal Mail tunnel for the Crossrail project, as shown in Fig. 6.1.7A. It was constructed in a parallel alignment using an

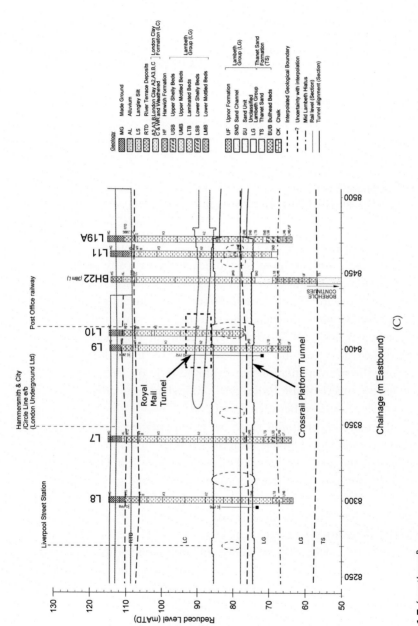

Figure 6.1.7 *(continued)*

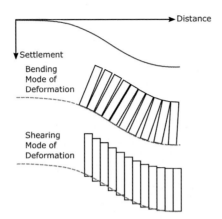

Figure 6.1.8 Bending and shear deformation modes for the existing cast iron lining tunnel in the longitudinal direction. (After Alhaddad et al., 2014 with permission from ASCE)

open face sprayed concrete lining (SCL) method. As mentioned in Section 6.1.3, Crossrail is one of the most significant infrastructure projects in the UK, delivering a new railway line linking Reading and Heathrow in the west of London with Shenfield and Abbey Wood in the east. A 6 m diameter pilot tunnel was constructed prior to the full enlargement to the final 11 m diameter. Figs. 6.1.7B and 6.1.7C show the typical soil profile along with the cross-section and longitudinal view of the tunnels. At the location where the instrumentation was deployed (Fig. 6.1.7A), the clearance between the invert of the Royal Mail Tunnel and the crown of the new Crossrail tunnel ranged between 1 and 2 m.

One of the main uncertainties faced by design engineers when a new tunnel is being constructed underneath an existing tunnel is the deformation mode of the existing tunnel linings (cast iron segments) in the longitudinal direction as it experiences tunneling-induced ground movements. In pure bending mode, the segmental linings deform in a continuous manner while in pure shear mode, the segmental linings are expected to displace vertically relative to each other as shown in Fig. 6.1.8. Induced stresses in the bolts and linings and hence the assessment of the integrity of the tunnel linings will differ significantly depending on the deformation mode.

Conventional instrumentation relies predominantly on discrete point displacement measurements that are carried out both manually and/or via automated total stations (ATS). The aim of measuring these displacements is to infer the amount of strain and stresses that are induced in the tunnel linings using interpolation between the measurement points. However, with limited measurement points and targets, assessment of the resultant stresses and strains can be very challenging. Distributed fiber-optic strain sensing, on the other hand, can provide the full strain profile of the tunnel lining as described below.

Distributed Fiber-Optic Strain Sensor Installation

The fiber-optic instrumentation layout of the Royal Mail tunnel consisted of a closed loop with 5 cross-sections at 10 m intervals and a 40 m long longitudinal section along the crown of the tunnel as shown in Fig. 6.1.9A. A Hitachi single core and single mode tight buffered cable (Fig. 6.1.2) was selected as the main strain sensing cable within the monitored section while Excel single mode loose tube cables were used to extend the connection to the top of the access shaft where a Yokogawa AQ8603 Brillouin Optical Time Domain Reflectometry (BOTDR) analyzer was located. All splices were kept in an enclosed splice holder for protection. The direction of tunneling for the new Crossrail platform tunnel within the monitored section was from ring R2950 in decreasing order to R2870, as shown in Fig. 6.1.9A (Gue et al., 2015).

The cables along the circumferential cross-sections were fully bonded to the flanges of the instrumented rings along the top half of the tunnel in order to obtain full strain profiles throughout the cable. Araldite 2021 epoxy was selected as the adhesive. The surfaces of the linings were cleaned to remove any grout, precipitate and dirt to ensure a strong bond between the sensing cable and cast iron lining. Low pull magnets were used to seat the cable temporarily to the cast iron flange while the epoxy set. Free running cables, unattached to the lining, were laid to the sides of the tunnel to connect the cross-sectional cables together.

A strain cable running along the tunnel crown was used to monitor the longitudinal behavior of the tunnel. However, flanges of the cast iron segments prevented the continuous gluing of the cable. A series of 10 mm thick aluminum L-brackets, measuring 110 (H) $\times$ 50 (W) $\times$ 80 (D) mm, were bolted to the mid-section of each pan to clear the flanges. The fiber-optic cable was then prestrained and glued onto each L-brackets with Araldite 2021 epoxy. Therefore, each section of fiber-optic cable located between two brackets acted as a gauge length. The strain measured in the middle of a gauge length represents the average strain on the structure between two brackets.

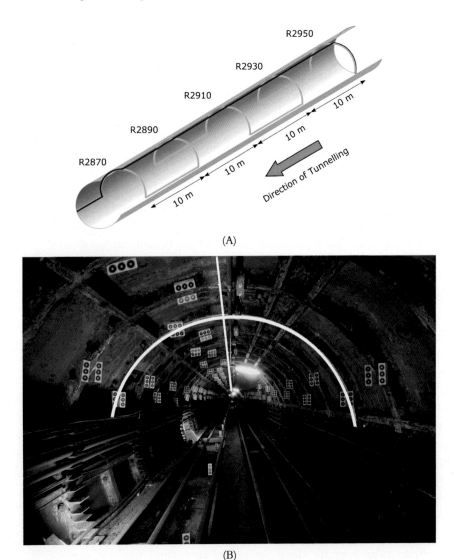

(A)

(B)

Figure 6.1.9 (A) Layout of fiber-optic cable in Royal Mail tunnel (Gue et al., 2015). (B) Photo of completed installation with fiber-optic cables highlighted in white

Fig. 6.1.9B shows the completed fiber-optic installation within the Royal Mail tunnel. It is crucial to note that the above mentioned setup for longitudinal strains will only detect strains from the bending mode deformation as the shearing mode deformation was expected to impose only negligible strains on the cable running along the crown.

As the duration of the monitoring scheme was significant, the analyzer had to be left on site continuously for months. For added security and protection from the harsh construction site environment, the analyzer was secured in a weather-proof strongbox at a safe predetermined location above the access shaft. This provided shelter and prevented accidental damage from site related activities. A filtered ventilation system was fitted in the strong box to provide a dust free cooling system for the analyzer. To prevent data loss in case of any accidental power failure, an uninterruptible power supply (UPS) was installed as a short-term backup system.

Results and Discussion: Cross-Sectional Behavior

Monitoring was carried out over a period of 11 months from April 16, 2013, to March 14, 2014. For clarity, only the data from the initial construction stage, the pilot tunnel construction stage and fully enlarged tunnel stage are presented and discussed in this case study. It is important to note that, as with any instrumentation system that is installed in an existing structure, only relative changes in strains and not absolute strains are measured. The changes in strain were computed from the baseline data collected on April 16, 2013, when there were no significant construction activities within the influence zone of the monitored section.

As the new Crossrail platform tunnel is excavated underneath the existing Royal Mail tunnel, stress is relieved around the new tunnel, causing the ground surrounding the existing tunnel above to move towards the newly created cavity, thus deforming the cast iron linings. Fig. 6.1.10A shows theoretical strain patterns that would develop along the top half of a tunnel when a new tunnel is constructed beneath it in two different configurations (directly underneath and to the side). The x-axes of the two plots refer to the distance along the fiber-optic cable around the intrados of the tunnel.

Fig. 6.1.10B presents the measured strain data for ring R2950 from A–A' throughout the three stages (the position of A–A' can be seen in Fig. 6.1.7A). The baseline is denoted by the horizontal black line at zero strain; in the initial construction stage (green dashed line) before the pilot tunnel was constructed within the influence zone, a passenger access tunnel was excavated at an offset to the lower side of Royal Mail tunnel in late July 2013. This caused the cast iron linings to deform in a skewed ellipse towards the passenger access tunnel (i.e., left depiction in Fig. 6.1.10A). The resulting strain profile was asymmetrical across the crown where it shows significant tensile strains on the side of the tunnel (from A to crown)

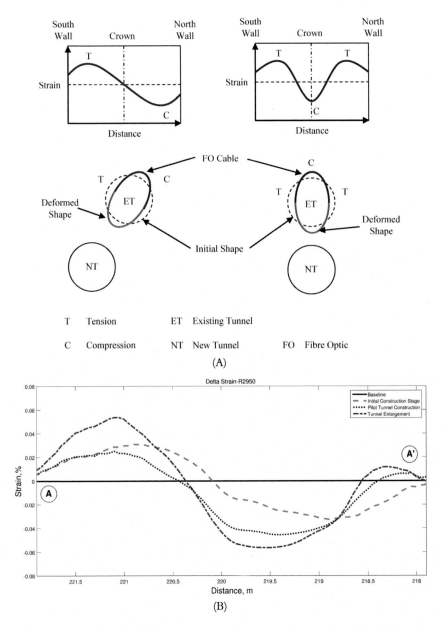

Figure 6.1.10 (A) Conceptualization of tunnel cross-section deformation mechanism based on strain distribution (Gue et al., 2015). (B) Cumulative change in strain for ring R2950 at various construction stages (Gue et al., 2015). (C) Change in strain at four locations along the longitudinal section. (D) Longitudinal deformation mechanism during pilot and tunnel enlargement (not to scale)

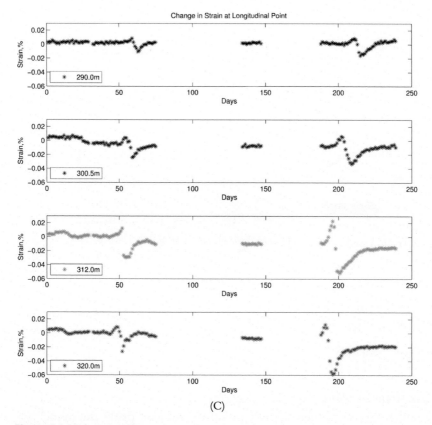

Figure 6.1.10 (continued)

nearest to the passenger access tunnel and compression on the far side (from crown to A′).

As the pilot tunnel enters the influence zone of ring R2950 in mid-September 2013, the cross-sectional profile changes to the expected shape for vertical ovalization (i.e., right depiction in Fig. 6.1.10A) where tensile strains prevalent on the sides of the tunnel while compressive strains are found over the crown (blue dotted line). The magnitude of strains increased to a peak of 550 $\mu\varepsilon$ (0.055 %ε) for the full tunnel enlargement in January 2014 (red dashed–dotted line) without changing the shape of the strain profile as the horizontal alignment of new tunnel alignment remains constant.

Calculating the equivalent stress in the cast iron lining by using the elastic stress strain relation and a Young's modulus of about 120 GPa and

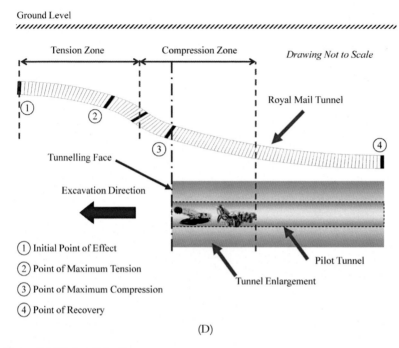

Ground Level

Tension Zone   Compression Zone   *Drawing Not to Scale*

Royal Mail Tunnel

① Initial Point of Effect

② Point of Maximum Tension

③ Point of Maximum Compression

④ Point of Recovery

Tunnelling Face

Excavation Direction

Pilot Tunnel

Tunnel Enlargement

(D)

Figure 6.1.10 *(continued)*

the peak strain of 550 $\mu\varepsilon$ would suggest an incremental stress of 66 MPa in the lining. However, there are two points to consider, firstly, the tunnel would have had a certain amount of stresses that was already in place before the instrumentation was carried out. If it is expected that the Royal Mail tunnel had an existing horizontal ovalization or squat, the induced vertical ovalization may actually reduce the stresses to some extent or vice versa. Moreover, the presence of wood packing across the joints of the segments would result in a zone with lower stiffness that would have been averaged out by the 1 m spatial resolution of the analyzer. Therefore, the strain values measured are influenced by the strain over the joints and that sustained by the cast iron panels themselves.

To quantify the actual amount of strains that developed in the cast iron segments, a deconvolution process would be needed. This would, however, require knowledge of the strain at each joint. Unfortunately, there were no local strain gauges installed over the joints in this field instrumentation. Hence, the maximum actual stresses within the cast iron segments would in fact be lower than the directly computed value of 66 MPa. Nonetheless, the measured strains would still be representative of the behavior of the cast iron ring as a whole (Gue et al., 2015).

Results and Discussion: Longitudinal Behavior

Based on empirical evidence, tunneling induced longitudinal green-field surface ground movements can be represented by a cumulative normal distribution curve (Attewell and Woodman, 1982; Mair and Taylor, 1997 and New and O'Reilly, 1991). It is expected that this "bow wave" movement would be transferred to the Royal Mail tunnel. Although strains were captured over the entire 40 m length along the crown, change in strains at 4 points, each located between adjacent cross-sectional cables are shown to facilitate understanding the behavior of the Royal Mail tunnel.

Point 320.0 m (red dots) was the first to encounter the excavation face, followed by point 312.0, 300.5, and 290.0 m in decreasing order. The first and second waves in Fig. 6.1.10C correspond to the construction effect from the pilot and tunnel enlargement respectively. Gaps in the data represent the time when no construction activities were taking place.

At day 44 in Fig. 6.1.10C, point 320.0 m experienced an increase in tensile strains peaking at day 49 before rapidly changing into compression by day 50 and recovers to approximately the same preconstruction values about 3 weeks later. Peak tensile strains were relatively low below 200 $\mu\varepsilon$ compared to the maximum compressive strains of about 300 $\mu\varepsilon$. Other points observed similar trends with slight delay in time due to their respective location from point 320.0 m. This is in line with the expected "bow wave" ground movement trend.

Tunnel enlargement entered the influence zone of point 320.0 m by day 187 and similar strain profile behavior was observed across all points albeit with larger strain magnitude since the final enlargement is significantly larger than the size of the pilot tunnel, causing larger ground induced movements. Similar to the pilot tunnel, compressive strains recorded larger values, with a maximum of about 600 $\mu\varepsilon$ in comparison to the tensile strains which remain mostly within 200 $\mu\varepsilon$.

A recovery zone was also observed with the strains reducing close to the pre-tunnel enlargement values; nonetheless, the fiber-optic data shows that the recovery is incomplete with some minor locked in stresses that remained throughout the monitoring period. As mentioned with the cross-sectional measurements, the strain values obtained would have been affected by wood packing inside the circumferential joints but it would still be indicative of the deformation of the cast iron tunnel as a whole.

Observing the data across the points, it is noticeable that the magnitude of strains reduces for both pilot and tunnel enlargement from point 320.0 m to point 290.0 m. This damping effect could be due to a number of reasons. Firstly, the point 320.0 m was the closest point to the passenger access tunnel which was constructed during the early stages, which may have affected the soil around this localized area. Secondly, a step plate junction was situated 10 m away from point 290.0 m, which would have changed the condition of fixity of the tunnel rings nearer to it in comparison with point 320.0 m, which was at the far end. A combination of these two effects could have played a role in the observed strain damping effect.

Aligning the observed trends with the construction data, it was found that the trough of each strain wave (point of maximum compressive strains shown at four longitudinal points in Fig. 6.1.10C) matches closely with the time when the excavation face was directly below the point of measurement. The longitudinal fiber optic data suggest that the deformation of the Royal Mail tunnel was similar to an inverted S-shape during the construction of both the pilot and enlargement tunnel. As shown in Fig. 6.1.10D, this S-shaped profile had four main characteristics: (a) an initial point of effect, (b) maximum tension, (c) maximum compression, and (d) point of recovery.

Conclusions

The distributed fiber-optic sensing system deployed in this study allowed the measurement of full strain profiles of the tunnel lining. This showed clear deformation patterns for cross-sectional and longitudinal deformations. In the longitudinal direction, the relative changes in strain changed rapidly as the tunneling face approached the measured point but its effects are short lived as recovery is observed as soon as the tunneling face exits the influence zone. This proves that the existing Royal Mail tunnel was subjected to the bending mode deformation to a certain degree; however, it should be noted that a compound deformation mode of both bending and shearing is also possible. In contrast with this, direct strain measurements for the cross-section shows that the effects are more permanent. This case study demonstrates that the distributed nature of the fiber-optic strain sensing technology provides a significantly superior set of information about the behavior of existing tunnels when subjected to tunneling-induced ground movements.

6.1.5 CASE STUDY 3: MONITORING OF CERN TUNNELS

Project Background & Aim of Monitoring

The Centre for European Nuclear Research (CERN) uses large and complex scientific instruments to study the basic constituents of matter by operating a network of underground particle accelerators and appurtenant tunnels. Several tensile cracks have been observed within a section of a concrete lined tunnel (Fig. 6.1.11A) called TT10, as shown in Fig. 6.1.11B. CERN asset engineers were then concerned about the safety and structural health of their infrastructure and designed a long-term sensing plan to monitor the behavior of these tunnels. However, they faced two challenges: (a) during the operation of the experiments most of the underground spaces are not accessible because of high radiation levels, and (b) radioactivity could potentially affect the reliability of conventional electro-magnetic monitoring equipment. Therefore, there was a clear need for radiation-resistant monitoring instruments that could be operated remotely.

Distributed fiber-optic sensing (DFOS) proved to be the most appropriate monitoring method as it appeared to satisfy the above-mentioned requirements. The University of Cambridge and its Centre for Smart Infrastructure & Construction (CSIC) have been using the distributed fiber optic sensing technology for more than 10 years, have had applications in a wide range on engineering structures, including a number of underground tunnels and have successfully validated the monitoring results against other conventional instruments and structural/geotechnical analyses. Therefore, a single fiber-optic (FO) sensing cable installed within the underground tunnel on the concrete lining surface and routed up to the ground surface was the selected method of sensing scheme. The purpose of this work was to monitor the long-term strains and deformations experienced by the existing underground tunnel section TT10 and potentially locate any development of tension cracks in the concrete lining.

Installation of Fiber-Optic Sensors & Planned Monitoring Scheme

A section of TT10 tunnel of about 100 m length was chosen to be initially monitored. The FO cable runs in both the longitudinal and circumferential directions of the tunnel, forming several (eight) circumferential loops and one straight cable section in the longitudinal direction, with the cable sections of interest prestrained as shown in Fig. 6.1.12A (solid line refers to prestrained cable section, whereas dashed line refers to loose cable section). The prestrained cable allows for accommodation of any compressive

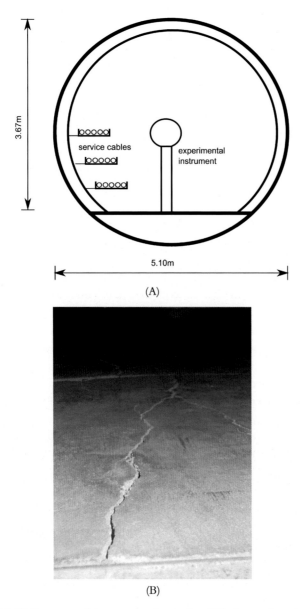

Figure 6.1.11 (A) Geometry of the examined tunnel and (B) observed cracks in the lining

strain without cable buckling. The cable is attached on the tunnel using ten hook-and-pulley systems thus forming a polygonal circumferential loop (Fig. 6.1.12B) and it is securely stuck on the hook-and-pulley system. At the bottom of the tunnel, the cable passes through a 5 mm cut section in

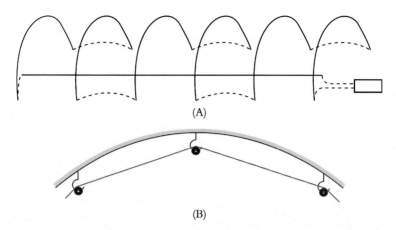

Figure 6.1.12 Installation scheme of fiber optic sensors: (A) layout of fiber-optic cable, and (B) attachment method

the floor concrete and then covered with epoxy glue and the section is then grouted.

At the end of the monitored section, the two ends of the cable are routed out of the tunnel section via a 50 m vertical shaft into a secure control room and no further access in the tunnel section is required, as long as there is no cable breakage requiring maintenance. Monitoring was planned to be carried out from outside of the tunnel, using an Omnisens BOTDA analyzer on which both ends of the FO cables are connected. The installation of the FO sensors took place and was completed in July 2014.

Since no major construction activity was taking place nearby and the aim is the long-term structural health monitoring of the tunnel, the BOTDA analyzer was not required to stay on site on a permanent basis. It was initially planned that monitoring would take place approximately every 2–3 months for the first year and then the long-term monitoring scheme would be designed depending on the outcome of the initial readings. Baseline readings were taken in July 2014 right after the sensors installation was completed and two further (progress) readings were taken in August 2014 and February 2015.

Current Monitoring Data

As discussed previously, the primary information obtained from a Brillouin based fiber-optic sensing is the characteristic Brillouin frequency of the cable and the secondary information is axial cable strain which is linearly proportional to the shift of the Brillouin frequency. Data for July 2014 is considered as baseline readings, whereas data for August 2014 and February 2015 is considered as progress monitoring readings.

The August 2014 and February 2015 progress monitoring data were then subtracted from the July 2014 baseline data to get the difference in Brillouin frequency, which correspond directly to the change in strain from the baseline measurement. Figs. 6.1.13A–B show the developed axial strain change profiles for two selected representative circumferential loops within the examined TT10 tunnel section (Loops 1 & 7) calculated from the FO cable Brillouin frequencies. The two ends of the section (i.e. start and end of the plot's x-axis) refer to the two sides of the tunnel cross-section, whereas the middle of the section refers to the crown of the tunnel.

As it may be observed from the latter figures, negligible strains develop within the first month (July 2014–August 2014) after the deployment of the sensors, as the recorded values of axial strain change do not seem to exceed 20 $\mu\varepsilon$. After seven months (July 2014–February 2015), the circumferential tunnel loops seem to show some noteworthy values of axial strain evolutions in the FO cable and a somewhat consistent profile of the strains for all the examined circumferential loops. The profile of the axial strains showed larger strain values at the lateral sides of the tunnel and smaller strains at the crown of the tunnel. There seems to be a consistent strain profile of strain value peaks at the two tunnel lateral sides and a strain value trough at the tunnel crown. Nevertheless, it should be noted that the absolute value of the observed strain evolution does not seem to exceed 100 $\mu\varepsilon$ which is not considered to be too significant. This outcome was expected, as the instruments were installed for long-term monitoring of the structures and higher values of strain are not likely to develop over a period of some months since no construction activity takes place at the vicinity of the tunnel. However, further monitoring is currently conducted to confirm this.

Conclusions & Future Work

Fiber-optic cable sensors have been installed at tunnel section TT10 at CERN for long-term strain and deformation measurements. A so-called "smart" installation was deployed with a number of spatially distributed monitoring sections: 8 circumferential tunnel loops with a longitudinal section within the tunnel.

Both ends of the cable have been brought back to the monitoring point in the control room above ground and plugged to the BOTDA machine offering a fully closed loop of the FO cable. Initial baseline readings showed a good signal transfer and a clear trend in defining the various sections of

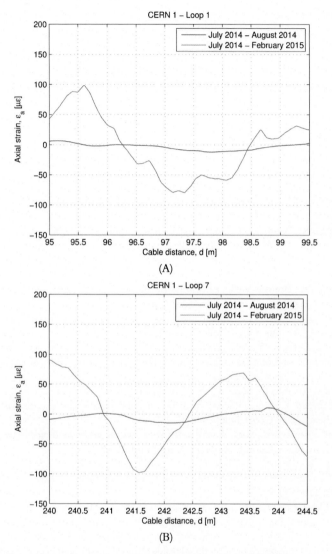

Figure 6.1.13 Fiber-optic axial strain readings: (A) loop 1 and (B) loop 7

interest. Three readings have been taken so far: the first set of baseline readings (July 2014) and two progress readings (August 2014 & February 2015).

In the progress readings taken for this tunnel area, the circumferential tunnel loops showed some minor values of axial strain development in the FO cable and a somewhat consistent profile of the strains for all circumferential loops. The values of all observed axial strains seem to be insignificant (i.e., less than 100 με), suggesting no real movement or deformation of the tunnel lining has taken place during the period of monitoring. This was expected as the purpose of this monitoring scheme is long-term tunnel deformations and no substantial strains would be anticipated over the relatively short period of six months.

REFERENCES

Alhaddad, M., Wilcock, M., Gue, C.Y., Bevan, H., Stent, S., Elshafie, M., Soga, K., Devriendt, M., Wright, P., Waterfall, P., 2014. Multi suite monitoring of an existing cast iron tunnel subjected to tunnelling induced ground movements. In: Proceedings of Geo-Shanghai International Conference 2014. Shanghai, China, May 26–28, pp. 357–361.

Amatya, B.L., Soga, K., Bennett, P.J., Uchimura, T., Ball, P., Lung, R., 2008. Installation of optical fibre strain sensors on soil nails used for stabilising steep highway cut slope. In: Proceedings of 1st ISSMGE International Conference on Transportation Geotechnics, pp. 276–282.

Amatya, B.L., Soga, K., Bourne-Webb, P.J., Amis, T., Laloui, L., 2012. Thermomechanical behaviour of energy piles. Géotechnique 62 (6), 503–519.

Attewell, P.B., Woodman, J.P., 1982. Predicting the dynamics of ground settlement and its derivatives caused by tunnelling in soil. Ground Eng. 15 (8), 13–22.

Bell, A., Soga, K., Ouyang, Y., Yan, J., Wang, F., 2013. The role of fibre optic instrumentation in the re-use of deep foundations. In: Proceedings of the 18th International Conference on Soil Mechanics and Geotechnical Engineering. Paris.

Bourne-Webb, P.J., Amatya, B., Soga, K., Amis, T., Davidson, C., Payne, P., 2009. Energy pile test at Lambeth College, London: geotechnical and thermodynamic aspects of pile response to heat cycles. Géotechnique 59 (3), 237–248.

Bourne-Webb, P.J., Amatya, B., Soga, K., 2013. A framework for understanding energy pile behaviour. Proc. Inst. Civil Eng.: Geotech. Eng. 166, 170–177.

Cheung, L., Soga, K., Bennett, P.J., Kobayashi, Y., Amatya, B., Wright, P., 2010. Optical fibre strain measurement for tunnel lining monitoring. Proc. Inst. Civil Eng.: Geotech. Eng. 163 (3), 119–130.

de Battista, N., Elshafie, M.Z.E.B., Soga, K., Williamson, M.G., Hazelden, G., Hsu, Y.S., 2015. Strain monitoring using embedded distributed fibre optic sensors in a sprayed concrete tunnel lining during the excavation of cross-passages. In: Proceedings of the 7th International Conference on Structural Health Monitoring of Intelligent Infrastructure. Torino, Italy, July 2015.

Galindez-Jamioy, C.A., López-Higuera, J.M., 2012. Brillouin distributed fiber sensors: an overview and applications. J. Sens. 2012, 1–17.

Gue, C.Y., Wilcock, M., Alhaddad, M.M., Elshafie, M.Z.E.B., Soga, K., Mair, R.J., 2015. The monitoring of an existing cast iron tunnel with distributed fiber optic sensing (DFOS). J. Civil Struct. Health Monitor. 5, 573–586.

Horiguchi, T., Kurashima, T., Tateda, M., 1989. Tensile strain dependence of Brillouin frequency shift in silica optical fibers. IEEE Photonics Technol. Lett. 1 (5), 107–108.

Iten, M., 2011. Novel Applications of Distributed Fibre Optic Sensing in Geotechnical Engineering. PhD Thesis. ETH, Zurich.

Janmonta, K., Uchimura, T., Amatya, B., Soga, K., Bennett, P., Lung, R., Robertson, I., 2008. Fibre optics monitoring of clay cuttings and embankments along London's ring motorway. In: Proceedings of GeoCongress 2008: Geosustainability and Geohazard Mitigation. New Orleans, Louisiana, March 9–12, pp. 509–516.

Kechavarzi, C., Soga, K., de Battista, N., Pelecanos, L., Elshafie, M., Mair, R.J., 2016. Distributed Fibre Optic Strain Sensing for Monitoring Civil Infrastructure – A Practical Guide. ICE Publishing, London, UK. ISBN 9780727760555.

Mair, R.J., Taylor, R.N., 1997. Theme lecture: bored tunnelling in the urban environment. In: Proceedings of the Fourteenth International Conference on Soil Mechanics and Foundation Engineering. Hamburg, September 6–12. Balkema, pp. 2353–2385.

Mohamad, H., 2012. Temperature and strain sensing techniques using Brillouin optical time domain reflectometry. In: Theodore, E. Matikas, Kara, J. Peters, Wolfgang, Ecke (Eds.), Smart Sensor Phenomena, Technology, Networks, and Systems Integration 2012. In: Proc. SPIE, vol. 8346.

Mohamad, H., Bennett, P.J., Soga, K., Mair, R.J., Bowers, K., 2010. Behaviour of an old masonry tunnel due to tunnelling-induced ground settlement. Géotechnique 60 (12), 927–938.

Mohamad, H., Soga, K., Pellow, A., 2011. Performance monitoring of a secant piled wall using distributed fibre optic strain sensing. J. Geotech. Geoenviron. Eng. 137 (12), 1236–1243.

Mohamad, H., Soga, K., Bennett, P.J., Mair, R.J., Air, R.J., Lim, C.S., 2012. Monitoring twin tunnel interactions using distributed optical fiber strain measurements. J. Geotech. Geoenviron. Eng., Am. Soc. Civil Eng. 138 (8), 957–967.

Mohamad, H., Soga, K., Amatya, B., 2014. Thermal strain sensing of concrete piles using Brillouin optical time domain reflectometry. ASTM Geotech. Test. J. 37 (2), 1–14.

New, B.M., O'Reilly, M.P., 1991. Tunnelling induced ground movements: predicting their magnitudes and effects. In: Proceedings of the 4th International Conference on Ground Movements and Structures. Cardiff, July 8–11. Pentech Press, pp. 671–697.

Nöther, N., 2010. Distributed Fiber Sensors in River Embankments: Advancing and Implementing the Brillouin Optical Frequency Domain Analysis. PhD Thesis. Technical University of Berlin.

Ohno, H., Naruse, H., Kihara, M., Shimada, A., 2001. Industrial applications of the BOTDR optical fiber strain sensor. Opt. Fiber Technol. 7, 45–64.

Ouyang, Y., Soga, K., Leung, Y.F., 2011. Numerical back-analysis of energy pile test at Lambeth College, London. In: Geotechnical Special Publication, pp. 440–449.

Pelecanos, L., Soga, K., Chunge, M.P.M., Ouyang, Y., Kwan, V., Kechavarzi, C., Nicholson, D., 2017. Distributed fibre-optic monitoring of an Osterberg-cell pile test in London. Geotech. Lett., 1–9.

Schwamb, T., Elshafie, M., Ouyang, Y., Janmonta, K., Soga, K., Fuentes, R., Ferreira, P., Swain, A., 2011. Monitoring of a secant pile wall using fibre optics sensors. In: Gatterman, J., Bruns, B. (Eds.), 8th International Symposium on Field Measurements and Geomechanics. 12–15 Sep. 2011, Berlin, Germany. TU Braunschweig. ISBN 3-927610-87-9, pp. 12–15.

Schwamb, T., Soga, K., Mair, R.J., Elshafie, M.Z., Sutherden, R., Boquet, C., Greenwood, J., 2014. Fibre optic monitoring of a deep circular excavation. Proc. ICE – Geotech. Eng. 167 (2), 144–154.

Soga, K., 2014. XII croce lecture: understanding the real performance of geotechnical structures using an innovative fibre optic distributed strain measurement technology. Riv. It. Geotech. 4, 7–48.

SUBCHAPTER 6.2

Fiber-Optic Sensor Networks: Environmental Applications

J.A. García*,†, D. Monzón-Hernández†, E. Bustos*
*Centro de Investigación y Desarrollo Tecnológico en Electroquímica S.C., Parque Tecnológico Querétaro s/n, Sanfandila, Pedro Escobedo, Querétaro, Mexico
†Centro de Investigaciones en Óptica A.C., León, Guanajuato, Mexico

6.2.1 INTRODUCTION

Optical techniques are powerful tools that gradually have gained a leading role in modern metrology. Most of the methods used in modern laboratory equipment for sensing and monitoring physical, chemical, or biological variables are based on optical methods. However, for outdoor applications such as environmental monitoring, most of this laboratory equipment is not suitable. Neither is it convenient for applications where continuous and real-time monitoring is a concern. In these situations optical sensors based on optical fibers constitute an alternative. The variety of optical fibers available today combined with the number of recently developed fiber devices, with improved characteristics, provides a countless potential for sensor configurations in environmental monitoring (García et al., 2014).

The inherent advantages of an optical fiber, such as lightweight, small size, resistance to electromagnetic interference, large bandwidth, have contributed to the development of sensors with enhanced performance. The distinguishing characteristics of these sensors are high sensitivity, resistance to environmental activity, flexibility and low-power consumption, and ease of use in normally inaccessible areas and interfacibility with data communication systems with no risk of electric shock on *in vivo* measurements. Thus, optical fiber sensors have been gradually replacing traditional sensors for rotation, acceleration, electric and magnetic field measurements, as well as temperature, pressure, acoustic signal, vibration, linear and angular positioning, strain, humidity, viscosity, chemical and pollutant content, pH, and gas phase measurements among many others (Calle et al., 2000; García et al., 2005; Castrellon-Uribe, 2012; García et al., 2014).

Fiber-optic sensors (FOS) work like most electronic sensors, except they use a glass fiber instead of copper wire for transmission of informa-

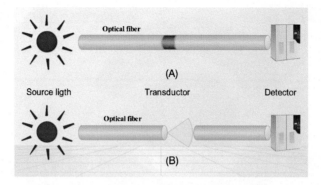

Figure 6.2.1 Basic fiber-optic sensor: intrinsic (A) and extrinsic (B) (Adapted from Chong et al., 2013)

tion, and photons instead of electrons to carry the information. In its most simple form an FOS system consists of a light source, optical fiber, a transducer or sensing element and a detector (Fig. 6.2.1). In these sensors an optical driving signal, whose characteristics are known, is continuously circulating along the optical fiber. When an external perturbation interacts with the sensor, at least one parameter of the circulating light, such as its intensity, phase, polarization, or wavelength is modulated (Chong et al., 2013). Although there are several proposals that classify FOSs, one of the simplest classifications describes FOSs as two types, intrinsic and extrinsic. Intrinsic sensors are based on a beam of light propagating through the fiber for which the sensory interaction occurs at the interface between the surrounding environment and the optical fiber. In extrinsic sensors, the optical fiber transmits the light beam, which can leave and return to the fiber so that the external environment affects the light beam. In these sensors the optical fiber does not play an important role in sensing but serves as a guide to drive the signal (Augousti and Grattan, 1999; Wolfbeis, 2004; McDonagh et al., 2008; García et al., 2014).

Extrinsic FOS is based mainly on the processes of light absorption-emission. The working mechanism of an FOS based on absorption is the same as spectrophotometers, allowing the detection of a wide range of organic and inorganic compounds in different phases, either liquid or gaseous. The fluorescence-based extrinsic FOSs require the use of light sources with specific wavelengths, since fluorescence processes depend on the excitation wavelength (Augousti and Grattan, 1999; Wolfbeis, 2004; McDonagh et al., 2008; García et al., 2014). On the other hand, among the intrinsic

Figure 6.2.2 FOS classification (adapted from Chong et al., 2013)

FOS types we find three variants of sensing mechanisms, as described in Fig. 6.2.2.

One remarkable feature of the intrinsic type FOS is the possibility of developing a spatially extended, position sensitive sensor system. Such a system is conventionally described as a "distributed system" if the position sensitivity extends over the entire active length of the fiber, and described as "quasi-distributed" if it consists of a series of sensitive regions spaced over the length of the fiber.

In this chapter, first a general background on the use of fiber optic sensor networks in environmental applications is presented. More focused discussion is offered on optical detection of gas leakage into atmosphere, water contamination, soil contamination, and their mapping using distributed fiber-optic sensors.

6.2.2 FIBER-OPTIC DEVICES FOR SENSING

Optical fibers are an efficient communication channel in which the optical signal, confined to the inner core of the fiber, is isolated from the influence of external changes. Therefore, one of the first challenges of the fiber-optic sensor research community was finding the way to access the optical signal in order to cause a perturbation in it. It is not surprising that the first chemical fiber sensors were based on cleaved optical fiber tips whose face was coated with a sensitive layer and in close contact with the sensed media. Although fiber tip sensors offer the advantage of being truly microscopic devices promising minimally invasive *in-situ* and real time monitoring of bio-chemical species, these are point sensors often capable of monitoring only a single parameter, therefore with limited applications. In the last two decades, the community has witnessed a significant increase in scientific research aimed to develop optical fiber devices capable of detecting external perturbations. In these devices, the fiber structure is modified by removing the fiber cladding by etching or polishing; reducing the fiber cross-section diameter by tapering; inducing a periodic change of the fiber core refractive index by means of the photorefractive effect, among others. In this regard optical fiber Bragg and long-period gratings are the most successful examples of devices whose response, codified in wavelength, is highly dependent on external parameters such as temperature or strain. Devices where evanescent wave interacts with the external medium such as fiber tapers, D-shape fibers, tilted Bragg gratings, or fibers with dissimilar core size are mostly used in chemical or biological sensing (Korposh and Lee, 2009; García et al., 2014).

Current research in the field of optical fiber sensors focuses on conception and development of new sensitive elements which can grow the sensing applications by increasing the number and range of analytes that can be measured by FOS. Two basic approaches are possible: (1) direct optical interaction with the analyte, or (2) indirect analysis using chemical indicators (i.e., compounds that change their optical properties by reaction with the analyte). These techniques are classifies as follows (Dakin and Culshaw, 1997; Ricciardi et al., 2013):

1. Direct optical interaction with the analyte
   1.1. Transmission Spectroscopy
   1.2. Absorption Measurements by Attenuated Total Reflection/Evanescent Field Absorption
   1.3. Absorption Measurement by Photoacoustic Spectroscopy
   1.4. Fluorescence (Luminescence) Spectroscopy
   1.5. Light Scattering
   1.6. Raman Spectroscopy
   1.7. Fourier Transform Spectroscopy
2. Indirect analysis using chemical indicators
   2.1. pH – Effects of Ionic Strength on pKa

2.2. Metal Chelators – Crown Ether Dyes (Chromoionophors); Chelators for Ions

2.3. Potential – Sensitive Dyes

2.4. Quenchable Fluorophores

2.5. Polymeric Supports and Coatings

2.6. Silicones

2.7. Hydrophobic Polymers

2.8. Silica Materials

2.9. Hydrophilic Supports

2.10. Diffusion and Permeation of Gases through Polymers

Sensors based on fiber-optic cable functions make use of the following important features of the cable to sense the environment:

(1) *Optical loss*, intrinsic and extrinsic energy loss properties

(2) *Refractive index*, index profile in radial direction and the reduction of index fluctuation along the axial direction

(3) *Shape*, cross-sectional shape and size, the surface finish and the fluctuation of the size along axial direction

Present fiber-optic sensors mostly use energy loss principles (i.e., changes in optical power in linearly positioned wave-guides) for chemical detection. These can be limited for distributed applications if energy depletes over a short stretch of the fiber sensor, or frequent sensor points are needed at a prohibitively expensive cost. Other sensors use the changes in refractive index and/or cross-sectional size of the fiber cable that change the light scattering property in optical fibers, known as Brillouin scattering (Horiguchi et al., 1995; Kee et al., 2000). Fiber-optic sensing based on Brillouin scattering has been used successfully in civil infrastructure for health monitoring (Bao et al., 2001; Ohno et al., 2001).

6.2.3 ENVIRONMENTAL APPLICATIONS OF FOS

Wide application of advanced chemical sensing in the environment may suffer from scaling issues. The real-world conditions often require self-referencing, spatially distributed, temporally continuous, and chemically selective sensors for monitoring regions spanning over long distances and large areas. When large area monitoring is required, use of point sensors can be cost prohibitive. Some of the non-point, distributed detection methods based on energy loss principles (Buerck et al., 2001) may be inadequate when scaled to wide area monitoring due to extensive energy input requirements. Having identified some of the shortcomings of wide area environmental sensing techniques, we present below the state-of-art FOS based technologies and devices available for select environmental monitoring and mapping, namely for sensing hazardous gases, water and soil contamination. It should be noted here that most of these technologies and devices may be adaptable to use above or underground, as needed.

6.2.3.1 FOS for Gas and Emission Sensing

In the last two decades, the air contamination, caused in part by the combustion of fossil fuels, has been intensifying. Therefore, research on development of sustainable energy sources as well as cost-effective, fast and reliable sensors to detect greenhouse gasses have become a priority. The vast quantity and many types of industrially produced organic gases have led many researchers to focus on the development of technologies for detection and prevention of their leakage into the environment (Mishra et al., 2012). Fiber-optic based gas sensors have been shown to possess important advantages over others. These sensors are particularly ideal for monitoring unstable or flammable gases, because they are inherently safe against sparks or electrical short circuits. Flares produced by the combustion of industrial gases have been used for many years as a major control mechanism for routine and emergency releases of gases generated by industrial processes. The accurate determination of gas emissions and the temperatures associated with their release are important for the understanding their impact on air quality. These measurements are often difficult because the emissions are directly released to the ambient air, and the gas temperatures are high, released form high stacks (up to several hundred meters). The emissions are variable by the process of combustion often complicated by factors such as meteorology, throughput and variability of constituents burned (Zwicker et al., 1998). Jin et al. (2013) reported fiber-optic gas sensors with photoacoustic spectroscopy (PAS), which has the potential for further development into cost-effective distributed and multiplexed fiber-optic gas detection systems with high detection sensitivity on the order of part-per-billion (ppb). One field where FOS shows great potential is in detection of volatile organic compounds (VOCs). These substances, including cleaning products, solvents, and other toxic agents that produce odors, are present daily in living and industrial enclosures. Although electronic devices do exist to detect hazardous concentrations of such substances in the atmosphere, they show practical drawbacks such as their large size and heavy weight (Elosua et al.,

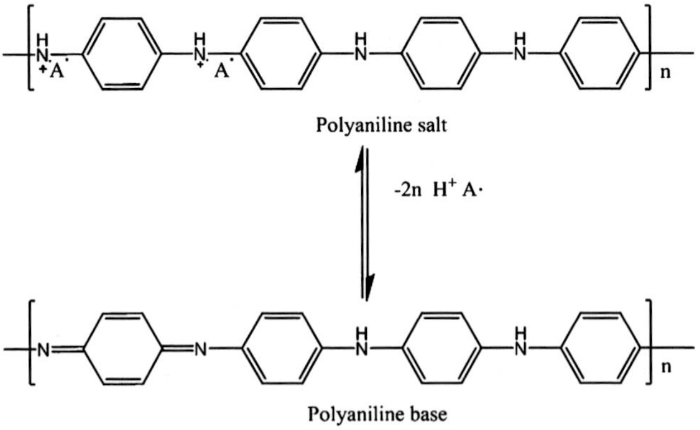

Figure 6.2.3 Chemical reaction of polyaniline in presence of ammonium (adapted from Mishra et al., 2012)

2014). The following sections discuss some of the advanced FOS systems developed to detect and quantify specific gases in the environment.

6.2.3.1.1 Fiber-Optic Coated With ITO and Polyaniline Using Plasmon Resonance for Monitoring Ammonia Gas

One of the gases widely used in industrial processes, whose leakage could present a serious risk to the environment and human health, is ammonia. Several fiber-optic ammonia sensors have been proposed, such as plasmon resonance-based FOS, which consists of un–cladded optical fiber coated with ITO ($In_2O_3 + SnO_2$) and a thin film of polyaniline (Mishra et al., 2012). This sensor can work with either signal strength (using the reversible reaction of polyaniline in the presence of acid media (Fig. 6.2.3)) or wavelength shift (due to plasmon ITO). The range of ammonia concentration measured with this device was from zero to 150 ppm.

6.2.3.1.2 Fiber-Optic Coated With Graphene Film Using Reflectivity for Monitoring Acetone Gas

Another example FOS development for detection of toxic compounds in air is the work carried out by Zhang et al. (2011). They developed a system which measures the changes of the reflectivity of an optical fiber tip coated with a 0.8 nm–thick graphene film as a function of the concentration of acetone exposure in the gas phase.

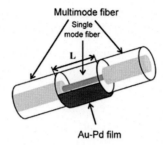

Figure 6.2.4 Schematic representation of heterocore device developed for sensing H_2 (adapted from Luna-Moreno et al., 2007)

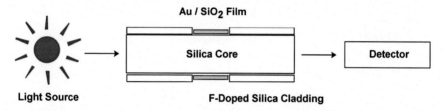

Figure 6.2.5 Schematic of the fabricated optical fiber sensor with an Au/SiO_2 film and a fiber doped silica cladding (adapted from Ohodnicki et al., 2013)

6.2.3.1.3 Fiber-Optic Heterocore Coated With Thin Film of Au–Pd for Monitoring Hydrogen Gas

Another one of gases of industrial and ecological origin, which has been considered the fuel of the future, but represents a danger by its high level of explosiveness, is hydrogen. Luna–Moreno et al. (2007) proposed a FOS using a fiber-optic hetero-core covered with a 10 nm thin film of Au–Pd, which allowed detection of H_2 (g) at concentrations ranging from 0.8% to 4% with a response time of 15 s (Fig. 6.2.4).

6.2.3.1.4 Fiber-Optic Coated With Au/SiO_2 for Monitoring Ambient Gases in Atmosphere

Ohodnicki et al. (2013) reported an FOS system of Au/SiO_2 deposited on silica-based optical fibers (Fig. 6.2.5) that remain stable under high temperature conditions of up to 1000°C and interactions with changing ambient gases in the atmosphere. They indicated that although additional investigation is required for full development of the device, the simplicity of the sensor design made it cost-effective for potentially widespread deployment.

6.2.3.1.5 Fiber-Optic Using Resonance Enhanced Multiphoton Ionization for Monitoring Volatile Organic Pollutants

Resonance enhanced multiphoton ionization (REMPI) is a useful technique for obtaining both qualitative and quantitative information about volatile organic pollutants. This technique has been used extensively for selective ionization in mass spectrometers (Shang et al., 1993; Clarck et al., 1993; Nesselrodth and Baer, 1994; Sin et al., 1984; Cornish and Baer, 1990; Nesselrodth et al., 1995; Buma et al., 1991; Cool, 1984; Tjossem and Cool, 1983; Mallard et al., 1982; Meier and Kohse-Hoinghaus, 1987), as well as in spectroscopic combustion analysis (Cool, 1984; Tjossem and Cool, 1983; Mallard et al., 1982; Meier and Kohse-Hoinghaus, 1987) reliably for the last two decades.

In the REMPI technique, a high powered, tunable pulsed laser is used to electronically excite a specific molecule with one or more photons. The absorption of a photon(s) then ionizes the molecule, releasing an electron(s). An electrode based at a high positive or negative potential is used for efficient collection of the electrons or ions. The current created by these electrons/ions flowing to the electrodes is then measured as a voltage drop across a "leak resistor." The resulting voltage is proportional to the concentration of the specific species being ionized. Due to the nonlinear nature of this technique, it is important to have high laser power densities near the collection electrode. A high power density increases the probability of ionizing the specific molecules of interest. Due to its sensitivity and qualitative capabilities, REMPI, is an excellent technique for remote analysis of various volatile organics in the environment. By using optical fibers as wave guide, it is possible to introduce high power laser pulses to a remote area (Marquardt et al., 1997). By placing a small electrode near the focus of the laser at the fiber output, a simple and remote probe can be constructed to conduct trace analysis down a well or at other inaccessible areas.

The REMPI system (Fig. 6.2.6) consists of a Q-switched Nd:YAG laser operating at either the second or the fourth harmonic with a high voltage power supply, an array of probe/collection electronics, and a fast digital oscilloscope (Marquardt et al., 1997). The high power laser pulse is first launched into the optical fiber and the laser light is transmitted through the fiber to the probe house where it is focused to a point, ionizing the gas vapor. Electrons created from this ionization are then collected on a platinum electrode biased at $+1000$ V. The current created by the collection of these electrons is then run through a 250 kΩ resistor and the voltage drop across

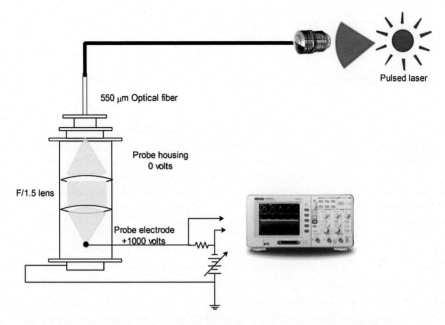

550 μm Optical fiber

Probe housing
0 volts

F/1.5 lens

Probe electrode
+1000 volts

Pulsed laser

Figure 6.2.6 Schematic diagram of REMPI apparatus showing launch and detection system with the lenses are 1 – inch diameter f/1.5 plano convex, and the platinum electrode is biased at +1000 V with respect to the aluminum housing (adapted from Marquardt et al., 1997)

the resistor is measured using a 1 $M\Omega$ coupled 500 MHz digital oscilloscope. The total acquisition time of average signal from 1000 laser shots at 10 Hz is approximately 100 seconds. The laser power at the probe tip is typically 5.4 mJ/pulse (266 nm) or approximately 20 mJ/pulse (532 nm), depending on the excitation wavelength used.

6.2.3.1.6 Fiber-Enhanced Raman Multigas Spectroscopy for Monitoring Atmospheric Gases

Versatile multigas analysis has high potential for environmental sensing of greenhouse gases and for noninvasive early stage diagnosis of certain disease states in human breath. Hanf et al. (2014) reported fiber-enhanced Raman spectroscopic (FERS) analysis of a suite of climate change relevant atmospheric gases, which allowed for reliable quantification of CH_4, CO_2, and N_2O alongside N_2 and O_2 with just a single measurement (Fig. 6.2.7). The technique was shown to achieve high analytical sensitivity down to a subpart per million detection limit with a 6 orders of

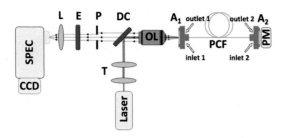

Figure 6.2.7 Design of the optical setup for fiber-enhanced Raman spectroscopy (FERS) consisting of laser, telescope (T), long-pass dichroic beam splitter (DC), objective lens (OL), fiber adapter assembly (A1 and A2), hollow-core photonic crystal fiber (PCF), powermeter (PM), pinhole (P), edge filter (E), aspheric lens (L), spectrometer (SPEC), and CCD detector (adapted from Hanf et al., 2014)

magnitude dynamic range, measured within one second. The natural isotopes $^{13}CO_2$ and $^{14}N^{15}N$ were quantified at low levels, simultaneously with major components such as N_2, O_2, and $^{12}CO_2$. Furthermore, there is great potential in the miniaturization efforts of FERS setups providing the advantage of portability and affordable prices while delivering high sensitivity and selectivity. A portable and robust FERS sensor, capable for real-time multicomponent quantification of biogenic gases is promising to be a powerful tool for the characterization of greenhouse gas fluxes as well as bedside clinical diagnosis of metabolic diseases.

6.2.3.1.7 Fiber-Optic Transmission Near-Infrared Spectroscopy for Monitoring Resin Curing and Humidity Ingress

Epoxy and amine based resin systems have been used extensively in the manufacture of advanced fiber reinforced composites (AFRC) (Afromowitz and Lam, 1989). For an AFRC, the stiffness and the strength of the material are dictated by the properties of the reinforcing fibers, but the inter-laminar shear and compressive strength and its sensitivity towards the ingress of fluids and temperature are influenced by the properties of its internal matrix. The integrity of the final matrix is often affected by the curing process of the resin composite. A number of cure monitoring techniques have been available, including dielectric analysis (Mopsik et al., 1989), ultrasonic velocity (Davis et al., 1991), nuclear magnetic resonance (Barton et al., 1994), Raman spectroscopy (Debakker et al., 1993) and optical fiber-based sensor systems (Druy et al., 1989 and 1992; Lam and Afromowitz, 1995a, 1995b).

A fiber-optic sensor based on near-infrared spectroscopy was developed and demonstrated to monitor the curing and subsequent moisture ingres-

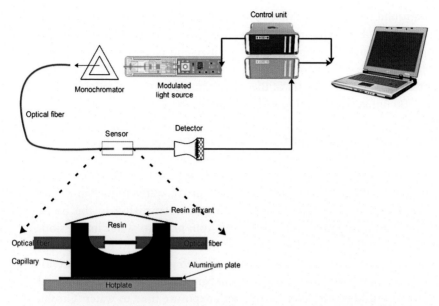

Figure 6.2.8 Schematic illustration of the optical fiber-based transmission cell for cure monitoring (adapted from Crosby et al., 1996)

sion of an epoxy–amine resin system (Crosby et al., 1996). In this sensor, first a slot was introduced at the center section of a 10 mm long precision bore metal capillary tube with an internal diameter of 250 μm. Two 50/125 μm multimode silica optical fibers were positioned within the capillary tube end-to end at the location of the slot, with a gap of approximately 1.5 mm between their end faces. The fibers were secured in position using an adhesive and the entire assembly was secure onto an aluminum plate, as shown in Fig. 6.2.8.

The experiments using this simple transmission sensor assembly were conducted using a fiber mono-chromator as the light source, which provided a stable, modulated, tunable light source over the wavelength range of 400–1700 nm. The input fiber of the transmission sensor was connected to the light source using a microscope objective lens. The output fiber of the sensor was connected to an InGaAs photodiode detector and lock-in amplifier. Output wavelength control of the spectrometer and the acquisition of the signal from the sensor were carried out using a PC (Crosby et al., 1996).

The isothermal cure monitoring experiments were carried out on a hotplate where the temperature was controlled at a precision of 1°C.

A background spectrum was first collected with the "air-gap" between the fiber-end faces. Following this, the epoxy-amine resin was placed over the sensor region filling the gap between the fiber-end faces. A thermocouple was place into the resin to monitor the actual temperature of the resin during the curing period. Spectra of the transmitting resin were collected over 1450–1700 nm at 5 minute intervals during the same period (Crosby et al., 1996).

This sensor was subsequently used to monitor moisture ingression into the resin. The cured resin was surrounded by a moat and the whole assembly immersed in distilled water. Spectra in region 1100–1700 nm were collected every hour over a period of 50 hours of water immersion. The spectra were converted to absorbance units by normalizing to the background spectrum obtained from the cured resin prior to water immersion. The negative absorbance values observed were attributed to the plasticization of the matrix, which in turn lowers the refractive index of the resin. The increase in the broad O–H absorption bands, centered at 1370 nm and at 1680 nm, could be observed clearly with exposure time to water (Crosby et al., 1996).

6.2.3.1.8 Fiber-Optic Using Optical Remote Sensing of Flare Emissions

Use of passive optical remote sensing (ORS) devices which utilize the emission spectra from the hot flare emissions have been studied many years (Haus et al., 1994). Passive systems seem ideal for these types of monitoring since all monitoring equipment remains at ground level and the optics is focused on the flare. However, there can be difficulty in obtaining accurate data because of the influence of varying background emissions and thermal effects from the flare (Zwicker et al., 1998).

Active ORS devices, which determine concentrations by absorption spectra, require a source of energy. Validating the accuracy of these measurements is easier particularly if the system is monostatic, that is, when there is a reflecting device at one end of the beam and the beam is modulated to remove background factors (Russwurm and Childers, 1996). Other types of ORS systems such as tunable diode lasers (TDLs) have also been used for studying flare emissions from petroleum related flares for much shorter flare stacks and monitoring for only methane emissions.

6.2.3.2 FOS for Water Contamination Sensing

The increasing concerns about drinking water quality and environmental protection of the natural water resources have significant impact on the development of advanced sensors for monitoring water. Point detection fiber-optic sensors have been developed for detection and quantification of chemical species (inorganic and organic), drugs, pollutants such as pesticides, and to monitor a wide variety of various chemical processes and biochemical reactions in water (Wolfbeis, 2000). In most of these devices the major component is the sensing or recognition element. Reactive and/or externally stimulated polymers have been utilized as chemical sensing materials. The interaction of an analyte with the polymer coating of the optical fiber has been the predominant mechanism of sensing of pollutants and chemicals in water (Philips et al., 2003). The following sections present some of the traditional and emerging fiber-optic based devices and techniques used to sense and monitor chemical and biological contamination in water.

6.2.3.2.1 *Fiber-Optic Using Fourier Transform Infrared Spectroscopy-Attenuated Total Reflectance and Near-Infrared for Monitoring Chlorinated Hydrocarbons in Water*

One of the main chemical species of large concern to environmental health are chlorinated hydrocarbons (CHCs) due to their high toxicity and novice effects on human health. For more than a couple of decades many researchers have been involved with developing fiber-optic based methods to detect CHCs in water, such as Walsh et al. (1995) who proposed a system using a silver halide coated optical fiber with polyisobutylene and an MCT (mercury–cadmium–telluride) detector. Using this device they could detect trichlorethylene (TCE) in the concentration range of 10 to 100 ppm with high sensitivity.

More recently, new approaches using advanced Fourier transform infrared spectroscopy-attenuated total reflectance (FTIR-ATR) sensor for *in-situ* and simultaneous detection of multiple CHCs have been proposed. The polycrystalline silver halide sensor fiber, which has a unique integrated planar-cylindric geometry, was coated with an ethylene/propylene copolymer membrane to act as a solid phase extractor. This construction resulted in a greatly amplified signal and contributed to a higher detection sensitivity compared to the previously reported sensors. The sensor system exhibited a high detection sensitivity when tested in a CHCs mixture of wide concentration range of 5–700 ppb, as shown in Fig. 6.2.9 (Lu et al., 2013).

Lavarioti et al. (2014) reported a series of experiments measuring the near-infrared optical absorption of alkanes, aromatics, and chlorinated hy-

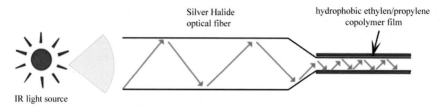

Figure 6.2.9 Schematic diagrams of the FTIR-ATR detection principle of CHCs (adapted from Lu et al., 2013)

drocarbons in liquid matrices. A spectral library was then developed to characterize the optical spectra of liquid hydrocarbons, where the near-infrared analysis was chosen due to its compatibility with optical fibers. Using this system they could differentiate between classes of hydrocarbons and discriminate between compounds within a class of similar molecular structure using statistical analyses, the principal component analysis (PCA) and the correlation coefficient (Spearman and Pearson methods). Regan (2014) reported that OP–FTIR can be used to illustrate the impact of sample type on environmental sensor measurements with a potential for real sample analysis of chlorinated hydrocarbons that can be verified with total organic contain (TOC) and carbon organic demand (COD) parameters used as indicators of the concentration of organic matter in the samples.

6.2.3.2.2 Fiber-Optic Using Evanescent Wave for Monitoring Nitrite in Water

Another type of fiber-optic based sensor utilize the interaction of the analyte and evanescent wave of the optical fiber for detection of contaminants in water. Kumar et al. (2002) employed the technique with an uncladded optical fiber to detect nitrite in water. Their system consisted of a 12 cm long uncladded plastic optical fiber of 200 μm core diameter inserted into a reaction cell. The surrounding water containing the nitrite ion acts as the cladding of the waveguide, absorbing the transmitted light at a particular wavelength that depend directly on the concentration of the nitrites. The system was able to detect nitrite ion at a concentration range of 1 to 1000 ppb (Fig. 6.2.10).

6.2.3.2.3 Fiber-Optic Using UV for Monitoring COD and BOD in Water

Chemical oxygen demand (COD) is an indicator of contamination that shows the amount of dissolved matter in water susceptible to being oxi-

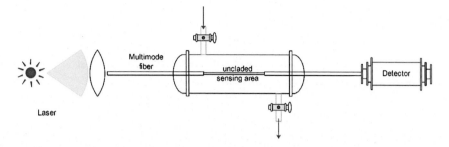

Figure 6.2.10 Experimental set-up for nitrite determination in water (adapted from Kumar et al., 2002)

dized. The most commonly used detection method of COD is the closed reflux or colorimetric method, which requires several hours obtaining a measurement and use of costly, hazardous reagents. Fang and Dai (2004) developed an online algorithm to construct adaptive models that predict COD in water samples. They analyzed samples with a spectrophotometer optical fiber with a response time of approximately 62 seconds without the need of sample preparation. The sample examinations were conducted at UV region of 258–380 nm, obtaining a measurement range between zero to 350 µL.

Biochemical oxygen demand (BOD) is an indicator of the amount of oxygen needed to oxidize biological organic matter present in water sample to CO_2, water, and ammonia. Because the current techniques to measure BOD can be time consuming, researchers have explored designing online sensors for this purpose. Chee et al. (1999) developed an optical sensor modified with a nylon membrane coated with *Pseudomona putida*. They measured the BOD in synthetic wastewater samples achieving good results compared with the traditional method, but reported some problems of reproducibility. Orellana and Moreno–Bondi (2002) developed another fiberoptic sensor by combining an oxygen-sensitive membrane and a porous membrane containing immobilized (innocuous) bacteria. This sensor was demonstrated successfully to measure the amount of oxygen relatable to the amount of organic matter in a wastewater sample in quantities of ppm.

6.2.3.2.4 Aptamer-Based Evanescent Wave Fiber Optic Bio-Sensor for Detection of Bisphenol-A in Water

One of the compounds present in surface waters that originate from the industrial production of epoxy resins, polysulfone resin, polyphenilene oxide

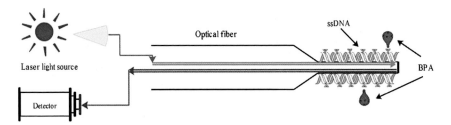

Figure 6.2.11 Schematic representation of FOS for BPA detection (adapted from Yildirim et al., 2014)

resin, polycarbonate, among others, is bisphenol-A (BPA). BPA is an endocrine disruptor compound that affects humans causing reproductive disorders, birth defects, various types of cancers and damage to the central nervous system. Among the most common methods for the detection of BPA in water are high performance liquid chromatography (HPLC), liquid chromatography coupled with electrochemical detection (LC-ED), liquid chromatography coupled with mass spectrometry (LC-MS), gas chromatography (GC), and gas chromatography coupled with mass spectrometry (GC-MS). These techniques are often time consuming, costly and require complex pretreatment processes of the test samples. Yildirim et al. (2014) developed a portable and easy-to-use aptamer-based evanescent wave fiber-optic biosensor for rapid and selective detection of BPA in environmental water samples, which showed good recovery, precision and accuracy, indicating that it was not susceptible to water matrix interferences even without the need of complicated sample pretreatments. A probe DNA molecule, which is the complementary sequence of a small part of the BPA aptamer, was covalently fixed onto the optical fiber sensor surface (Fig. 6.2.11). A fast indirect competitive detection mode was developed using samples containing different concentrations of BPA in the range of 2 to 100 nM at the detection limit of 0.45 ng m^{-1}, and the detection time of 400 s.

6.2.3.2.5 Fiber-Optic With a Bragg Grating for Monitoring pH in Water

One of the important chemical parameters of interest in water to measure on line is the pH, particularly in extreme environmental conditions, such as in nuclear reactors, wastewater treatment plants or in groundwater monitoring systems. Several researchers have developed systems or devices for measuring pH using fiber optic sensors, including Deboux et al. (1995) who created an on line sensor for pH ranges from 3 to 9 with high sensitivity (0.75 dB/pH) and immune to changes in ionic strength of the

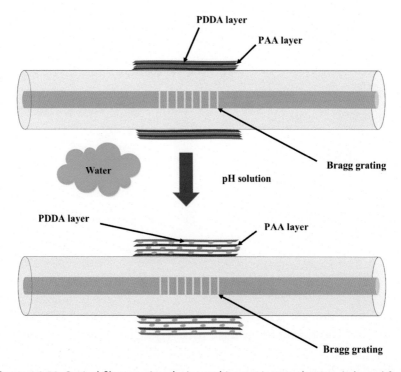

Figure 6.2.12 Optical fiber sensing device and its sensing mechanism (adapted from Shao et al., 2013)

medium. This sensor consisted of an optical fiber with the exposed core and coated with methylene blue and a protective membrane that allowed the process of diffusion and exchange with the surrounding environment. Research for development of a pH sensing *in vivo* systems was performed using a multimode optical fiber of 125-micron diameter. The tip of the fiber was tapered employing concentrated HF up to 50 microns and then it was coated with seminaphthorhodamine-1 1 mM carboxylate. This device was used to measure pH changes in the brains of rats to track the effects of trauma in living organisms (Grant et al., 2001). More recently Shao et al. (2013) used optical fiber with a Bragg grating coated with a self-assembled film of poly-diallyldimethylammonium chloride (PDDA) and polyacrylic acid (PAA) for detection of pH in water. Their device was able to detect pH within a range of 4.6 to 6.0, with 0.01 units of accuracy and response time of 10 s (Fig. 6.2.12).

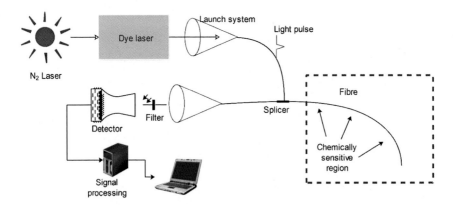

Figure 6.2.13 The experimental configuration for the distributed system (adapted from Wallace et al., 1996)

6.2.3.2.6 Fiber-Optic Using Optical Time Domain Reflectometry for Monitoring pH in Solution

The optical time domain reflectometry (OTDR) experimental arrangement (Fig. 6.2.13) consists of four main parts: (1) light source, (2) launch unit and splice, (3) sensing section, and (4) light detector and signal processing unit. The light source comprises a dye laser pumped by N_2. The dye laser produces short pulses at 490 nm which is launched into a 1×2 optical fiber coupler through a 40X microscope objective. Fiber, with sol–gel claddings at several discrete positions, was spliced to the coupler and a photomultiplier. The output of this photonic array was displayed on a storage oscilloscope before being transferred to a computer (Wallace et al., 1996).

In an experiment using this OTDR array, the sol–gel cladded sections of the fiber, linearly positioned several meters apart on the fiber were immersed sequentially in eight buffer solutions of pH ranging from 4 to 11.8. The sol–gel films were allowed to fully hydrate for one hour in the buffer solutions while a control section was left in air. The time domain responses of the immersed sensor sections of the fiber were compared to those of the control sections over 4 hours. The fluorescence signals from control sections remained constant whereas the signals form the active sensors exhibited a 3–4 fold increase as the pH was increased from 4 to 11.8. The response of the sensor was reproducible over the timescale of the experiment but when the sensor was left soaking for a period of one week a significant change in the signal was observe due to the dye leaching from the sol–gel matrix.

Covalently binding the dye to the fiber in subsequent trials was thought to help resolve this problem. The position resolution of the distributed system was dependent upon the width of the signal pulses (i.e., a function of the base electronics used) but also the intrinsic decay time of the fluorescence signal as it travels from the sensing position to the signal recorder. In the experiments reported, the sample pulses of 15 ns width were captured on a 500 MHz oscilloscope, which corresponded to a position resolution of about 1 meter (Wallace et al., 1996).

6.2.3.2.7 Fiber-Optic Using Hetero-Core Fiber Coated With an Acrylic Polymer Doped with Prussian Blue for Monitoring pH in Solution

García et al. (2014) reported the development of a fiber-optic pH sensor based on hetero-core fiber structure coated using an acrylic polymer doped with Prussian blue (PB). In this design, the pH changes of the surrounding medium affected the PB present in the coating layer and produced a change in its refractive index. The pH changes were observed as changes in the hetero-core transmission signal. The hetero-core fibers were constructed using two different length and two different types of optical fibers. In this case, two types of single-mode fibers (SMA and SMB) and two of multimode fibers (MMA and MMB) were used. First, two pieces of the MM fiber, stripped of its original polymer coating (3 cm section) were spliced to a stripped SM fiber on each of its ends to construct the basic hetero-core fiber. The hetero-core fibers were then treated with 0.1 mM PB, 4% polyvinyl alcohol (PVOH), 50% acrylic polymer emulsion (APE) at different combinations, such as PVOH + PB or APE + PB, to develop the reactive coatings over the appropriate stripped surfaces. The coating was accomplished using a small U-shape container made of a glass capillary fixed to a mechanical mount and filled with the mixture of the polymer support (PVOH or APE) and PB. Then the single-mode section of the hetero-core fiber was immersed for 5 minutes into this solution, after which it was removed and dried at room temperature. In this manner, the coating material was adhered to the single-mode section of the hetero-core fiber. One end of the hetero-core fiber was connected to a white light source and the other to the spectrum analyzer. This set-up was used to measure the transmission of light first during the modification process of the fiber and later to measure the response of the modified fiber to pH changes.

In order to test the device, the optical fiber sections modified with PVOH + PB or APE + PB were immersed into solutions where the pH was

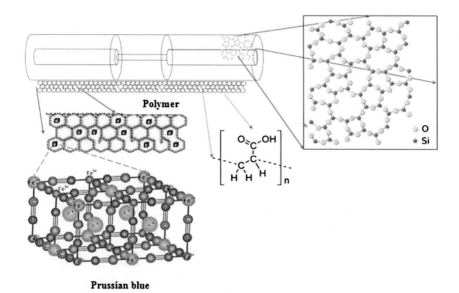

Figure 6.2.14 Schematic representation of FO modified with the polymer APE with Prussian blue for monitoring pH in solution

varied by adding 0.1 M NaOH or 0.1 M HCl. The transmission spectrum changes were recorded in the wavelength range from 350 to 1700 nm. The transmission intervals decreased with increasing wavelength, demonstrating good sensitivity of measurement. The pH changes were more evident at pH values less than 7, as shown in Fig. 6.2.14. This was attributed to the fact that at pH values higher than 7, the PB complex tends to hydrate causing the signal grow weaker with each incremental change in pH (García-Jareño et al., 1996). Based on the results reported, the longer length hetero-core device was recommended (10 mm rather than the 5 mm) due to the larger evanescent wave field and larger analysis spectrum gain (i.e., 2 dB) obtained, and higher measurement sensitivity.

6.2.3.3 FOS For Soil Indices and Soil Contamination Sensing

One of the unique features of the optical fiber technology is the possibility to construct distributed sensors for large volume or area applications. Rapid development in the area of smart sensor technologies over the last two decades has given way for exploring structurally integrated optical fiber to form the basis for smart structure technology both under and above ground. In these setups discrete measurements can be accomplished along a waveguide line with a given spatial resolution (Galindez-Jamioy and López-Higuera, 2012). Brillouin optical time domain analysis (BOTDA) and Brillouin optical time domain reflectometry (Bao et al., 2001; Pamukcu et al., 2006a and 2006b; Anastasio et al., 2007a and 2007b; Cui et al., 2009, 2010, 2011a and 2011b), and hetero-core long-period fiber gratings (LPFG) sensors are two such systems that can deliver distributed sensing along linear positioning of a single optical fiber placed over a large area or space. Other examples of sensor systems that can provide distributed measurements include multiplexed long gauge interferometric sensors, LPFGs with sensitive films, hetero-core devices, fiber Bragg gratings on doped fibers (i.e., germanium doped), for which the closer the ambient refractive index to that of the fiber cladding the stronger its sensitivity to externally stimulated refractive index changes. This high sensitivity of measurement has piqued the interest of many to develop various types of refractive index-based LPFG sensors which constitute most of the chemical sensing applications in soils and other geological media (Orellana and Haigh, 2008; Kasik et al., 2010).

6.2.3.3.1 Fiber-Optic Using BOTDR for Monitoring Water Content in Soil

A BOTDR sensing system to detect water content changes in soil was reported (Texier et al., 2005; García et al., 2014). Hydrophilic polymers, used as water transducers, were integrated with optical fiber to correlate fiber Brillouin strain response to the water content of the surrounding soil environment. In these experiments, AEP60 hydrophilic polymer discs (2 cm thick × 5 cm diameter) were stringed along 100-m continuous optical fiber by wounding the fiber about the discs to secure them at discrete spacing along the continuous fiber. The AEP60 polymer typically expands up to 400% over dry volume when exposed to water. Full expansion was reproducible over many wetting and drying cycles, and was consistent over a wide range of pH and dissolved solid concentrations. The diameter of the polymer disc used in these set of experiments was selected particularly to accommodate the minimum curvature of bending of the fiber, as shown in Fig. 6.2.15A. Each disc was embedded and allowed to swell in a clay soil sample of different water content, packed in watertight equal volume cylindrical cells of 14-cm diameter and 28-cm height. The sensing experiments were conducted in a temperature-controlled environment, at 25°C so that Brillouin measurements were not influenced by thermal expansion or contraction of the fiber.

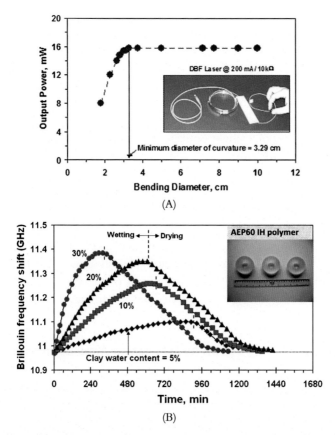

Figure 6.2.15 Assembly and test results for minimum fiber bending radius assessment (A), Brillouin frequency shift variation with time of optical fiber integrated with swelling AEP60 polymer discs embedded in clay with different water contents (B) in a BOTDR system (adapted from Texier et al., 2005)

The Brillouin shift of the wound fiber over the swelling polymer disc was measured at 20 minute intervals up to the maximum observable swelling. When no significant change in Brillouin shift was recorded for three consecutive measurements, the polymer transducers were removed from the clay chambers and left open for air-drying. In the experiments described above, the Brillouin signal variation due to the straining of the wound fiber was accurately correlated to the maximum swelling of the polymer disc, which was correlated to the original water content of the clay media, as shown in Fig. 6.2.15B.

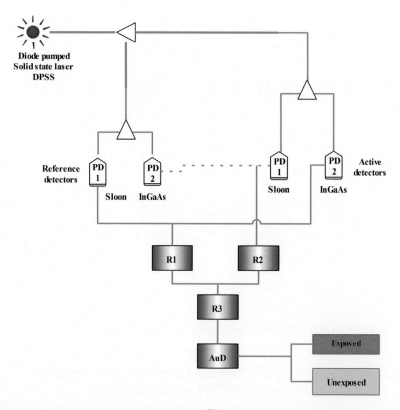

Figure 6.2.16 Schematic testing of HySense™ sensor (adapted from Mendoza et al., 2013)

6.2.3.3.2 Fiber-Optic Using Fluorescent Coating for Monitoring the Fuel Leaks

Mendoza et al. (2013) reported the development of a fast response, high sensitivity, distributed fiber-optic fuel leak detection (HySense™, Fig. 6.2.16) system for buried or above ground pipelines, tanks, pumps, and valves. This device was constructed using an optical fiber with a hydrocarbon sensitive fluorescent coating to detect the presence of fuel leaks present in close proximity along the length of the sensing fiber. It was reported that the HySense™ system operates in two modes, leak detection and leak localization, and will trigger an alarm within seconds of exposure contact.

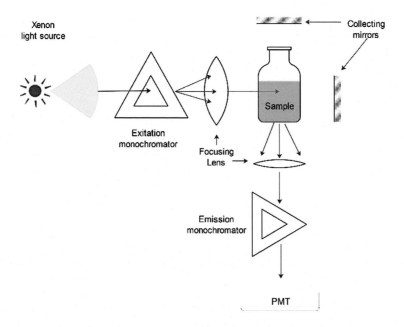

Figure 6.2.17 Optical Schematic of SSL (adapted from Hyfantins et al., 1998)

6.2.3.3.3 Fiber-Optic Using Synchronous Scanning Luminoscope for Monitoring Contaminated Soil and Ground Water

The synchronous scanning luminoscope (SSL) is a field-portable, synchronous luminescence spectrofluorometer developed for on-site analysis of contaminated soil and ground water. The portable optical components consist of a xenon flash lamp, two monochromators, focusing lenses, a sample holder, and a photomultiplier tube, as shown in Fig. 6.2.17. In addition, the SSL can utilize up to 5 meters of optical fiber for remote monitoring. A laptop computer provides for the SSL control functions, spectral display, quantifications, and data storage. The SSL has been reported capable of accurately performing quantitative analysis of polyaromatic hydrocarbons (PAHs), creosotes, and polychlorinated biphenyls (PCBs) in complex mixtures (Vo-Dinh, 1978, 1989; Vo-Dinh et al., 1979). Synchronous fluorescence was shown to reduce the complexity of the fluorescence spectra and allow for rapid field assessment of different types of contaminants at hazardous waste sites with a high correlation to laboratory data. Synchronous fluorescence is obtained when a constant wavelength interval is maintained between the excitation and the emission monochromators throughout the spectrum. This procedure takes advantage of the overlap between the ab-

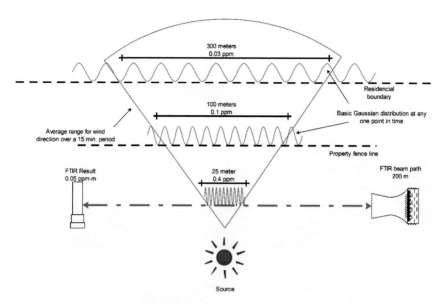

Figure 6.2.18 Simplified approach to estimate downwind concentration using OP-FTIR (adapted from Virag and Kricks, 1998)

sorption spectra and emission spectra for given compounds and provides greater selectivity when analyzing mixtures of compounds (Hyfantins et al., 1998).

6.2.3.3.4 Fiber-Optic With Open-Path Fourier Transform Infrared Spectroscopy Monitoring Remediation of Polluted Sites

Open-path Fourier transform infrared spectroscopy (OP-FTIR) has been used to monitor the perimeter of remediated polluted sites during the earth excavation and stockpiling operations. The technique has been shown capable of providing compound differentiation at atmosphere measurement levels that would ensure the protection of the public health and the environment. OP-FTIR is capable of monitoring a large number of chemical compounds along its entire beam path (Fig. 6.2.18). Therefore, it would be unlikely that pollutants would go undetected especially where there are several emission sources and wind directions are less than favorable (Virag and Kricks, 1998).

Using OP-FTIR, the air monitoring data is collected and screened on near real-time basis. This system could use an ETG FTIR remote sensor, cube reflector, flat mirrors, WS-12 anemometer, meteorological remote

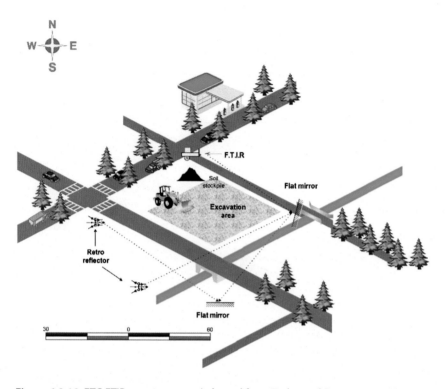

Figure 6.2.19 ETG FTIR remote sensor (adapted from Taylor and Brunnert, 1998)

sensor connected to a computer (Fig. 6.2.19). The device was shown to detect and quantify volatile organic compounds such as toluene, ethyl benzene, o-xylene, m-xylene, and p-xylene during remediation operations at a hazardous waste site (Taylor and Brunnert, 1998).

The usefulness of OP-FTIR technology was tested in an urban-industrial setting by monitoring the ambient air quality over a seven-day period (Kagann et al., 1998). Several of the ambient species, which are ozone precursors and referred to as criteria pollutants in Title I of the Clean Air Act Amendment of 1990, showed poor OP-FTIR detection limits because of the overlap of their infrared absorption by very strong water vapor lines in such an environment. Due to the high detection limit requirements for sulfur dioxide and nitrogen dioxide, OP-FTIR has not been the method of choice for measuring air quality in urban environments. However, OP-FTIR can be used in the detection of ozone, carbon monoxide, sulfur dioxide, nitrogen dioxide among many others for ambient air quality monitoring of urban environments.

6.2.3.4 Mapping With Array-Based Distributed Fiber-Optic Sensors

Spatially resolved mapping of chemical and biological constituents is an important requirement in a variety of practical applications. Spatially resolved analyte monitoring can simultaneously indicate and locate when an accepted level of exposure to toxic species has been exceeded and may help track the exposure to the source.

For example, DNA and antibody based sensors provide biologically relevant response with selective sensitive materials, but they may lack the capacity to provide reliable real-time analyses due to their slow response times. Furthermore, many such sensors, which are capable of high sensitivity and high selectivity detection, have been fashioned for single analyte detection (i.e., small molecule detection and immunological assays for toxins and pathogens). Only a few selected cases employ array sensors, which display multianalyte detection, such as oligomeric receptors that can yield vast molecular diversity for chemical libraries (Savoy et al., 1998).

Another example where multianalyte detection may be important is the Latex Agglutination Test (LAT) one of the most commonly used techniques for clinical analyses. Commercially available LATs for more than 60 analytes are used routinely for the detection of infectious disease, illegal drugs, and early pregnancy tests. The vast majority of these tests operate on the principle of agglutination of latex particles (polymer microspheres) which occurs when the antibody-derived microspheres become effectively "crosslinked" by a foreign antigen resulting in the attachment to, or the inability to pass through a filter. The dye-doped microspheres are then detected by colorimetric analysis upon removal of the antigen carrying solution. However, LATs cannot be utilized real time for multianalyte detection schemes as the nature of the response intrinsically depends on a cooperative effect of the entire collection of microspheres (Savoy et al., 1998).

Similarly, array sensors of tin oxide, conductive organic polymers and carbon black/polymer composites use surface acoustic wave (SAW) signals for identifying varieties of organic vapors (Afromowitz and Lam, 1989; Nakamoto et al., 1990, 1991; Shurmer et al., 1991; Logergan et al., 1996). When these sensors are exposed to volatile compounds, some of the chemical components adsorbed onto the surfaces of different elements lead either to small changes in the electrical resistance or mass of the composite layers. These small differences in their behavior are recognized as a pattern and the overall "finger print" response for the array serves as the basis for an olfaction of the analyte species. Although SAW sensors yield extremely sensitive responses to vapor, engineering challenges have prevented the creation of large arrays having multiple sensing sites (Savoy et al., 1998).

Advances in localized analyte recognition have been developed by covalently fixing polymeric "cones", grown by photopolymerization, onto the remote face of fiber-optic bundles. These sensor probes are designed with the goal of obtaining unique, continuous, and reproducible responses from sample-localized regions of dye-doped polymers. Here, the polymer serves as a solid support for "indicator" molecules that provide information about test solutions through changes in their optical properties. These polymer-supported sensors have been used for multianalyte detection, including pH, metals, and specific biological entities (Dickinson et al., 1997).

6.2.4 GENERAL CONCLUSIONS

The fiber-optic sensors can used effectively to create a truly distributed chemical sensing capability for selectively detecting and quantifying pollutants in air, water, and soil by real-time acquisition over large distances in the environment. The fiber-optic sensor networks are conducive to most environmental applications owing to their versatility by small size, lightweight, environmental ruggedness, immunity to electromagnetic and high temperature interference, and also their functionality by offering large bandwidth, high sensitivity and accuracy. Most current fiber-optic technologies, as reviewed in here, are capable of point or semidistributed measurement of contaminants or environmental indices by strategic placement of the sensing or monitoring devices at locations of concern. These include those used for gas and emissions detection, sensing and quantifying various indices and toxic chemicals and biological substances in water and soil.

ACKNOWLEDGMENTS

The authors want to acknowledge Consejo Nacional de Ciencia y Tecnología (CONACyT), Centro de Investigaciones en Óptica (CIO) and Centro de Investigación y Desarrollo Tecnológico en Electroquímica (CIDETEQ) for providing the facilities to conduct some of the research reported in here.

REFERENCES

Afromowitz, M.A., Lam, K.Y., 1989. Fiber optic cure sensor for thermoset composites. Fiber optic smart structures and skins. Proc. SPIE 986 (25), 135–139.

Anastasio, S., Pamukcu, S., Pervizpour, M., 2007a. BOTDR detection of chemical & liquid content. In: Proc. of the 7th FMGM, GSP 175 Int. Sym. on Field Measurements in Geomechanics. ASCE, Boston, MA, pp. 1–12.

Anastasio, S., Pamukcu, S., Pervizpour, M., 2007b. Chemical selective BOTDR sensing for corrosion detection on structural systems. In: Proc. of the 7th Int. Workshop on Structural Health Monitoring (IWSHM 2007). Stanford, CA, pp. 1701–1708.

Augousti, A.T., Grattan, K.T.V., 1999. Optical Fiber Sensor Technology. Chemical and Environmental Sensing. Optoelectronics, Imaging and Sensing. Springer, Netherlands.

Bao, X., DeMerchant, M., Brown, A., Bremner, T., 2001. Tensile and compressive strain measurement in the lab and field with the distributed Brillouin scattering sensor. J. Lightwave Technol. 19, 1698.

Barton, J.M., Buist, G.J., Hamerton, I., Howlin, B.J., Jones, J.R., Liu, S., 1994. High temperature 1H NMR studies of epoxy cure: a neglected technique. Polym. Bull. 33, 215–219.

Buerck, J., Roth, S., Kraemer, K., Mathieu, H., 2001. OTDR distributed sensing of liquid hydrocarbons using polymer-clad optical fibers. In: Dowding, C.H. (Ed.), Proceedings of the Second International Symposium and Workshop on Time Domain Reflectometry for Innovative Geotechnical Applications. Academic, Evanston, IL, pp. 496–509.

Buma, W.J., Kohler, B.E., Song, K., 1991. Lowest energy excited singlet states of isomers of alkyl-substituted hexatienes. J. Chem. Phys. 94, 4691–4698.

Calle, Martín A., Lechuga, L.M., Prieto, F., 2000. Optical Sensors Based on Evanescent Field Sensing Part 1. Surface Plasmon Resonance Sensors, USA.

Castrellon-Uribe, J., 2012. Optical fiber sensors: an overview. In: Yasin, Moh, Harun, Sulaiman W., Hamzah, Arof (Eds.), Fiber Optic Sensors. InTech, Croatia. Chapter 1.

Chee, G.-J., Nomura, Y., Ikebukuro, K., Karube, I., 1999. Development of highly sensitive BOD sensor and its evaluation using preozonation. Anal. Chim. Acta 394 (1), 65–71.

Chong, S., Aziz, A., Harun, S., 2013. Fibre optic sensors for selected wastewater characteristics. Sensors 13 (7), 8640–8668.

Clarck, A., Ledingham, K.W.D., Marshall, A., Sander, J., Singhal, R.P., 1993. Attomole detection of nitroaromatic vapours using resonance enhanced multiphoton ionization mass spectrometry. Analyst 118, 601–607.

Cool, T.A., 1984. Quantitative measurement of NO density by resonance three – photon ionization. Appl. Opt. 23, 1559–1572.

Cornish, T.J., Baer, T., 1990. Identification of conformational isomers of methyl – substituted cyclohexanonoe and tetrahydropyran frozen in a molecular beam. J. Phys. Chem. 94, 2852–2857.

Crosby, P.A., Powell, G.R., Liu, T., Xu, X., Fernando, G.F., 1996. In-situ core monitoring of an epoxy/amine resin system using an optical fibre transmission sensor. In: Kim, V., Bennett, D., Kim, Byoung Yoon, Liao, Yanbiao (Eds.), Proceedings of SPIE: Fiber Optic Sensors, vol. 2895. Beijing, China, pp. 109–115.

Cui, Q., Pamukcu, S., Xiao, W., Guintrand, C., Toulouse, J., Pervizpour, M., 2009. Distributed fiber sensor based on modulated pulse base reflection and Brillouin gain spectrum analysis. Appl. Opt. 48 (30), 5823–5828.

Cui, Q., Pamukcu, S., Lin, A., Xiao, W., Toulouse, J., 2010. Performance of double side band modulated probe wave in BOTDA distributed fiber sensor. Microw. Opt. Technol. Lett. 52, 2713–2717.

Cui, Q., Pamukcu, S., Lin, A., Xiao, W., Herr, D., Toulouse, J., Pervizpour, M., 2011a. Distributed temperature-sensing system based on Rayleigh scattering BOTDA. IEEE Sens. J. 11 (2), 399–403.

Cui, Q., Pamukcu, S., Xiao, W., Pervizpour, M., 2011b. Truly distributed fiber vibration sensor using pulse base BOTDA with wide dynamic range. IEEE Photonics Technol. Lett. 3 (24), 1887–1889.

Dakin, J., Culshaw, B., 1997. Optical Fiber Sensors Volume for Applications, Analysis and Future Trends. Artech House Inc., Boston–London, USA.

Davis, A., Ohn, M.M., Liu, K.X., Measures, R.M., 1991. A study on an opto-ultrasonic technique for cure monitoring. In: Fiber Optic Smart Structures and Skins IV. Proc. SPIE 1588 (35), 264–274.

Debakker, C.J., George, G.A., John, N.A., Fredericks, P.M., 1993. The kinetics of the cure of an advanced epoxy-resin by Fourier transform Raman and near-IR spectroscopy. Spectrochim. Acta, Part A, Mol. Biomol. Spectrosc. 49 (5–6), 739–752.

Deboux, B.J.C., Lewis, E., Scully, P.J., Edwards, R., 1995. A novel technique for optical fiber pH sensing based on methylene blue adsorption. J. Lightwave Technol. 13 (7), 1407–1414.

Dickinson, T.A., Walt, D.R., White, J., Kauer, J.S., 1997. Generating sensor diversity through combinatorial polymer synthesis. Anal. Chem. 69, 3413–3418.

Druy, M.A., Elandjian, L., Stevenson, W.A., 1989. Composite cure monitoring with infrared transmitting optical fibers. Fiber optic smart structures and skins. Proc. SPIE 986 (25), 130–134.

Druy, M.A., Glatkowski, P.J., Stevenson, W.A., 1992. Embedded optical fiber sensors for monitoring cure cycles of composites. Active materials and adaptive structures. Proc. SPIE 163, 805–808.

Elosua, C., Bariain, C., Matias, I.R., 2014. Optical fiber sensing applications: detection and identification of gases and volatile organic compounds. In: Yasin, Moh, Harun, Sulaiman W., Hamzah, Arof (Eds.), Fiber Optic Sensors. InTech, Croatia. Chapter 2.

Fang, J., Dai, L., 2004. Rapid detection of chemical oxygen demand using least square support vector machines. In: Proceedings of the Fifth World Congress on Intelligent Control and Automation (WCICA 2004), vol. 15–19. Hangzhou, China, pp. 3810–3813.

Galindez-Jamioy, C.A., López-Higuera, J.M., 2012. Brillouin distributed fiber sensors: an overview and applications. J. Sens. 204121, 17.

García, M.G., Álvarez, E., García, A., Solana, G., Vilchez, F., González, I., 2005. Applications of analytical chemistry in environmental research. In: Palomar, Manuel (Ed.), Research Signpost. Kerala, p. 273.

García, J.A., Monzón, D., Martínez, A., Pamukcu, S., García, R., Bustos, E., 2014. Optical fibers to detect heavy metals in environment: generalities and case studies. In: InTech, pp. 427–457. Chapter 14.

García-Jareño, J.J., Navarro-Laboulais, J., Vicente, F., 1996. Electrochemical study of nafion membranes/Prussian blue films on ITO electrodes. Electrochim. Acta 41 (17), 2675–2682.

Grant, S.A., Bettencourt, K., Krulevitch, P., Hamilton, J., Glass, R., 2001. In vitro and in vivo measurements of fiber optic and electrochemical sensors to monitor brain tissue pH. Sens. Actuators B, Chem. 72 (2), 174–179.

Hanf, S., Keiner, R., Yan, D., Popp, J., Frosch, T., 2014. Fiber-enhanced Raman multigas spectroscopy: a versatile tool for environmental gas sensing and breath analysis. Anal. Chem. 86, 5278–5285.

Haus, R., Schaefer, K., Bautzer, W., Heland, J., Moseback, H., Bittner, H., Eisermann, T., 1994. Mobile Fourier transform infrared spectroscopy monitoring of air pollution. Appl. Opt. 33 (24), 5682–5689.

Horiguchi, T., Shimizu, K., Kurashima, T., Tateda, M., Koyamada, Y., 1995. Development of a distributed sensing technique using Brillouin scattering. J. Lightwave Technol. 13, 1296–1302.

Hyfantins, G.J., Watts, J.W., Finnegan, T.P., 1998. Field applications of a portable luminoscope for hazardous screening. In: Vo-Dinh, Tuan, Spellicy, Robert L. (Eds.), The International Society for Optical Engineering: Environmental Monitoring and Remediation Technologies. In: Proceedings of SPIE, vol. 3534, Boston, Massachusetts, pp. 92–99.

Jin, W., Ho, H.L., Cao, Y.C., Ju, J., Qi, L.F., 2013. Gas detection with micro- and nano-engineered optical fibers. Opt. Fiber Techol. 19 (6), 741–759.

Kagann, R.H., Wang, C.D., Chang, K.-L., Lu, Ch.-H., 1998. Open-path FTIR measurement of criteria pollutants and other ambient species in an industrial city. In: Vo-Dinh, Tuan, Spellicy, Robert L. (Eds.), The International Society for Optical Engineering: Environmental Monitoring and Remediation Technologies. In: Proceedings of SPIE, vol. 3534, Boston, Massachusetts, pp. 140–149.

Kasik, I., Mrazek, J., Martan, T., Pospisilova, M., Podrazky, O., Matejec, V., Hoyerova, K., Kaminek, M., 2010. Fiber-optic pH detection in small volumes of biosamples. Anal. Bioanal. Chem. 398, 1883–1889.

Kee, H.H., Lees, G.P., Newson, T.P., 2000. All-fiber system for simultaneous interrogation of distributed strain and temperature sensing by spontaneous Brillouin scattering. Opt. Lett. 25, 695.

Korposh, S., Lee, S.-W., 2009. Fabrication of sensitive fibre-optic gas sensors based on nano-assembled thin films. In: Optical Fiber New Developments. In: InTech, Croatia.

Kumar, P.S., Vallabhan, C.P.G., Nampoori, V.P.N., Pillai, V.N.S., Radhakrishnan, P., 2002. A fibre optic evanescent wave sensor used for the detection of trace nitrites in water. J. Opt. A, Pure Appl. Opt. 4 (3), 247.

Lam, K.Y., Afromowitz, M.A., 1995a. Fiber optic epoxy composite cure sensor. 1. Dependence of refractive index of an autocatalytic reacton epoxy system at 850 nm on temperature and extend of cure. Appl. Opt. 34 (25), 5635–5638.

Lam, K.Y., Afromowitz, M.A., 1995b. Fiber optic epoxy composite cure sensor. 2. Performance characteristics. Appl. Opt. 34 (25), 5639–5644.

Lavarioti, M., Kostarelos, K., Pourjabbar, A., Ghandehari, M., 2014. In situ sensing of subsurface contamination – part I: near-infrared spectral characterization of alkanes, aromatics, and chlorinated hydrocarbons. Environ. Sci. Pollut. Res. 21, 5849–5860.

Logergan, M.C., Severin, E.J., Doleman, B.J., Beaber, S.A., Grubbs, R.H., Lewis, N.S., 1996. Array – based vapor sensing using chemically sensitive. Carbon black – polymer resistors. Chem. Mater. 8, 2298.

Lu, R., Mizaikoff, B., Li, W.W., Qian, C., Katzir, A., Raichlin, Y., Sheng, G.P., Yu, H.Q., 2013. Determination of chlorinated hydrocarbons in water using highly sensitive mid-infrared sensor technology. Sci. Rep. 3, 2525.

Luna-Moreno, D., Monzón-Hernández, D., Villatoro, J., Badenes, G., 2007. Optical fiber hydrogen sensor based on core diameter mismatch and annealed Pd–Au thin films. Sens. Actuators B, Chem. 125 (1), 66–71.

Mallard, W.G., Miller, J.H., Smyth, K.C., 1982. Resonantly enhanced two-photon photoionization of NO in an atmospheric flame. J. Chem. Phys. 76, 3483–3492.

Marquardt, B.J., Cullum, B.M., Saw, T.J., Angel, S.M., 1997. Fiber-optic probe for determining heavy metals in solids based on laser – induced plasmas. Proc. SPIE 3105, 203–212.

McDonagh, C., Burke, C., MacCraith, B., 2008. Optical chemical sensors. Chem. Rev. 108 (2), 400–422.

Meier, U., Kohse-Hoinghaus, K., 1987. REMPI detection of CH_3 in low-pressure flames. Chem. Phys. Lett. 142, 498–502.

Mendoza, E., Kempen, C., Esterkin, Y., Sun, S., 2013. Distributed fiber optic fuel leak detection system. Photonic applications for aerospace, commercial, and harsh environments IV. In: Kazemi, Alex A., Kress, Bernard C., Thibault, Simon (Eds.), Proc. of SPIE, vol. 8720, 872005.

Mishra, S.K., Kumari, D., Gupta, B.D., 2012. Surface plasmon resonance based fiber optic ammonia gas sensor using ITO and polyaniline. Sens. Actuators B, Chem. 171–172, 976–983.

Mopsik, F.I., Cang, S.S., Hunston, D.L., 1989. Dielectric measurements for cure monitoring. Mater. Eval. 47 (4), 448.

Nakamoto, T., Fukunishi, K., Moriizumi, T., 1990. Identification capability of odor sensor using quartz-resonator array and neural-network pattern recognition. Sens. Actuators B, Chem. 1, 473.

Nakamoto, T., Fukuda, A., Moriizumi, T., Asakura, Y., 1991. Improvement of identification capability in an odor-sensing system. Sens. Actuators B, Chem. 3, 221.

Nesselrodth, D.R., Baer, T., 1994. Cyclic ketone mixture analysis using $2 + 1$ resonance enhanced multiphoton ionization mass spectrometry. Anal. Chem. 66, 2497–2504.

Nesselrodth, D.R., Potts, A.R., Baer, T., 1995. Observation of ethyl-substituted cyclohexanone and spectroscopy. J. Phys. Chem. 99, 4458–4465.

Ohno, H., Naruse, H., Kihara, M., Shimada, A., 2001. Industrial applications of the BOTDR optical fiber strain sensor. Opt. Fiber Technol. 7, 45–64.

Ohodnicki, P.R., Buric, M.P., Brown, Th.D., Matranga, Ch., Wang, C., Baltrus, J., Andio, M., 2013. Plasmonic nanocomposite thin film enabled fiber optic sensors for simultaneous gas and temperature sensing at extreme temperatures. Nanoscale 5, 9030–9039.

Orellana, G., Haigh, D., 2008. New trends in fiber-optic chemical and biological sensors. Current Anal. Chem. 4, 273–295.

Orellana, G., Moreno-Bondi, M.C., 2002. From molecular engineering of luminescent indicators to environmental analytical chemistry in the field with fiber-optic (bio)sensors. In: 15th Optical Fiber Sensors Conference Technical Digest.

Pamukcu, S., Cetisli, F., Texier, S., Naito, C., Toulouse, J., 2006a. Dynamic strains with Brillouin scattering distributed fiber optic sensor. GeoCongress 187, ASCE, 31–36.

Pamukcu, S., Texier, S., Toulouse, J., 2006b. Advances in water content measurement with distributed fiber optic sensor. GeoCongress 187, ASCE, 7–12.

Phillips, C., Jakusch, M., Steiner, H., Mizaikoff, B., Fedorov, A.G., 2003. Model-based optimal design of polymer-coated chemical sensors. Anal. Chem. 75, 1106–1115.

Regan, F., 2014. Sample matrix effects on measurements of chlorinated hydrocarbons using a fiber-optic infrared sensor. In: Instrumentation Science and Technology, vol. 42. Taylor & Francis Group, pp. 1–14.

Ricciardi, A., Consales, M., Quero, G., Crescitelli, A., Esposito, E., Cusano, A., 2013. Lab-on-fiber devices as an all-around platform for sensing. Opt. Fiber Technol. 19, 772–784.

Russwurm, G.M., Childers, J.W., 1996. FT-IR open-path monitoring guidance document. EPA/600/R-96/040.

Savoy, S., Lavigne, J.L., Yoo, J.S.-J., Wright, J., Rodriguez, M., Goodey, A., McDoniel, B., McDevitt, J.T., Anslyn, E.V., Shear, J.B., Ellington, A., Neikirk, D.P., 1998. Solution – based analysis of multiple analytes by a sensor array: toward the development of an "electronic tongue". In: Büttgenbach, Stephanus (Ed.), The International Society for Optical Engineering: Chemical Microsensors and Applications, vol. 3539, Boston, Massachusetts, pp. 17–26.

Shang, Q.Y., Moreno, R.D., Disselkamp, R., Bernstein, E.R., 1993. Rydberg spectra of diethyl ether, diisopropyl ether, and methyl vinyl ether: analysis of the torsional motion. J. Chem. Phys. 98, 3703–3712.

Shao, L.-Y., Yin, M.-J., Tam, H.-Y., Albert, J., 2013. Fiber optic pH sensor with self-assembled polymer multilayer nanocoatings. Sensors 13 (2), 1425–1434.

Shurmer, H.V., Corcoran, P., Gardner, J.W., 1991. Integrated arrays of gas sensors using conducting polymers with molecular sieves. Sens. Actuators B, Chem. 4, 29.

Sin, C.H., Tembreull, R., Lubman, D.M., 1984. Resonant two-photon ionization spectroscopy in supersonic beams for discrimination of disubstituted benzenes in mass spectrometry. Anal. Chem. 56, 2776–2781.

Taylor, R., Brunnert, J., 1998. Fugitive emission monitoring with open-path FTIR at times beach, Missouri, City Park. In: Vo-Dinh, Tuan, Spellicy, Robert L. (Eds.), The International Society for Optical Engineering: Environmental Monitoring and Remediation Technologies. In: Proceedings of SPIE, vol. 3534, pp. 2–8.

Texier, S., Pamukcu, S., Toulouse, J., 2005. Advances in subsurface water-content measurement with a distributed Brillouin scattering fibre-optic sensor. In: Proc. of SPIE, vol. 5855, OFS-17, Bruges, Belgium, pp. 555–558.

Tjossem, P.J.H., Cool, T.A., 1983. Detection of atomic hydrogen in flames by resonance four-photon ionization at 365 nm. Chem. Phys. Lett. 100, 479–483.

Virag, P., Kricks, R., 1998. In: Vo-Dinh, Tuan, Spellicy, Robert L. (Eds.), Environmental Monitoring and Remediation Technologies. In: Proceedings of SPIE, vol. 3534, Boston, Massachusetts, pp. 187–193.

Vo-Dinh, T., 1978. Multicomponent analysis by synchronous luminiscence spectrometer. Anal. Chem. 50, 396–401.

Vo-Dinh, Tuan, 1989. Chemical Analysis of Polycyclic Aromatic Compounds. Wiley, New York.

Vo-Dinh, T., Gammage, R.B., Hawthorne, A.R., Thorngate, J.H., 1979. Recent advances in synchronous luminescence spectroscopy for trace organic analysis. In: Proceedings of the 9th Materials Research Symposium. National Bureau of Standards Special Publications, Gaitherburg, MD, pp. 679–684.

Wallace, P.A., Yang, Y., Uttamlal, M., Campbell, M., Holmes-Smith, A.Sh., 1996. Characterisation of a quasi-distributed optical fibre chemical sensor. In: Proceedings of SPIE: Fiber Optic Sensors V, Beijing, China, pp. 103–108.

Walsh, J.E., MacCraith, B.D., Meaney, M., Vos, J.G., Regan, F., Lancia, A., Artioushenko, V.G., 1995. Midinfrared Fiber Sensor for the In-Situ Detection of Chlorinated Hydrocarbons.

Wolfbeis, O.S., 2000. Fiber-optic chemical sensors and biosensors. Anal. Chem. 72, 81R.

Wolfbeis, O., 2004. Optical Sensors. Industrial Environmental and Diagnostic Applications. Springer Series on Chemical Sensors and Biosensors. Springer-Verlag, Berlin, Heidelberg.

Yildirim, N., Long, F., He, M., Shi, H.-C., Gu, A.Z., 2014. A portable optic fiber aptasensor for sensitive, specific and rapid detection of bisphenol-A in water samples. Environ. Sci.: Proc. Impacts 16 (6), 1379–1386.

Zhang, H., Kulkarni, A., Kim, H., Woo, D., Kim, Y.-J., Hong, B.H., Choi, J.-B., Kim, T., 2011. Detection of acetone vapor using graphene on polymer optical fiber. J. Nanosci. Nanotechnol. 11 (7), 5939–5943.

Zwicker, J.O., Ringler, E., Kagann, R., 1998. Monitoring of CO flare emissions using open-path Fourier transform infrared technology. In: Vo-Dinh, Tuan, Spellicy, Robert L. (Eds.), Proceedings of SPIE – The International Society for Optical Engineering: Environmental Monitoring and Remediation Technologies.

CHAPTER 7

Advances and Challenges in Underground Sensing

Suk-Un Yoon*, Andrew Markham[†], Niki Trigoni[†], Traian E. Abrudan[†], Orfeas Kypris[†], Christian Wietfeld[‡]

*Samsung Electronics, Suwon, South Korea
[†]University of Oxford, Oxford, United Kingdom
[‡]Communication Networks Institute (CNI), TU Dortmund University, Dortmund, Germany

SUBCHAPTER 7.1

Wireless Signal Networks for Global Underground Sensing

Suk-Un Yoon
Samsung Electronics, Suwon, South Korea

7.1.1 INTRODUCTION

Electromagnetic signal propagation has been widely used in soil science as a means of geo-sensing and characterization of soil properties. For example, moisture content and salinity of soil have been measured using different techniques such as remote electromagnetic induction sensors, four-electrode sensors, time domain reflectometric (TDR) sensors, and ground penetrating radar (GPR) (Yoon et al., 2012). The subsurface monitoring has been achieved through destructive and nondestructive techniques including direct soil sampling, probing and soundings, and using geophysical mapping tools. Even though the aforementioned techniques have been successfully applied to characterize the state of subsurface media, there are challenges associated with them, including difficulties in the real-time event data reporting, tracking geo–hazard such as landslides and slope failures, and deployment challenges, specifically the requirement of wired connections in some techniques.

Underground Sensing.
DOI: http://dx.doi.org/10.1016/B978-0-12-803139-1.00007-2
Copyright © 2018 Elsevier Inc. All rights reserved.

To resolve the challenges associated with the existing techniques, wireless sensors have been used recently for subsurface monitoring application. Wireless sensor nodes in general underground monitoring applications, however, provide only point measurements and are incapable of providing global measurement for characterization of subsurface medium. This chapter presents a novel concept of wireless signal networks (WSiNs) to bridge the gap between real-time monitoring and global measurements. The proposed wireless signal networks use the signal strength variation in the soil medium as the main indicator of an underground event or a physical change in soil properties. The wireless signal networks are based on the received radio signal strength variations on the distributed underground sensor nodes to monitor and characterize the sensed area. This chapter introduces potential wireless signal networks applications. Due to the location of underground sensor nodes, the installation and management of underground signal networks has more constraints than aboveground deployment. This chapter presents challenges and potential solutions on the deployment and management of wireless signal networks with underground sensor nodes.

In the underground sensing applications, the monitored events can be characterized by the localization of the deployed sensors. Based on the localization, sensors estimate their locations using the received signal strength from a set of seed sensors. The soil properties and conditions such as degree of compaction, gradation, and salinity level can potentially be monitored based on the received signal strength (RSS) information. The received signal strength with respect to distance from the source node is used to classify different subsurface events. The received signal strength variation can be classified by a minimum distance classifier based on Bayesian decision rule. To apply the Bayesian theory on wireless underground sensor networks, the computational power is an important factor due to the limited computational power of wireless sensor nodes. In the chapter, a *window-based minimum distance classifier* is introduced as an efficient classifier for wireless signal networks with high accuracy and less computation.

7.1.2 WIRELESS SIGNAL NETWORKS

7.1.2.1 Concept of Wireless Signal Networks

The proposed wireless signal networks (WSiNs) are based on the received radio signal strength variation as the main indicator of an event in the physical domain (subsurface environments) of the wireless signal network. Soil

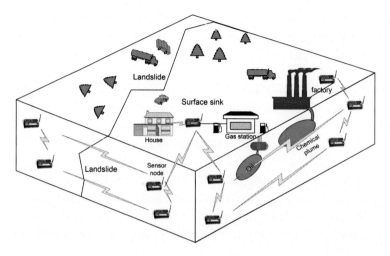

Figure 7.1.1 The concept of wireless signal networks (WSiNs) for subsurface monitoring

properties such as density, water, and mineral content are known to affect electromagnetic signal propagation. If these physical properties of the host soil change during the evolution of a subsurface event, the soil properties will affect the radio transmission and the received signal strength on the event area. The concept of subsurface monitoring with wireless signal networks for detection of landslide and chemical plume/oil leakage is shown in Fig. 7.1.1. The proposed wireless signal networks can provide both point sensing and global sensing between the underground sensor nodes, and real time measurements which can be used for subsurface monitoring applications.

7.1.2.2 Subsurface Monitoring Applications

To monitor and characterize subsurface geo-events, wired or wireless sensor nodes are widely used. The proposed WSiNs can be used may applications accomplished by existing wired and wireless sensor networks. The potential applications with wireless signal network based on the existing monitoring applications with wired/wireless sensor networks are listed as follows:

- Mine Application. A mine requires underground monitoring system and a good place to deploy wireless underground sensor nodes. On the mine application, the radio signal propagation can be either in soil or in a tunnel of a mine structure. Wireless signal networks can assist rescuers in searching and detecting survivors in a collapsed mine. In a mining case study, safety assurance and rescue communication systems in high-stress environments are presented in Misra et al. (2010). A fixed infrastructure deployed above ground in systems can provide stable long-distance communication in harsh mine environments. The data can be sent over wired cables such as twisted pair and coaxial cables, or via optical fiber cables. A coal mine personnel global positioning system based on the location information of sensor nodes is presented in Liu et al. (2010).

- Agriculture Application. To monitor subsurface soil conditions, such as water and mineral content, and to provide the collected data for appropriate irrigation and fertilization, WSiNs can be used for the agriculture application. With the locally collected data from sensor nodes, each sprinkler could be activated based on local sensors rather than irrigating the whole field in response to local sensing data. In Hu et al. (2010), applications and experiences of deploying sensor networks in a sugar farm are introduced.

- Structure Health Monitoring. Wireless signal networks can be used to monitor the structural health of underground components of an integrated structure such as building foundation, bridge pier, or dam's foundation. A customized wireless sensor network for monitoring structural health is introduced and called DuraNode in Park et al. (2005). To monitor the structural health of a civil structure, a wireless sensor network is used in monitoring theater's structure in Cho et al. (2008).

- Underground Infrastructure Monitoring. Monitoring pipes, electrical wiring, and liquid underground storage tanks is a good practical application of wireless signal networks. Monitoring the water pipe under the road is a good example of subsurface infrastructure monitoring (Trinchero et al., 2010, 2009a, 2009b; Trinchero and Stefanelli, 2009). As a water distribution monitoring system, a wireless sensor network is configured in Lin et al. (2008).

- To estimate the underground fault location, an automatic underground distribution fault location system with a one-line diagram from another intranet web application for the estimated fault location feeder is introduced in Sabin et al. (2009). Wireless signal networks can be used for underground oil tank monitoring as well. An automatic oil leak detection system of an underground tank is introduced in Fujita et al. (1999), which could be a good potential application of wireless signal networks.

- Natural Disaster. In the natural disaster applications, the subsurface hazards include a landslide, an earthquake, and monitoring an active fault zone that involve shifting and moving of earth masses and are characterized by the localization of underground sensor nodes. WSiNs can detect landslides, where the concept is introduced as Senslide in Sheth et al. (2005). For slope monitoring applications, the design, development, and evaluation of a tilt and soil moisture sensor network are introduced in de Dios et al. (2009). A volcano monitoring sensor network architecture, where the network consists of 16 sensor nodes, each with a microphone and seismometer, collecting seismic and acoustic data on volcanic activity, is introduced in Werner-Allen et al. (2006) to monitor volcanic eruptions.

- Environmental Monitoring. Near rivers and aquifers where chemical runoff could contaminate drinking water supplies, monitoring the presence and concentration of various toxic substances on soil are important. Sensor nodes can be used for monitoring outdoor water quality such as salinity and the level of the underground water table, and the deployment of a wireless sensor network is investigated for the water quality monitoring (Dinh et al., 2007).

- Subway security monitoring. Rail-based mass transit systems are vulnerable to criminal acts such as vandalism and terrorism, and wireless sensor/signal networks can be used for subway security monitoring and underground tunnel and rail monitoring applications. Sensor networks as integrated surveillance systems for the physical security of metro railways are introduced in Bocchetti et al. (2009).

- Sport Field Application. Monitoring sports fields, where they can be used to monitor soil conditions at golf courses, soccer fields, baseball fields, and grass tennis courts, can be a good practical application of wireless signal networks.

- Security Application. For home and commercial security, where sensors could be deployed underground around the perimeter of a building in order to detect intruders, the data collected from sensor nodes can be useful information for the security application. Wireless pressure sensors deployed at a shallow depth along the length of a border could provide information to alert authorities to illegal crossings.

- Home Application. Wireless signal networks can be used for home security and gardening applications. A sensor deployed in shallow underground around house can detect intruder by the pressure sensor or received signal variations and generate alarm signal. Monitoring both local moisture level from moisture sensor and global soil moisture level from received signal strength can be used to control sprinklers. Monitoring and preventing overflow of a septic tank in home is also a good potential application of wireless signal networks, where the leakage from the tank can be detected from received signal strength variation.

7.1.2.3 Subsurface Monitoring of WSiNs

When wireless signal networks adopt existing sensing technologies and devices such as light sensor, accelerometer, moisture sensor, ammonia/gas detection sensor, pressure sensor, temperature sensor, humidity sensor, seismic sensor, fiber optical sensor, etc., the wireless signal networks can accomplish more accurate subsurface monitoring applications from point and global sensing. The sensing devices of the traditional sensor node such as moisture sensor or temperature sensor could provide point sensing from the subsurface sensing. Thus, the traditional sensing devices have difficulties to monitor global subsurface hazard and characteristics due to the limited subsurface sensing capability of the sensor's equipment. With the proposed wireless signal networks, the real-time global subsurface monitoring can be accomplished. The wireless signal network uses the received signal strength variation as the main indicator of an event in the physical domain as a global medium sensing. Soil properties such as density, water, and mineral content are known playing important roles in the received signal attenuation. When the physical properties of the host soil change during the evolution of a geo-event, the changed properties affect the transmission quality and received signal strength within the region of the event. With the monitoring of soil properties and conditions such as compaction, gradation, pH level, and salinity, potential applications can be subsurface monitoring for oil leakage from subsurface reservoirs, water leakage from underground pipelines, seepage in earth dams, and sea water intrusion near sea shore. With the monitoring of soil movement such as perturbation of earth masses, potential applications can be landslide, land subsidence, earthquake, and active fault zone monitoring.

7.1.3 DEPLOYMENT CHALLENGES OF WSiNs

The deployment of wireless sensor nodes for WSiNs in challenging environments such as underground incurs new issues due to the install location. This chapter provides the challenging issues and potential solutions of WSiNs deployment.

7.1.3.1 Installation and Management

Digging a hole to install sensor nodes and finding underground sensor nodes are much more difficult compared to aboveground networks installation and management. When managing the underground sensor networks, the sensor can be easily damaged when digging even when the sensor position is traceable. Thus, the network and topology should be designed to minimize installation and management costs before deploying underground sensor networks. For example, the sensor nodes having high energy consumption can be deployed in shallow depth which will be easier for the management. Replacing batteries of underground sensors is a challenging issue due to the underground environments. To avoid replacing a battery, high capacity battery should be used and power saving sensor operation such as sleep mode should be applied. When sensor nodes have high capacity batteries such as D batteries (15 000 mAh) comparing to with the widely used AA batteries (2000~2400 mAh), the sensor nodes' expected lifetime can be extended to about 7 times longer. To minimize energy consumption on underground sensor nodes' operation, sleep mode and long data reporting interval can be helpful. To save energy, the sensor nodes sleep during the sniff period and then wake up at the end of the sniff period to detect clear channel and initiate data transmission or listen for data (Crossbow Technology a). It is mathematically estimated that the sensors' expected lifetime is about 4 years when two D batteries are used, packets are sent once or twice per hour, with 1 second sniff period. The average power consumption is based on Crossbow's power calculation spreadsheet (Crossbow Technology b; Crossbow Technology c) for the estimation. If the batteries are attached in parallel more and more, the expected sensors' lifetime can be extended as well, depending on the number of batteries.

7.1.3.2 Underground Radio Propagation and Communication Distance

In air communication, the received signal strength is decaying with the square of the distance, which is described by the Friis equation. In geomedia, the signal strength attenuation is much faster than in air communication due to the transmitted power loss in the propagation medium. For lossy dielectrics such as soil medium, the permittivity and electrical conductivity are main signal attenuation factors and dependent on the operating frequency. These two properties (permittivity and electrical conductivity) represent the displacement (polarization) current and the

conduction current, respectively. These currents incur the power losses on
the electromagnetic signal propagation in the soil medium. An accurate
and simple subsurface wireless signal propagation model for low-power devices such as wireless sensor nodes at a frequency of 2.4 GHz is developed
and its performance is evaluated using real MICAz wireless sensor nodes
(Yoon et al., 2011a). The underground medium in the underground radio
propagation model has permittivity $\varepsilon = \varepsilon_r\varepsilon_0$, permeability μ_0, and electrical conductivity σ, where ε_0 (8.85×10^{-12} F/m) is the permittivity and μ_0
($4\pi \times 10^{-7}$ H/m) is the permeability of air. In the derivation of the underground radio propagation model, the source is imagined to be a vertical
electric dipole of length ds and carrying a current I. The received signal
strength in geo-media (soils) is defined as follows (Yoon et al., 2011a):

$$P_r(d) = \frac{A_{eff}\cos\theta_\eta}{2|\eta|}\left(\frac{Ids\mu_0\omega}{4\pi}\right)^2\frac{e^{-2\alpha d}}{d^2}$$

$$= K\frac{e^{-2\alpha d}}{d^2} \tag{7.1.1}$$

where A_{eff} is the effective antenna area of the receiver, $\eta = \sqrt{\frac{i\omega\mu}{\sigma+i\omega\varepsilon}}$ is
the intrinsic wave impedance of the medium, θ_η is the phase angle of
the intrinsic impedance $\eta = |\eta|e^{j\theta_\eta}$, ω is angular frequency, d is the dis-
tance from the source, α is the attenuation constant of the medium, and
$K = \frac{A_{eff}\cos\theta_\eta}{2|\eta|}(\frac{Ids\omega\mu_0}{4\pi})^2$. The attenuation constant α characterizing the soil
properties is defined as follows:

$$\alpha = \text{Re}(\gamma) = \omega\sqrt{\frac{\mu\varepsilon}{2}\left[\sqrt{a+\left(\frac{\sigma}{\omega\varepsilon}\right)^2}-1\right]} \tag{7.1.2}$$

where γ is the propagation constant, ω is angular frequency, ε is the permittivity, and μ is the magnetic permeability of the soil, and σ is the electric
conductivity of the medium. In case of 2.4 GHz MICAz and 433 MHz
MICA2, the communication radii are relatively short, namely 20 and 30 cm
in wet clay type soil (electrical conductivity = 780 mS, relative permittivity
= 30). If the underground sensors adopt a low frequency, the communication radius can be extended more effectively. There are candidates of low
frequency bands for underground communication such as low frequency
ISM (industrial, scientific, and medical) bands of 125 and 134 kHz, as well
as 6.78 and 13.56 MHz. ISM bands are reserved radio bands internationally

for the use of radio frequency for industrial, scientific, and medical purposes. Based on the theoretical estimation with Eq. (7.1.1), wireless sensor nodes can achieve 1.6~2.3 m communication with 13.56 and 6.78 MHz bands, and 31~32 m communication with 134 and 125 kHz bands even in wet clay underground considering 1 W transmission power and the same MICAz's antenna gain.

7.1.4 SUBSURFACE EVENT DETECTION AND CLASSIFICATION

To detect and classify subsurface events, the received signal strength information can be used and a minimum distance classifier based on Bayesian decision theory is proposed for the subsurface event classification (Yoon et al., 2012). To apply Bayesian theory to runtime wireless underground sensor networks, the computational power should be considered because wireless sensor nodes have limited computational power. In case of a MICAz sensor node, it has a low-power microcontroller (ATmega128L) which speed is 4 or 7 MHz (Crossbow Technology d). The heavy computational works on the low-power CPU cannot be performed, when it sends or receives data due to its hardware and software architecture. In wireless signal networks, the received signal strength information is collected while sensor nodes send and receive data. The MICAz wireless sensor consumes power in each mode as follows: 0 dBm TX, 17.4 mA; −10 dBm TX, 11 mA; and receive mode 19.7 mA (Crossbow Technology d). In our subsurface event detection and classification experiments, 0 dBm TX power is used. In this chapter, a window-based minimum distance classifier is designed for wireless sensor networks considering the energy and computational efficiency while maintaining high accuracy with less computation. The proposed window-based classifier has two steps: event detection (based on the deviation criterion) and event classification (minimum distance classification). When a subsurface event is detected, the window-based classifier classifies the geo-events on the event occurring regions called a classification window. The proposed event detection and window-based classification is evaluated with a water leakage experiment.

7.1.4.1 Event Detection and Window Selection

Due to the limited battery and computational power on wireless sensor node, the event classification at every sensing time is not reasonable. When the wireless sensor node sends or receives data, it cannot perform the event classification based on the probabilities. So, if the received signal strength does not change from previously collected data at a specific location, there is no new geo-hazard or event in subsurface environments and the event classification is not required. Therefore, it is important to detect the event occurring region for a simple classification such as two-category case (ω_1, event; ω_2, no-event) where ω_1 is assigned to binary value 1 and ω_2 to binary value 0. The subsurface event ω_1 at kth position ($1 \leq k \leq N$) is detected by the received signal strength deviation from previous M sample average at the nth sensing time t_n as follows:

$$\left| x_k(t_n) - \left[\sum_{j=n-1}^{n-M} x_k(t_j) \right] / M \right| \geq \varsigma \tag{7.1.3}$$

where ς is the deviation criterion. The deviation criterion can be empirically determined (e.g., 3~5 dBm) with the measured data where RSS variation in soil is small in normal case.

7.1.4.2 Event Classification on Selected Window

For the subsurface event classification in underground wireless signal networks, there are N positions to sense geo-events (Yoon et al., 2012). Let $\{\omega_1, \omega_2, ..., \omega_c\}$ be the finite set of c states of geo-events. $P(\omega_j)$ represents the prior probability where the event is in state ω_j. The measurement variability is expressed in probabilistic terms as x that is considered a random variable the distribution of which depends on the event state expressed as $p(x|\omega_j)$. With x where $P(\omega_i|x)$ is greater than $P(\omega_j|x)$, there is a higher possibility that the true state of the event is ω_i. Then, choosing ω_i will minimize the error probability. The scalar random variable x is replaced by the feature vector $\vec{x}$ with more than one measurement, where $\vec{x}$ is in an N-dimensional Euclidean space R^N. The posterior probability $P(\omega_i|\vec{x})$ can be calculated from $p(\vec{x}|\omega_i)$ by Bayes equation:

$$P(\omega_i|\vec{x}) = \frac{p(\vec{x}|\omega_i)P(\omega_i)}{p(\vec{x})} \tag{7.1.4}$$

where $p(\vec{x}) = \sum_{j=1}^{c} p(\vec{x}|\omega_j)P(\omega_j)$.

The Bayes formula represents the fact that the prior probability $P(\omega_j)$ can be converted to *a posteriori* probability $P(\omega_i|\vec{x})$ by observing the value of $\vec{x}$. In the Bayes decision rule, the role of the posterior probabilities is emphasized and the evidence factor $p(\vec{x})$ is unimportant when making a decision. In event classifications, a set of discriminant functions $g_i(\vec{x})$ is introduced, where $i = 1, ..., c$. The classifier is said to assign a vector $\vec{x}$ to class ω_i if $g_i(\vec{x}) > g_j(\vec{x})$ for all $j \neq i$. For the minimum error classification, discriminant functions are expressed as follows (Duda et al., 2000):

$$g_i(\vec{x}) \triangleq P(\omega_i|\vec{x}) = \frac{p(\vec{x}|\omega_i)P(\omega_i)}{\sum_{j=1}^{c} p(\vec{x}|\omega_j)P(\omega_j)}$$

$$\triangleq p(\vec{x}|\omega_i)P(\omega_i)$$

$$\triangleq \ln p(\vec{x}|\omega_i) + \ln P(\omega_i) \tag{7.1.5}$$

where $\sum_{j=1}^{c} p(\vec{x}|\omega_j)P(\omega_j)$ does not affect on the decision and can be ignored.

The log-scale expression of discriminate functions will be used to develop a minimum distance classifier because it does not affect the decision as well. The measured received signal strength from wireless sensor nodes is assumed to be independent and normally distributed as in Locher et al. (2005). That represents that each measurement is statistically independent and its probability density function is normally distributed. Then, the discriminant function can be expressed with the densities $p(\vec{x}|\omega_i)$ which are multivariate normal, $p(\vec{x}|\omega_i) \sim N(\vec{\mu}_i, \Sigma_i)$, where $\vec{\mu}_i$ is the mean vector and Σ_i is the covariance matrix. The general multivariate normal density in N dimensional is expressed as:

$$p(\vec{x}) = \frac{1}{(2\pi)^{N/2}|\Sigma|^{1/2}} \exp\left[-\frac{1}{2}(\vec{x} - \vec{\mu})^t \Sigma^{-1}(\vec{x} - \vec{\mu})\right] \tag{7.1.6}$$

where $\vec{x}$ is N-dimensional column vector, $\vec{\mu}$ is the N-dimensional mean vector, Σ is the N-by-N covariance matrix, and $|\Sigma|$ and Σ^{-1} are its determinant and inverse, respectively (Duda et al., 2000). With multivariate normal density, the discriminant function can be written as follows:

$$g_i(\vec{x}) = -\frac{1}{2}(\vec{x} - \vec{\mu}_i)^t \Sigma^{-1}(\vec{x} - \vec{\mu}_i) - \frac{M}{2}\ln 2\pi - \frac{1}{2}\ln(\Sigma_i) + \ln P(\omega_i). \tag{7.1.7}$$

When the features are statistical independent and each feature has the same variance σ^2, the covariance matrix is expressed as $\Sigma_i = \sigma^2 I$ where I is the identity matrix. Then, the discriminant functions can be simplified as follows (Locher et al., 2005):

$$g_i(\vec{x}) = -\frac{\|\vec{x} - \vec{\mu}_i\|^2}{2\sigma^2} + \ln P(\omega_i) \tag{7.1.8}$$

where $\| \ \|$ denotes the Euclidean norm, that is,

$$\|\vec{x} - \vec{\mu}_i\|^2 = (\vec{x} - \vec{\mu}_i)^t(\vec{x} - \vec{\mu}_i). \tag{7.1.9}$$

If the prior probabilities $P(\omega_i)$ in the log-scale expression are the same for all c classes, the $\ln P(\omega_i)$ term becomes an unimportant additive constant which can be ignored and the optimum decision rule can be a minimum distance classifier. To classify a feature vector $\vec{x}$, the Euclidean distance $\|\vec{x} - \vec{\mu}_i\|$ from each $\vec{x}$ to each of the c mean vectors is measured, and $\vec{x}$ is assigned to the category of the nearest mean. Then, the subsurface events can be classified according to the training data $(\vec{\mu}_i)$ and measured received signal strengths $(\vec{x})$. The training data can be taken from empirical data or theoretically estimated data such that the event ω_i has training data $\vec{\mu}_i$. The measurement $\vec{x}$ can be classified to the event ω_i that has the highest probability that the true state of the event is ω_i or the minimum distance between $\vec{x}$ and $\vec{\mu}_i$. Based on the minimum distance classifier, the decision boundary for the subsurface events can be determined.

After the subsurface event detection as described in Section 7.1.4.1, the subsurface events are classified on the selected classification window. The minimum error subsurface events classification can be accomplished with the discriminant functions with $\vec{x} = \{x_k\} \in \Psi$ where Ψ is a set of W features in classification window which size is W. The window based selection can reduce the classification error rate. Consider that a particular x_k is observed and decision α_i is taken. When the true state of the event is ω_j, the classification loss is expressed as $\lambda(\alpha_i|\omega_j)$. Because $P(\omega_j|x_k)$ is the probability of the true state with the event ω_j, the expected loss by taking decision α_i on selected window, R_w, is expressed as follows:

$$R_W(\alpha_i|\vec{x}) = \sum_{k=1}^{W} \sum_{j=i}^{c} \lambda(\alpha_i|\omega_j) P(\omega_j|x_k)$$

$$\leq \sum_{k=1}^{N} \sum_{j=i}^{c} \lambda(\alpha_i|\omega_j) P(\omega_j|x_k) = R_N(\alpha_i|\vec{x}_N) \qquad (7.1.10)$$

where R_N is the expected loss on whole range in which $W \leq N$. As shown in Eq. (7.1.10), the expected loss on window based classifier R_W is lower than the one on whole range classifier R_N, where the received signal strength variations in no event region are ignored on the window based classifier.

7.1.5 EVALUATIONS OF WIRELESS SIGNAL NETWORKS

The proposed wireless signal networks for subsurface event detection and classification is evaluated with widely used sensor nodes (i.e., MICAz) due to their stability for the measurements of received signal strength information. The MICAz sensor nodes are consist of the CC2420 RF transceiver that includes a digital direct sequence spread spectrum baseband modem. To avoid interference from Wi-Fi channels, the operating frequency was configured to be 2.48 GHz that is ZigBee channel 26 and nonoverlapping with Wi-Fi 802.11b. Before the evaluations, all wireless sensor nodes are calibrated and selected working in 1~2 dBm error bounds on the received signal strength measurement. A wireless sensor node sends a packet periodically, and the multiple receivers calculate the received signal strength from the received packets and store the date on the flash memory. The save data from buried sensor nodes is retrieved in the laboratory with serial or Ethernet programming boards connected to the desktop.

7.1.5.1 Experiments of Subsurface Event Detection

To demonstrate the functionality of the WSiNs detecting transient changes in host soil within the network domain, three experiments were designed. Based on the event detection rule introduced in Section 7.1.4.1, the subsurface events are detected with the empirically or theoretically determined deviation criterion ζ which is 3–5 dBm in our experiments. By generating transient changes within host soil mass (coarse sand, $D_{50} = 3.3$ mm where D_x is the diameter of the soil particles for which x of the particles are finer) in the large soil box/tank, three subsurface event detection experiments were conducted, where these events were (1) water intrusion, (2) relative density change, and (3) relative motion.

7.1.5.1.1 Evaluation of Water Intrusion Detection

By gradually injecting water between the buried wireless sensor nodes, water intrusion events were simulated. Figs. 7.1.2A–B show the sensor nodes configuration and water intrusion experiment results, respectively. All transceivers were buried at a depth of 20 cm at the locations as shown in Fig. 7.1.2A. The initial water content of the soil was 6% and the first event was started at 17 minutes (elapsed time), where water was injected between the sender and node R2 (Receiver 2). The next event, water injected between the sender and node R4, was simulated at 19 minutes. The last event was initiated at 32 minutes by injecting water between the sender and node R5. There are significant decreases in the received signal strength right after the water injection in all three cases as shown in Fig. 7.1.2B. There is no change in the received signal strength at node R3 (fairly constant), however, where there was no water inject event. After the water

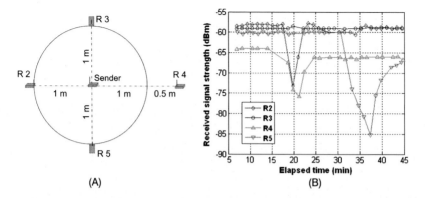

Figure 7.1.2 Evaluation of water intrusion detection: (A) configuration of wireless sensor nodes; (B) the measured received signal strength on each node for water intrusion detection

injection events, the average water content of soil between the sender and R2/R4/R5 was measured as 12%, 11%, and 16%, respectively. The results based on the water content and receive signal strength measurements show that there is the more received signal strength decreases with the higher water content on the propagation medium. The increased water content of soil increased the permittivity and electrical conductivity of the medium which result in the received signal strength decrease. Thus, the magnitude of received signal decrease is proportional to the magnitude of change in the water content comparing nodes R2 and R5. The node R4 has lower received signal strengths comparing with R2/R3/R5 since the node R4 was located at farther distance than the other nodes.

7.1.5.1.2 Evaluation of Relative Density Change Detection

By applying vibration at the soil surface with magnetic vibration equipment, relative density change experiment through soil compaction was conducted. All transceivers were buried at a depth of 20 cm with the sensor node configuration as shown in Fig. 7.1.3A. Each shaded area, between the sender and receivers from node R2 to node R5 in Fig. 7.1.3A, was compacted about 2 minutes starting at 27, 32, 36, and 38 minutes. In the relative density change experiment, the water content of the soil was maintained as 5%. The compactness change of soil is detected by marked drops in received signal strengths at all nodes as shown in Fig. 7.1.3B. The received signal strength decreases may be attributed to the increase in electrical conductivity of the compacted soil medium. Compacted soil has more surface

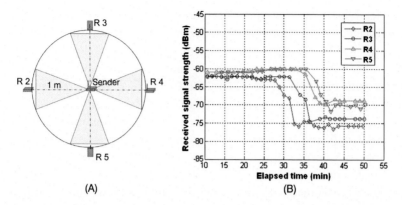

(A)   (B)

Figure 7.1.3 Evaluation of relative density change detection: (A) configuration of wireless sensor nodes for relative density change; (B) the measured received signal strength on each node for relative density change detection

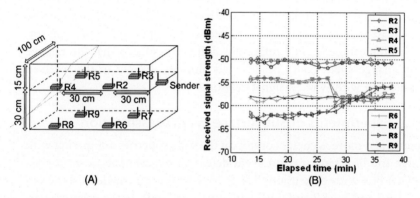

(A)   (B)

Figure 7.1.4 Evaluation of relative motion detection: (A) configuration of wireless sensor nodes for relative motion; (B) the measured received signal strength on each node for relative motion detection

particle contacts compared to the loose soil, and provides more electron flow resulting in higher electrical conductivity of the compacted media. Thus, wireless signal networks can detect the relative density changes, based on that the received signal strength decreases as the soil compactness (relative density) increases.

7.1.5.1.3 Evaluation of Relative Motion Detection

In the experiment of relative motion detection as shown in Fig. 7.1.4A, the sender and the nodes R2, R3, R4, and R5 were installed in the same plane at a depth of 15 cm from the surface, and the nodes R6, R7, R8, and R9 were installed at the same plane at a depth of 45 cm from the surface. The soil mass was held using a wooden plate near nodes R4/R5/R8/R9 while the other sides near the sender node remained continuous. Around 27 minutes from the measurement start, the event triggered by removing the wooden plate. The results of received signal strength measurements are shown in Fig. 7.1.4B. Even after the event, the received signal strength at nodes R2, R3, R6, and R7 remained fairly constant. Because neither the physical properties of the soil nor the distance between the sender and receivers changed within the area, the received signal strength has no change. At nodes R4 and R5 that were directly affected by the event, the received signal strength decreased and remained constant as shown in Fig. 7.1.4B. The received signal strength decease is attributed to a combination of multiple factors such as change in the distance between the sender and receivers, antenna orientation, and density of soil between the sender and receivers. In contrast, nodes R8 and R9 have the received signal strength increases. This increase may be attributed to the reduced mass of soil located on top of these nodes initially. The important point to notice is that the relative motion change event could be detected through changes in the received signal strength with wireless signal networks.

7.1.5.2 Experiments of Subsurface Event Classification

For better and more accurate control on the soil properties comparing with the large soil box/tank used in event detection experiments, a small plastic (PVC) box with dimensions $118 \times 13 \times 13$ cm was made as shown in Figs. 7.1.5A and 7.1.5B and filled with controlled soils. In the event classification experiment, physical soil properties (e.g., soil density) and temporally dynamic variables (e.g., soil water content, salinity, soil temperature, etc.), which affect the electric conductivity and permittivity of the radio propagation medium, were controlled. On the evaluation of the proposed classifier, the classifier was applied into a water leakage experiment where soil properties were controlled where the soil is fine sand ($D_{50} = 0.58$ mm).

In the experiment of subsurface event classification, sensor nodes (S2–S6) sent a packet at every 60 seconds and the receiver node (S1) measured the received signal strength from the received packet. The node S1, which could be a sink node or an intermediate node that can send or receive data, can estimate the location of water leakage event and classify the water contents on the soil media between the transmitters and receiver based on the collected received signal strength information. In the water

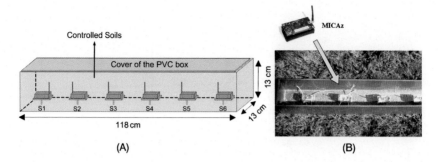

Figure 7.1.5 Details of subsurface event detection and classification experiments: (A) designed PVC box to install sensors (one transmitter and receivers); (B) PVC box with sensors before soil filling

leakage experiment, the water content of the soil in the box was 12% before the water leakage event. The water leakage event was initiated at 26 minutes from the experiment start by injecting water into the regions between 40 and 60 cm from the first node (S1). After the water leakage event, the average volumetric water content of the soil in the box was measured as 5%.

7.1.5.2.1 Reference Data Generation

In the evaluation of event classification, the geo-events can be classified based on training data which could be the measured data or theoretically estimated data. With the underground radio propagation model proposed in Yoon et al. (2011a), the received signal strength with different water contents can be estimated as the reference data (training data for the event classification) which are shown in Fig. 7.1.6. On the estimation based on Eq. (7.1.1), the electric conductivity value was determined from the measured data of the soil samples and the relative permittivity of the soil is estimated between 19 and 30 based on Hubbard et al. (1997).

7.1.5.2.2 Event Classification of Water Leakage Experiment

The time evolution of the received signal strength is presented in Fig. 7.1.7, when the water leakage event was conducted. The nodes at 55 and 95 cm have decreased received signal strength due to high signal attenuation from increased water content after the water leakage event at 26 minutes. Thus, the node at 55 cm generates event detection signal at 26, 27, and 28 minutes and the node at 95 cm generates event detection signal at 28 and 29 minutes based on the event detection rule. In the experiment, the deviation

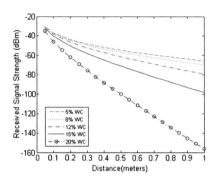

Figure 7.1.6 Received signal strength estimation with different water content

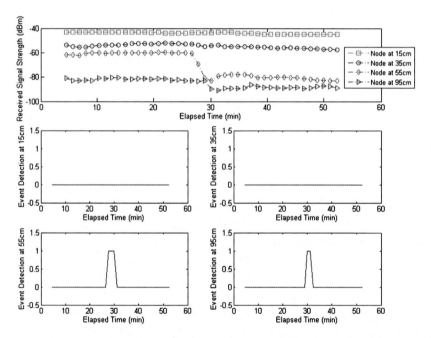

Figure 7.1.7 The received signal strength measurements of water leakage experiment and event detection signals from different sensor nodes located 55 and 95 cm from the sender

criterion ζ is chosen as 3 (dBm) and the previous sensing time M is used as 3 (minutes). As a result, the minimum distance classification will be conducted only from 26th to 29th minute based on the event detection signals.

When water leakage events are detected, the window-based minimum distance classifier classifies the detected event based on the measured data

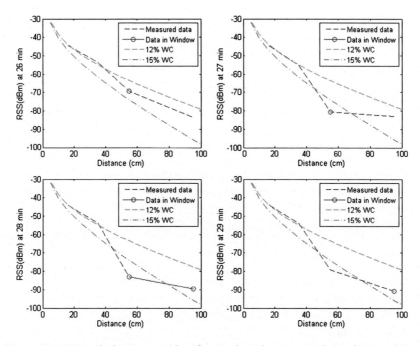

Figure 7.1.8 Water leakage event classification based on received signal strength between 26 and 29 minutes (detected events at 26, 27, and 28 minutes are classified as 15% water leakage event)

by calculating the minimum distance between the detected event and the reference data using Eq. (7.1.8). The measured received signal strength between 26th and 29th minute is shown in Fig. 7.1.8, where the event detection signals were generated. The red circle in the figure represents the data in the classification window that can be used in the minimum distance classification. Based on the window-based minimum distance classifier, the detected events at 26, 27, 28, and 29 minutes are classified as 15% water leakage event as shown in Fig. 7.1.8. If a whole-range minimum distance classification is applied, the detected events are classified as 12% water leakage event at 26 and 27 minutes and 15% water leakage event at 28 and 29 minutes. The window-based minimum distance classification and the whole-range minimum distance classification are compared in Fig. 7.1.9. The proposed event detection method generates four event detections during the water leakage experiment.

When the subsurface events are detected, the window-based minimum distance classifier can detect and classify the water leakage event with

Time (min.)	Event Detection	Window-based MDC		Whole Range MDC	
		Generated Event	Classified Event	Generated Event	Classified Event
26	Node at 55 cm	15% WC	15% WC	15% WC	12% WC
27	Node at 55 cm	15% WC	15% WC	15% WC	12% WC
28	Nodes at 55, 95 cm	15% WC	15% WC	15% WC	15% WC
29	Node at 95 cm	15% WC	15% WC	15% WC	15% WC

Figure 7.1.9 Comparison results of Minimum Distance Classifier (MDC)

100% accuracy; while the whole range minimum distance classifier has 50% classification accuracy. In the computational power comparison on the experiment, the window-based minimum distance classification has 68% less computation than the whole range minimum distance classification.

7.1.6 CONCLUSION

The chapter introduces a novel concept of wireless signal networks (WSiNs) that can be used for real-time and global subsurface monitoring and characterization based on the received signal strength variations between wireless sensor nodes. The received signal strength data from underground sensor nodes are used to characterize the subsurface events with the proposed wireless signal network. On the subsurface characterization, the soil properties affecting the signal propagation on the underground communications are investigated with experiments of real wireless sensor nodes. For evaluations of wireless signal networks to detect subsurface events, three subsurface experiments such as water intrusion detection, relative density change detection, and relative motion detection were conducted using MICAz sensor nodes. For the subsurface events classification, a window based minimum distance classifier is introduced and evaluated. With the water leakage experiment, the proposed window-based minimum distance classifier detects the subsurface water leakage event and classifies the event into different water contents. Based on the theoretical analysis and simulated experiments, the chapter shows the proof of concept of the wireless signal networks to monitor global subsurface environment.

REFERENCES

Bocchetti, G., Flammini, F., Pragliola, C., Pappalardo, A., 2009. Dependable integrated surveillance systems for the physical security of metro railways. In: Proc. of IEEE International Conference on Distributed Smart Cameras.

Cho, S., et al., 2008. Smart wireless sensor technology for structural health monitoring of civil structures. Steel Struct. 8, 267–275.

Crossbow Technology a. Guidelines for wsn design and deployment. http://www.xbow.com.

Crossbow Technology b. http://www.xbow.com/support/Support_pdf_files/XMesh_Avg_Current.xls.

Crossbow Technology c. MICA2 AA battery pack service life test. http://www.xbow.com.

Crossbow Technology d. Products of MICAz, MICA2, and MTS sensor boards. http://www.xbow.com.

de Dios, R.J., Enriquez, J., Mendoza, F.G.V.E.A., Talampas, M.C., Marciano, J.J., 2009. Design, development, and evaluation of a tilt and soil moisture sensor network for slope monitoring applications. In: Proc. of IEEE TENCON 2009.

Dinh, T.L., et al., 2007. Design and deployment of a remote robust sensor network: experiences from an outdoor water quality monitoring network. In: Proc. of 32nd IEEE Conference on Local Computer Networks.

Duda, R.O., Hart, P.E., Stork, D.G., 2000. Pattern Classification, 2nd edn. Wiley-Interscience, New York, NY, USA.

Fujita, S., Fujiwara, K., Sumita, J., Konya, Y., Muroyama, S., 1999. Automatic oil leak detection system for an underground tank. In: Proc. of IEEE Telecommunication Energy Conference.

Hu, W., Le, T.D., Corke, P., Jha, S., 2010. Design and deployment of long-term outdoor sensornets: experiences from a sugar farm. In: Proc. of EEE Pervasive Computing.

Hubbard, S.S., Peterson, J.E., Majer, E.L., Zawislanski, P.T., Williams, K.H., Roberts, J., Wobber, F., 1997. Estimation of permeable pathways and water content using tomographic radar data. Lead. Edge 16, 1623–1628.

Lin, M., Wu, Y., Wassell, I., 2008. Wireless sensor network: water distribution monitoring system. In: Proc. of IEEE RWS 2008.

Liu, Z., Li, C., Ding, Q., Wu, D., 2010. A coal mine personnel global positioning system based on wireless sensor networks. In: Proc. of the 8th World Congress on Intelligent Control and Automation.

Locher, T., Wattenhofer, R., Zollinger, A., 2005. Received-signal-strength-based logical positioning resilient to signal fluctuation. In: Proceedings of the Sixth International Conference on Software Engineering, Artificial Intelligence, Networking and Parallel/Distributed Computing, 2005 and First ACIS International Workshop on Self-Assembling Wireless Networks. Towson, MD, USA, 23–25 May, pp. 396–402.

Misra, P., Kanhere, S., Ostry, D., Jha, S., 2010. Safety assurance and rescue communication systems in high-stress environments: a mining case study. IEEE Commun. Mag..

Park, C., Xie, Q., Chou, P., Shinozuka, M., 2005. DuraNode: wireless networked sensor for structural health monitoring. In: Proc. of IEEE Sensors, pp. 277–280.

Sabin, D.D., Dimitriu, C., Santiago, D., Baroudi, G., 2009. Overview of an automatic underground distribution fault location system. In: Proc. of IEEE Power and Energy Society General Meeting.

Sheth, A., et al., 2005. Senslide: a sensor network based landslide prediction system. In: Proc. of SenSys'05, pp. 280–281.

Trinchero, D., Stefanelli, R., 2009. Microwave architectures for wireless mobile monitoring networks inside water distribution conduits. IEEE Trans. Microw. Theory Tech. 57.

Trinchero, D., Galardini, A., Stefanelli, R., Fiorelli, B., 2009a. Microwave acoustic sensors as an efficient means to monitor water infrastructures. In: Proc. of IEEE IMS 2009.

Trinchero, D., Stefanelli, R., Galardini, A., Fiorelli, B., 2009b. Wireless sensors for a wire-independent analysis of fluid networks. In: Proc. of IEEE Sensors Applications Symposium.

Trinchero, D., Stefanelli, R., Ricardo, M., Cerquera, P., 2010. Design and optimization of the electromagnetic front-end for wireless sensors floating in dissipative media. In: Proc. of IEEE RWS 2010.

Werner-Allen, G., et al., 2006. Deploying a wireless sensor network on an active volcano. IEEE Internet Comput. 10, 18–25.

Yoon, S.-U., Cheng, L., Ghazanfari, E., Pamukcu, S., Suleiman, M.T., 2011a. A radio propagation model for wireless underground sensor networks. In: Proc. of the IEEE Globecom 2011.

Yoon, Suk-Un, Ghazanfari, Ehsan, Cheng, Liang, Wang, Zi, Pamukcu, Sibel, Suleiman, Muhannad T., 2012. Subsurface event detection and classification using wireless signal networks. Sensors 12 (11), 14862–14886.

SUBCHAPTER 7.2

Magneto-Inductive Tracking in Underground Environments

Traian E. Abrudan, Orfeas Kypris, Niki Trigoni, Andrew Markham
University of Oxford, Oxford, United Kingdom

7.2.1 INTRODUCTION

GPS has revolutionized outdoor navigation and tracking, providing highly accurate positioning anywhere in the world. However, as it uses high frequency radio waves, it suffers heavily from multipath and attenuation effects that are caused by buildings as well as thick vegetation, and is completely unusable underground. There is a strong need to be able to position people, equipment, and assets in underground environments without resorting to manual surveying. In this chapter, we discuss how positioning in GPS-denied environments can be accomplished by using low-frequency magnetic fields. These fields have the special property of being able to penetrate most natural media without further attenuation and due to their long wavelengths, do not suffer from multipath effects. However, the generation and detection of magneto-inductive fields is challenging which has limited their uptake. This chapter discusses applications where they have been successfully used and provides an overview of the fundamental approaches to positioning in single and multihop scenarios. Finally, it concludes by outlining challenges and limitations.

7.2.1.1 Approaches to GPS-Denied Tracking

High frequency radio propagation suffers from multipath and attenuation in underground tunnels. In particular, RF cannot penetrate more than a few cm of soil/rock, limiting its use to line-of-sight (LOS) scenarios. This in turn means that a network for positioning in underground mines using RF typically needs to be relatively dense to maintain a high proportion of LOS paths. A number of researchers (e.g., Huang et al., 2010) have attempted to transfer localization techniques which work relatively well in above-ground scenarios to underground environments. One of the most common techniques is the use of ZigBee or WiFi based RSSI measurements. Cypriani et al. (2015) used Wi-Fi measurements in a 70 m underground mine to demonstrate that mean errors in the order of 10 m were possible using RSSI measurements. Their innovations included the use of "capture points" to continually measure channel properties and variations. However, the major limitation with this approach is that the 90% positioning errors exceeded 25 m. ZigBee based time-of-flight measurements in underground tunnels were presented in Bedford and Kennedy (2012), which showed that acceptable ranging errors in the order of a 10 m could be obtained in linear tunnels of lengths up to 200 m. However, the trials were conducted in straight tunnels – performance around corners was not tested. Ultra wideband (UWB) has also been proposed as a means for positioning in underground mines, as it can achieve excellent (cm-level) accuracy in many indoor environments. Nkakanou et al. (2011) obtained UWB measurements in an underground mine and showed that for the LOS case, path loss was around 1.8, whereas for no line of sight, path loss increased greatly to an exponent of 4.0. Similarly the RMS delay spread increased from 1 to 4 ns over a range of only 10 m. The delay spread is caused by arrival of multiple delayed impulses, caused by the rough tunnel surface and reflections.

Other options that have been explored include geomagnetic positioning (Haverinen and Kemppainen, 2011; Park and Myung, 2014), laser map-matching (Shaffer et al., 1992), seismic waves (Squire et al., 2009), acoustic techniques (Pfeil et al., 2015) and dead-reckoning (Hawkins et al., 2006). Acoustic positioning has been used for precise underground positioning next to large machinery, as in Pfeil et al. (2015). Again, however, this technique is LOS based and requires signals from four noncollocated sources to determine the location of the receiver. Optical based techniques, both based on natural mine features and deployed landmarks can provide superior positioning accuracy in the order of a few cm. Apart from being LOS, their main limitation is continuous operation in dusty environments. Geomagnetic positioning is a form of spatial fingerprinting that relies on distortions of the Earth's magnetic field that are caused by metal objects and rocks. With the aid of a good fingerprinting database, it was shown that accuracies in the order of a few meters were obtainable in an extremely deep (1400 m) mine (for details, refer to Haverinen and Kemppainen, 2011). The major advantage of this modality is that it is infrastructureless and relies simply on the preexisting magnetic field distortions. However, its drawbacks of requiring an accurately surveyed fingerprint map, and the need to travel a relatively long distance (ca. 35 m) to localize the target limit its widespread adoption.

7.2.1.2 Magneto Inductive Technology

There is one wireless technology that has the property of being able to penetrate most naturally found materials (soil, water, rock, etc.) with minimal loss, namely magneto-inductive (MI) technology. MI at the scale in underground mines refers to very low frequency (VLF) fields (wavelengths of hundreds of meters or even kilometers). At these wavelengths, the energy is virtually entirely contained within the quasistatic magnetic field and inductive coupling occurs between source and sensor. The field itself is nonpropagating and nonradiative. As such, it cannot suffer from multipath, and attenuation of signal is primarily due to the magnetic permeability of the media through which it passes.

MI communication has had a long history of use in underground environments (see Yarkan et al., 2009 for a comprehensive overview of different underground communication technologies). This was first proposed by Tesla in 1896, although the first real through-the-earth (TTE) communication system was presented by Wadley (1946) in 1946 in a mine in South Africa. A VLF based system was shown by Barkand et al. (2006) which allowed for audio transmission through approximately 100 m of overburden. The drawback of this system was the large size of the antennas, requiring a 20 m loop to be laid out to perform communication. An interesting approach to mine communication is to exploit the large amount of metal rails and wiring to promote signal propagation in the medium field (MF) band (Stolarczyk, 1991). These conductors act as waveguides, allowing for far more efficient propagation than through solid media. MI has also been used as a communication modality for wireless underground sensor networks (e.g., see Sun and Akyildiz, 2010) for applications such as precision agriculture, as opposed to high-frequency RF based propagation (e.g., see Silva and Vuran, 2009; Stuntebeck et al., 2006). To overcome the issues with high path loss, a novel passive waveguide scheme was proposed in Sun and Akyildiz (2010) where intermediate passive resonant loops act as flux concentrators. This helps to reduce the overall path loss, at the cost of a loss in omnidirectionality. A multihop MI based communication system, which helps to address some of the scale limitations of single hop devices was presented in Markham and Trigoni (2012a). This employed triaxial antennas, in contrast to earlier work which used single axis antennas. The advantages of using triaxial antennas in a communication system are two-fold: firstly, antennas do not need to be oriented carefully with respect to one another, and secondly, a novel encoding scheme, termed magnetic vector modulation, can be used to nearly triple the achievable data rate without incurring a power or bandwidth penalty.

MI positioning, which involves measuring characteristics of the established field, has been relatively well researched in a number of areas, including virtual reality and medical positioning. Typically, the focus of these developments has been on very high accuracy (mm) over short ranges (3 m). The seminal paper by Raab et al. (1979) presented a 3D positioning system that could use a single three-axis (triaxial) source to determine both the position and orientation of a triaxial sensor. The key innovation was treating the received signals as three-dimensional vectors with both sign and magnitude. By doing so, they demonstrated that by using spatial transformations and coordinate rotations that the position and orientation of the receiver could be determined by treating the source as three orthogonal dipoles. This was a key insight that provided one of the first motion-capture systems. In particular, MI positioning has found widespread use in medical applications, as it can provide mm level accuracy for guiding catheters and other instruments inside the human body (Hummel et al., 2005). One of the issues that faces MI positioning systems in medical applications is the large amount of distorting (metallic) material that is typically found in an operating theatre. This in itself has led to a number of studies investigating calibration techniques (e.g., see Kindratenko, 2000) to reduce or mitigate distorted positions. Underground MI positioning is a relatively new area of exploration – in this chapter, we outline some recent work in this field and its applications, ranging from positioning in caves to tracking animals underground.

7.2.2 CHANNEL MODEL

The channel relates how transmitted signals appear at a spatially displaced and possibly rotated receiving antenna. It relates how much energy from the transmitter arrives at the receiver and characterizes the impact of the media where the quasistatic signal has been established.

7.2.2.1 Source

A magneto-inductive source is created by modulating the current passing through a coil, generating a magnetic moment. The magnitude of the magnetic dipole moment of a magneto-inductive source is given by

$$m = NIA, \tag{7.2.1}$$

where N is the coil number of turns, I is the current passing through the coil, and A is the coil cross-sectional area. Transmitting coils are typically air-cored as most permeability enhancing materials (e.g., ferrite) saturate under high magnetic fields. It can be seen from Eq. (7.2.1) that for a given cross-sectional former area, the moment can be increased either by adding more turns or increasing the current. Note, however, that for a given wire gauge, these are coupled parameters – increasing the number of turns increases the coil resistance, which makes it more challenging to drive the same amount of current through the coil. Fundamentally, the magnetic moment is thus related to the mass of copper (or other conductor) used to make the transmitting antenna.

This can be shown as follows: consider the DC resistance R of a coil with N turns and circumference l,

$$R = \frac{Nl}{\sigma A}, \tag{7.2.2}$$

where A is the cross-sectional area of the wire. At higher frequencies, where the skin effect becomes significant, one must replace A with the effective cross-sectional area $A_{\text{eff}} \simeq \pi \delta d$, where δ is the skin depth, and d is the wire diameter; this substitution gives the AC resistance. The complex impedance

Z of the coil is defined as

$$Z = R + j\omega L - \frac{j}{\omega C} \qquad (7.2.3)$$

where L and C are the inductance and capacitance of the coil, respectively. When the coil is tuned to resonance such that $\omega = 1/\sqrt{LC}$, the imaginary part disappears and we are only left with the real part, R. At low frequencies, the current I flowing through the circuit due to an applied voltage V_p will thus be equal to

$$I = \frac{V_p}{R} = \frac{V_p \sigma A}{Nl}. \qquad (7.2.4)$$

The loop area of a circular coil is given by $A_l = l^2/2\pi$. Thus, the generated magnetic moment in a coil of N turns with a loop area A_l due to a current I will be

$$m = NIA_l = \frac{V_p \sigma A l}{4\pi} \qquad (7.2.5)$$

which simplifies to

$$m = \frac{V_p \sigma}{4\pi} V = \frac{V_p \sigma \rho_m}{4\pi} m_1 = km_1 \qquad (7.2.6)$$

where V is the volume of one turn of wire, ρ_m is the mass density of the conducting material, $k = \frac{V_p \sigma \rho_m}{4\pi}$ is a constant that incorporates applied voltage, coil geometry and conductor properties, and m_1 is the mass of one turn of wire. Therefore, given a constant applied voltage, and provided the frequency is low enough such that $\delta \gg d/2$, *the generated magnetic moment is linearly dependent on the mass of one turn of conducting material.* The reason why the moment does not increase with N is because increasing the turns also increases the total resistance of the coil, which has a counterbalancing effect.

A simple calculation for determining when this relationship is true can be performed as follows: consider a copper coil with electrical conductivity $\sigma = 5.96 \times 10^7$ S/m, magnetic permeability $\mu = \mu_0 = 4\pi \times 10^{-7}$ H/m, operating at a frequency of $f = 100$ kHz, and with a wire diameter d of 0.2032 mm (corresponding to AWG 32 wire). The skin depth will be

$$\delta = \sqrt{\frac{2}{\sigma \omega \mu}} = 0.20616 \text{ mm}. \qquad (7.2.7)$$

Table 7.2.1 Reference levels for the electric field strength E, magnetic field strength H, and magnetic flux density B, for general public exposure to time-varying electric and magnetic fields (unperturbed rms values). Transmitter magnetic moment m was calculated from B, assuming that the reference level is set at a 1 m and 10 m distance from the transmitting magnetic dipole. Adapted from International Commission on Non-Ionizing Radiation Protection (2010)

Frequency	$E\,(\mathrm{kV\,m^{-1}})$	$H\,(\mathrm{A\,m^{-1}})$	$B\,(\mathrm{T})$	$m\,(\mathrm{A\,m^2})$ @ 1 m	$m\,(\mathrm{A\,m^2})$ @ 10 m
50 Hz–400 Hz	$2.5 \times 10^2/f$	1.6×10^2	2×10^{-4}	2×10^3	2×10^6
400 Hz–3 kHz	$2.5 \times 10^2/f$	$6.4 \times 10^4/f$	$8 \times 10^{-2}/f$	$8 \times 10^5/f$	$8 \times 10^8/f$
3 kHz–10 MHz	8.3×10^{-2}	21	2.7×10^{-5}	2.7×10^2	2.7×10^5

Since $\delta \gg d/2$, we conclude that the relationship of (7.2.6) holds. In fact, to make efficient use of the mass of conducting material, *we must have* $\delta \gg d/2$, to ensure that at the chosen operating frequency the current density is as uniform as possible over the cross-sectional area of the wire.

The turns of the antenna cause it to act as an inductor, with a reactive impedance at the driving frequency. To increase efficiency, the coil is coupled with a tuned capacitor to make it resonate at the desired carrier frequency. At resonance, the complex impedance of the loop inductor and tuning capacitor cancel each other out, leaving the coil resistance as the effective impedance of the loop. To yield a high magnetic moment, the coil resistance should be as low as possible, which makes the antenna quality factor (Q) high. However, a very high Q leads to a limited transmitter bandwidth and high sensitivity to component tolerances and variation. It can be seen from this that the optimal design of a transmitting antenna depends on a number of factors and great gains in performance can be made by matching the requirements of the system to the constraints on the antenna.

Limits are set on the maximum allowable magnetic moment, both for health and safety of humans and also for compliance to intrinsic safety standards. The allowable limits depend on the frequency, as at higher frequencies, currents are induced more easily in conductive tissue. Guidelines for maximum magnetic moment set by ICNIRP are shown in Table 7.2.1, which are set by the reference distance. Using a reference distance of 1 m, it can be seen that at a frequency of 2.5 kHz, the maximum (safe for human exposure) magnetic moment is 320 A m^2.

7.2.2.2 Impact of Media

A quasistatic magnetic field is established in the surrounding material by the source. When operating in air, the relative coefficient of permeability of air, μ_r, is virtually unity, so there is insignificant deviation between the free-space model and this setup. However, in dense media, such as rock and soil, there are two factors that can alter the strength of the field sensed at the receiver. The first is the relative permeability and the second is the conductivity of the material. The relative permeability for most nonferromagnetic materials is very close to 1, so this effect is typically negligible. However, the conductivity of the material can have a major impact on the establishment of the field as the time-varying magnetic field induces a current in the media which acts to oppose the primary field. This can cause rapid attenuation of the magnetic field and consequent reduction in range.

The relative permeability μ_r of biological mass is very close to one, and the same is valid for air and water. This fact is important, as absorption by people is not only detrimental to positioning, it can also carry safety risks. Water is frequently encountered in underground mines, and thus the ability to penetrate water with little attenuation is essential. Most rocks, silt, and clay have a relative permeability is very close to one. The work by Telford et al. (1990) addresses the magnetism of rocks and minerals, and point out that magnetically important minerals are surprisingly few in number. Except for magnetite, pyrrhotite, and titanomagnetite, all the other minerals have $\mu_r \approx 1$. In conclusion, since most soils in nature do not contain magnetite, we can safely assume that the relative permeability of most underground environments is close to one.

Concerning conductivity, most dry rocks have very low conductivity. However, when the water content increases, the conductivity may also increase. Most minerals possess high conductivity, except for some compounds that contain iron, copper, and other metals. In order to quantify the penetration capabilities of the magnetic field, we provide in Table 7.2.2 skin depth calculations for various classes of rocks and minerals. These are pessimistic, assuming very high water concentrations, which is not typically the case in underground environments. Two different frequencies are considered: 2.5 kHz (used for localization in, e.g., Abrudan et al., 2015; Markham and Trigoni, 2012a), and 10 MHz (used for communication in Sun and Akyildiz, 2010). The skin effect is substantially higher at higher frequencies due to the contribution of the conductivity, and therefore, the secondary field will be significant, which leads to deviations from the magnetic dipole, and to higher attenuation with the distance. The table demonstrates that although the permeability of most media is close to unity, the conductivity can have a much greater impact on achievable range.

Table 7.2.2 The minimum skin depth for some different rocks and minerals at two different frequencies: 2.5 kHz and 10 MHz

Material	Skin depth [m] ($f = 2.5$ kHz)	Skin depth [m] ($f = 10$ MHz)
Natural waters (ign. Rocks)	30.20	0.48
Natural waters (sediments)	3.90	0.06
Sea water	10.07	0.16
Ice	≥ 100	≥ 100
Surface waters (ign. Rocks)	3.18	0.05
Surface waters (sediments)	31.83	0.62
Clays	10.07	0.16
Schists (calcareous and mica)	45.02	0.75
Conglomerates	≥ 100	11.50
Sandstones	10.07	0.16
Limestones	71.18	1.25
Granite	≥ 100	56.29
Slates (various)	≥ 100	8.03
Bitum. Coal	7.80	0.12
Hematite	0.58	0.01
Pyrite	0.05	0.00
Pyrolusite	0.71	0.01
Pyrrhotite	0.45	0.01
Titanomagnetite	0.08	0.00
Magnetite	0.45	0.01

7.2.2.3 Receiver

Although magnetic sources typically comprise coils of conductive wire, there are many more approaches to sensing magnetic fields, with different size, weight, power, and cost implications. The simplest sensor is the humble coil, which senses alternating magnetic fields by their induced voltage. Tumanski (2007) shows that the SNR of the air cored coil is governed primarily by the diameter of the core. This intuitively makes sense from the fact that a larger cross-sectional area will intersect more flux lines, creating a higher signal compared to the self-noise of the sensor. To reduce the size of receiving antennas, a material with high permeability (e.g., metglas) can be used as a core, at the cost of linearity and increased noise. The voltage output from the coil is related to the rate of change of the magnetic flux, so inherently has a frequency dependence. This can be integrated (either with an analog circuit or digitally) to provide a flat, wideband response.

Solid state sensors based on hall or magneto-resistive properties of semi-conductors can provide reasonable sensitivity, but in small packages. The gold-standard for ultra-high sensitivity remains the SQUID (supercooled quantum inference device) sensor. However, such sensors are expensive and bulky. For a very comprehensive overview of magnetic sensing technology, refer to Tumanski (2007).

7.2.2.4 Channel Model

In this section, we provider a simple model to predict the spatial distribution of the magnetic field in an undistorted environment. Consider a triaxial transmitter located at $(x, y, z) = (0, 0, 0)$ and a receiver located at $\mathbf{r} = (x_r, y_r, z_r)$. The range between the transmitter and receiver is $r = \|\mathbf{r}\|_2$. We consider generating a signal on three mutually orthogonal antenna, which we term a *triaxial* transmitter. In practice, this can be accomplished by sequentially energizing each coil in turn (time-multiplexing) or using a different frequency for each coil (frequency-multiplexing). As will be shown later, the use of triaxial antennas at both transmitter and receiver conveys unique advantages for positioning and tracking. Let us consider a triaxial whose X, Y, Z coils are subsequently energized. The magnetic moments corresponding to the three axes are:

$$\mathbf{m}_i = N_{TX} I_{TX} A_{TX} \mathbf{e}_i, \tag{7.2.8}$$

where N_{TX} is the coil number of turns, I_{TX} is the current through the coil, A_{TX} is the cross-sectional area of the transmitter coil, and $\mathbf{e}_i$, $i = 1, 2, 3$ are the excitation versors corresponding to the standard Euclidean basis vectors.

Considering a generic magnetic moment vector $\mathbf{m}$, the magnetic field flux density at position $\mathbf{r}$ may be expressed using the magnetic dipole equations as

$$\mathbf{B}(\mathbf{r}, \mathbf{m}) = \frac{\mu_{TX}}{4\pi}\left[\frac{3\mathbf{r}(\mathbf{m}^T\mathbf{r})}{r^5} - \frac{\mathbf{m}}{r^3}\right] = \frac{\mu_{TX}}{4\pi r^3}\left[\frac{3\mathbf{r}\mathbf{r}^T}{r^2} - \mathbf{I}_3\right]\mathbf{m}, \tag{7.2.9}$$

where μ_{TX} is the transmitter coil core magnetic permeability, $\mathbf{I}_3$ is the 3×3 identity matrix, and $(\cdot)^T$ denotes the matrix transpose operation. Given the three mutually perpendicular the magnetic moments $\mathbf{m}_i$, $i = 1, 2, 3$ in Eq. (7.2.8), we obtain the corresponding magnetic field flux density vectors $\mathbf{B}_i = \mathbf{B}(\mathbf{r}, \mathbf{m}_i)$.

Let $\mathbf{\Omega}_t$ be a 3×3 orthogonal matrix whose columns represent the axes of the triaxial receiver in the transmitter frame. Grouping the $\mathbf{B}_i$ column vectors in a matrix $\mathbf{B}_{1,2,3} = [\mathbf{B}_1, \mathbf{B}_2, \mathbf{B}_3]$, the magnetic field can be expressed as

$$\mathbf{\Omega}_t^T \mathbf{B}_{1,2,3} = \frac{\mu_{TX} N_{TX} I_{TX} A_{TX}}{4\pi r^3} \mathbf{\Omega}_t^T \left[\frac{3\mathbf{r}\mathbf{r}^T}{r^2} - \mathbf{I}_3 \right] \underbrace{[\mathbf{e}_1, \mathbf{e}_2, \mathbf{e}_3]}_{\mathbf{I}_3}. \qquad (7.2.10)$$

Each excitation $\mathbf{e}_i$ at TX will induce a voltage on the ith receiver axis, which may be represented by a 3×1 vector

$$\mathbf{v}_i \propto 2\pi f \mu_{RX} N_{RX} A_{RX} \mathbf{\Omega}_t^T \mathbf{B}_i, \qquad (7.2.11)$$

where f is the excitation carrier frequency, N_{RX} is the receiver coil number of turns, and A_{RX} is the receiver coil cross-sectional area. The energy transfer between transmitter and receiver triaxial coils can be modeled as a 3×3 MIMO channel matrix $\mathbf{S}$ given by

$$\mathbf{S} \triangleq [\mathbf{v}_1, \mathbf{v}_2, \mathbf{v}_3], \qquad (7.2.12)$$

whose ith column $\mathbf{v}_i$ is given in Eq. (7.2.11). The three excitations correspond to a space-time code whose coding matrix is $\mathbf{I}_3$ (see Eq. (7.2.10)). Combining Eqs. (7.2.10)–(7.2.12), we can write

$$\mathbf{S} = c\mathbf{\Omega}_t^T \left[\frac{3\mathbf{r}\mathbf{r}^T}{r^2} - \mathbf{I}_3 \right]. \qquad (7.2.13)$$

The elements $s_{j,i}$ of the MIMO channel matrix $\mathbf{S}$ are proportional to the voltages induced from the ith TX coil to the jth RX coil. The scaling constant decays with the third power of range, i.e., $c \propto 1/r^3$. Note that the matrix $\mathbf{S}$ contains the 3D RX position $\mathbf{r}$ and orientation $\mathbf{\Omega}_t$ which we are interested in.

7.2.3 SINGLE HOP LOCALIZATION

In this section, we describe how, given the channel model, we can derive the positions and orientations of a receiver. A major advantage of using triaxial antennas at source and receiver is that a single source can be used to localize a receiver in 3D. This is in stark contrast to conventional range based systems which require four or more noncollocated anchors to estimate 3D position. This result arises from the property that the magnetic field is a vector field, i.e., it has both magnitude and direction. Fundamentally, if one considers the equation for the flux Φ passing through a coil with area A,

$$\Phi = \mathbf{BA} = BA\cos(\theta), \tag{7.2.14}$$

it can be seen that there is an angular dependence between the voltage sensed and the generated magnetic field. This property is exploited to provide positioning information as described below.

The channel matrix $\mathbf{S}$ may be used to estimate the position the receiver in 3D, subject to a hemispherical ambiguity, which is inherent due to the symmetry of the dipole magnetic field. Therefore, the estimated position corresponds to either (x_r, y_r, z_r), or $(-x_r, -y_r, -z_r)$. This ambiguity may be using map information, inertial measurements, or multiple transmitters. The rest of this section addresses the single transmitter magneto-inductive 3D positioning. First, we estimate the range from the total received power, which corresponds to the Frobenius norm of the MIMO channel matrix. Then, we can then compute the bearing angles (azimuth and elevation) from the eigendecomposition of channel matrix.

Range Estimation

The Frobenius norm of $\mathbf{S}$ defines the total received power, and in undistorted environments, it follows an inverse-cube law

$$\|\mathbf{S}\|_F \propto r^{-3}. \tag{7.2.15}$$

Since Frobenius norm is invariant to orthogonal transforms, the range is also invariant with respect to the relative orientation of the triaxial transmitter and receiver. This is an important factor, as the orientation of the receiver is difficult or impossible to keep constant in a number of underground applications. Let us define the total received signal strength indicator (RSSI) measured in dB as

$$\rho \triangleq 20\lg \|\mathbf{S}\|_F. \tag{7.2.16}$$

Due to the inverse cube law of the magnetic filed decay, the RSSI at distance r can be expressed as

$$\rho = \rho_0 - 60\lg(r/r_0), \tag{7.2.17}$$

given the RSSI ρ_0 measured at reference distance r_0 (determined by calibration). Thus, the transmitter–receiver distance may be estimated as

$$r = r_0 10^{(\rho_0 - \rho)/60}. \tag{7.2.18}$$

Position Estimation

Knowing the range, we still need to determine the direction of the position vector $\mathbf{r}$ in 3D, i.e., the position versor $\mathbf{u} = \mathbf{r}/r$. We first need to get rid of the orientation information. Let us define the matrix

$$\mathbf{C} = \mathbf{S}^T \mathbf{S}. \tag{7.2.19}$$

According to Eq. (7.2.13), the orientation cancels out in Eq. (7.2.19). Therefore, $\mathbf{C}$ is orientation invariant and may be used to estimate the 3D position versor $\mathbf{u} = \mathbf{r}/r$. We propose an elegant algebraic technique exploiting the rank-one term $\mathbf{r}\mathbf{r}^T/r^2$ in (7.2.13). Let $\mathbf{C} = \mathbf{U}\mathbf{D}\mathbf{U}^T$ be the eigendecomposition of $\mathbf{C}$. We obtain

$$\frac{\mathbf{r}}{\|\mathbf{r}\|}\frac{\mathbf{r}^T}{\|\mathbf{r}\|} = \mathbf{U}\left[\frac{1}{3c}\mathbf{D}^{1/2} + \mathbf{I}_3\right]\mathbf{U}^T \tag{7.2.20}$$

which is also a rank-one matrix. Therefore, the position versor we are interested in is the eigenvector $\mathbf{u}_{\max}$ corresponding to the maximum eigenvalue of $\mathbf{C}$, i.e., $\mathbf{r}/\|\mathbf{r}\| = \mathbf{u}_{\max}$. Finally, the 3D position vector can be expressed as

$$\mathbf{r} = r\mathbf{u}_{\max}. \tag{7.2.21}$$

Orientation Estimation

The orientation matrix $\boldsymbol{\Omega}_t$ can be estimated as

$$\mathbf{S} = c\boldsymbol{\Omega}_t^T \underbrace{\left[3\mathbf{u}_{\max}\mathbf{u}_{\max}^T - \mathbf{I}_3\right]}_{\mathbf{P}}, \tag{7.2.22}$$

where $\mathbf{P} \triangleq 3\mathbf{u}_{\max}\mathbf{u}_{\max}^T - \mathbf{I}_3$ is always an invertible matrix (its eigenvalues are $\{2, -1, -1\}$). The orthogonal Procrustes problem in Eq. (7.2.22) can be solved for orthogonal $\boldsymbol{\Omega}_t$ by polar decomposition. The polar decomposition of $c\mathbf{P}\mathbf{S}^T = \mathbf{U}\mathbf{H}$ is a product of an orthogonal matrix $\mathbf{U}$ and a positive semidefinite symmetric matrix $\mathbf{H}$. The nearest matrix satisfying the equality (7.2.22) under Frobenius norm is the orthogonal polar factor of $\mathbf{P}\mathbf{S}^T, \forall c > 0$ (for details, see Higham, 1989). Therefore, the matrix describing the orientation of the RX frame in the TX frame is

$$\hat{\boldsymbol{\Omega}}_t = s\,\mathrm{pf}\{\hat{\mathbf{P}}\hat{\mathbf{S}}^T\}, \tag{7.2.23}$$

where $\hat{\mathbf{P}} = 3\hat{\mathbf{u}}_{\max}\hat{\mathbf{u}}_{\max}^T - \mathbf{I}_3$ and $\hat{\mathbf{S}}$ is the estimated channel matrix. For a full-rank matrix $\mathbf{A}$, the polar factor can be calculated as $\mathrm{pf}\{\mathbf{A}\} \triangleq \mathbf{A}(\mathbf{A}^T\mathbf{A})^{-1/2}$. The multiplication by $s \triangleq \mathrm{sign}[\det(\mathrm{pf}\{\hat{\mathbf{P}}\hat{\mathbf{S}}^T\})]$ ensures that $\hat{\boldsymbol{\Omega}}_t$ is a proper rotation.

Using these above relationships, the position and orientation of the receiver can be determined in 3D, using signals from a single transmitter.

7.2.4 MULTIHOP LOCALIZATION

In the previous section, we showed how the receiver position and orientation can be determined for a single transmitter–receiver pair. In this section, we show how information from multiple transmitters can be fused by a single receiver for improved estimation, following the derivation presented in Abrudan et al. (2016). Assume that there are N magneto-inductive infrastructure transmitters and one receiver carried by the user. Again, we consider position and orientation estimation independently. Note, that as the system is receiver-centric, the number of receivers can be arbitrarily large.

We express the 3D position of the receiver with respect to a transmitter in terms of a range (distance) and bearing versor (a unit vector pointing from a transmitter in the direction of the receiver). By decoupling range from bearing angle, the maximum likelihood formulation becomes conceptually simpler.

Range Likelihood

From Eq. (7.2.17), it follows that the likelihood of the range corresponding to the nth TX is $l_r(r_n) = \mathcal{N}(\rho_n; \tilde{\rho}_n(r_n), \sigma_n^2)$, where ρ_n is the measured RSSI, and $\tilde{\rho}_n(r_n) = \rho_{0,n} - 60\log_{10}(r_n/r_{0,n})$ is the RSSI predicted by the path-loss model in Eq. (7.2.17) for the range r_n.

Let $\mathbf{p}$ and $\mathbf{t}_n$ denote the position vectors in the world frame corresponding to the RX and the nth TX, respectively. Then, the range can be expressed as $r_n = \|\mathbf{p} - \mathbf{t}_n\|$. Assuming that the RSSI uncertainties are mutually independent for all N transmitters, the joint log-likelihood function including the corresponding ranges is

$$L_\tau(\mathbf{p}) = c_1 - \sum_{n=1}^{N} \frac{1}{2\sigma_n^2} \left(\rho_n - \rho_{0,n} + 60 \log_{10} \frac{\|\mathbf{p} - \mathbf{t}_n\|}{r_0} \right)^2, \qquad (7.2.24)$$

where c_1 is a constant that does not depend on $\mathbf{p}$.

Bearing Versor Likelihood

Unlike conventional range-based multilateration, we are able to estimate angles between transceivers in 3D. In order to deal with angular uncertainties, we need probability density functions defined in a proper parameter space, that exhibits 2π-periodicity. We model the bearing versor error using von Mises–Fisher distribution, which is defined for the random 3-dimensional unit vector $\mathbf{x} \in \mathbb{R}^3$ as follows:

$$p(\mathbf{x}; \boldsymbol{\mu}, \kappa) = \frac{\kappa}{4\pi \sinh \kappa} \exp\left(\kappa \boldsymbol{\mu}^T \mathbf{x}\right), \qquad (7.2.25)$$

where $\boldsymbol{\mu}$ is a unit vector pointing in the mean direction, and κ is the concentration parameter. In order to deal with the hemispherical ambiguity in position estimation which magneto-inductive positioning is subject to due to the field symmetry, we use abalanced mixture of two Fisher distributions Costa et al. (2014) whose means correspond to the estimated position versor of the nth TX $\hat{\mathbf{u}}_n = \hat{\mathbf{r}}_n/r$, and its antipodal point $-\hat{\mathbf{u}}_n$, respectively. The corresponding bimodal bearing likelihood for the nth TX is $l_\angle(\mathbf{r}) = [p(\mathbf{r}; \hat{\mathbf{u}}_n, \kappa_n) + p(\mathbf{r}; -\hat{\mathbf{u}}_n, \kappa_n)]/2$. For N transmitters, the joint log-likelihood function that includes the estimated bearing versors $\hat{\mathbf{u}}_n = (\hat{\mathbf{p}} - \mathbf{t}_n)/\|\hat{\mathbf{p}} - \mathbf{t}_n\|$ is

$$L_\angle(\mathbf{p}) = c_2 + \sum_{n=1}^{N} \ln \, \cosh\left[\kappa_n \frac{(\mathbf{p} - \mathbf{t}_n)^T}{\|\mathbf{p} - \mathbf{t}_n\|} \frac{(\hat{\mathbf{p}} - \mathbf{t}_n)}{\|\hat{\mathbf{p}} - \mathbf{t}_n\|} \right], \qquad (7.2.26)$$

where c_2 is a constant that does not depend on $\mathbf{p}$. In order to set the concentration parameter κ_n which characterizes the spread of von Mises–Fisher p.d.f., we use the eigenvalue criterion in Abrudan et al. (2015).

Joint Log-Likelihood Maximization

Our goal is to maximize the joint log-likelihood corresponding to the N TXs that includes range and 3D bearing information $L(\mathbf{p}) = L_\tau(\mathbf{r}) + L_\angle(\mathbf{r})$, and this is achieved by using a steepest ascent algorithm. Fig. 7.2.1 shows

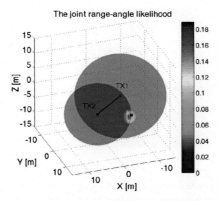

Figure 7.2.1 Illustration of the joint range-bearing likelihood function for two transmitters TX1 and TX2. The two spheres correspond to the most likely ranges. The joint likelihood is maximized close to the true RX position (shown by the black dot)

that the joint range and bearing likelihood function has a well-defined maxima in the vicinity of the true position.

Using the derived maximum likelihood, it is possible now to find the best position estimate of the receiver. Using a multihop configuration allows for scalable and more precise positioning, as the range limitations caused by the inverse cube law path loss are overcome.

7.2.5 APPLICATIONS OF UNDERGROUND MI POSITIONING

In this section, we present two examples of how MI can be used to provide through-earth positioning capability. We discuss how MI can be used to form an iteratively deployable multihop positioning infrastructure, suitable for rescue situations and how MI can be used to precisely track the locations of burrowing animals underground.

7.2.5.1 Iteratively Deployable Positioning Architecture

In an underground rescue situation, such as a mining collapse or caving accident, it is necessary to have accurate positioning of rescue team members (which could be people or even robotic agents). However, valuable time cannot be wasted setting up a positioning system. A novel, iterative, positioning system is presented in Abrudan et al. (2016) as a proof-of-concept. The key idea is that a single anchor node is placed at a known location (such as a tunnel entrance) to act as the origin. The receiver is localized relative to the transmitter. However, this faces the issue that positioning range

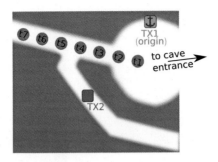

Figure 7.2.2 2D schematic layout of cave system, showing surveyed anchor transmitter (TX1), secondary transmitter at unknown position (TX2), and test points (t1–t7) where the receiver is localized. Note that there are no line-of-sight paths between either transmitter and the test points, or between the two transmitters

is limited by the coverage of the first transmitter. To extend the range, additional transmitters can be deployed. The receiver, using the first transmitter as the origin, can localize the additional transmitters and subsequently use them as anchors when outside the range of the original anchor. In this way, a positioning system can be iteratively deployed, without requiring any additional surveying effort or worrying about anchor geometry as would be the case in range based systems.

The outline of the main phases in this approach is discussed as follows:

Phase I. In this initial phase, receivers localize themselves relative to the transmitter at the origin (TX1).

Phase II. In this phase, to maintain coverage, receivers derive the location of the additional transmitter(s) (TX2 in this case) with respect to the origin.

Phase III. Positions of the receiver is then estimated with all the transmitters using a maximum likelihood approach.

We carried out experiments in a man-made cave within a space of $15 \times 15 \times 2$ m^3, with a schematic layout shown in Fig. 7.2.2. It is well-known that accurately mapping underground caves in 3D is an extremely difficult and time-consuming task. Ground truth locations were surveyed using laser-based range and inclination meters. The origin was set closer to the cave entrance, such that its global position can be inferred easily. The transmitter TX1 was placed at the origin and acted as a primary anchor, and secondary anchor TX2 was placed in a different tunnel. Note that the tunnels were not planar, they exhibit a difference in elevation of approximately 2 m.

The two transmitters were equipped with square triaxial coils (30 cm side and 80 turns), similar to Markham and Trigoni (2012a). The receiver

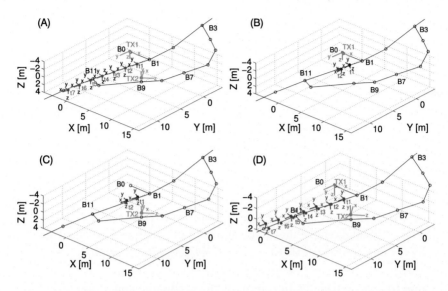

Figure 7.2.3 Linear displacement experiment. The surveyed positions and orientation of TX1 and TX2 are shown in green. (A) Surveyed RX positions and orientations t1–t7 are shown in magenta color. (B) Phase I: Two estimated RX 3D positions and orientations using TX1 as anchor node. (C) Phase II: Estimated TX2 position and orientation using the two previously estimated RX positions and orientations as virtual anchors. (D) Phase III: Estimated RX positions and orientations using ML and both TX1 and TX2

contains small triaxial coils (around 20 mm long) and an IMU for gravity vector estimation. The system transmitted vector modulated signals at a frequency of 2.5 kHz, with a bandwidth of 31.25 Hz and a code length of 36 symbols.

The experiment was carried out close to the origin (see Fig. 7.2.3A). A measuring tape was laid on the ground and magneto-inductive measurements were collected from two different transmitters (TX1 acting as single anchor, and TX2 shown in green), every two meters along the tape (points t1 to t7 in magenta color). The surveyed receiver 3D positions and orientations are shown in Fig. 7.2.3A by the magenta dots and frames, respectively. The subsequent relative receiver orientations were with the x-axis parallel to the tape, pointing in the direction opposite to the origin, and the z-axis pointing down. Fig. 7.2.3 shows the three phases of the network deployment and discovery algorithm.

Phase I is shown in Fig. 7.2.3B, where the receiver 3D positions and orientations are estimated using TX1 only as an anchor. The hemispherical ambiguity was removed with the basic knowledge that the measured locations are on the positive side of the y-axis of the anchor TX1. The orientation ambiguity was removed based on the gravity vector, which was estimated from the RX inertial measurement unit.

Phase II is depicted in Fig. 7.2.3C, where two estimated receiver 3D positions and orientations are used as virtual anchors in order to estimate the position and the orientation of TX2.

Phase III is shown in Fig. 7.2.3D. The newly discovered anchor TX2 is used to estimate the receiver 3D positions and orientations, in tandem with TX1 when available.

In terms of receiver positioning accuracy, the estimation approach provides a very reliable estimate, the positioning errors are well below 1 m for most of the locations and orientation estimation errors remain below 10 degrees. Although the scale of the proof-of-concept trial is small, it demonstrates the fundamentals of accurate 3D positioning through solid rock. In addition, the use of multiple deployable secondary anchors allows the system to scale gracefully. In summary, this technique could find great use in underground rescue, autonomous mining and cave exploration.

7.2.5.2 Revealing Underground Animal Behavior

Many animals spend a large proportion of their lives underground in burrows or dens, to provide safety and refuge. Due to a lack of noninvasive technologies which can determine where they precisely are underground, little has been known about their detailed activity. This is especially important for animals such as badgers (*Meles meles*) which live in large, communal underground burrows which can house up to 20 animals over a few hundred square meters (see Noonan et al., 2015). To address this problem, a low power MI tracking technology was developed. This comprised three core components: the transmitting antennas, placed above the area to be monitored; the tracking collars, worn by the study animals; and the aboveground basestation which wirelessly received buffered data when animals emerged to forage every night, as described in Markham et al. (2010).

The transmitting antennas are time multiplexed and emit unique digital codes, modulated with amplitude-shift-keying. They are large (2 m × 15 m) rectangular loops of cable that are dug into the ground to provide long-term robustness (for details, see Markham and Trigoni, 2012b). Typically, 8 antennas arranged in an X–Y matrix are used to provide sufficient spatial coverage.

The collars comprise a low power microcontroller, an analog triaxial MI sensor, a flash memory, 802.15.4 wireless transceiver, and are powered by a small lithium cell. The MI sensor listens for an encoded signal from each antenna, and converts this measurement to a single RSSI value. By using three orthogonal sensors, the orientation of the collar, which is unknown, becomes an invariant. The RSSI values from all the transmitting antennas are stored and time-stamped in the onboard flash memory. The radio transceiver periodically tries to establish a link with the basestation, which typically occurs when an animal emerges every night to forage. The device then reliably transfers the stored data to the basestation, ready for post-processing.

To determine the animal's trajectory, a model of the magnetic field generated by each transmitting antenna is used. These are simple geometric structures, so application of Biot–Savart's law allows for the magnetic field strength to be readily calculated using the method presented by Misakian (2000). A particle filter is then used to estimate the animal's location, given the MI field readings and knowledge of the antenna geometry. Typical accuracies of 30 cm RMS error in 3D are obtained, allowing for extremely precise tracking of underground animal movement, as shown in Fig. 7.2.4. With a high temporal sampling rate of 0.3 Hz, it is possible to obtain detailed trajectories of animal movements, both above and below ground, as can be seen in Fig. 7.2.5. Lastly, one of the major advantages of such a system is that over time it not only tracks animal movements, but as animals are constrained to move within their burrows, also reveals their underground tunnel structure, shown in Fig. 7.2.6.

In summary, a low-power MI positioning system was used to provide highly accurate GPS-like positioning of a number of animals simultaneously. This in turn has helped to uncover new biological findings about how animals interact underground and what the precise function of their burrows is (see Noonan et al., 2015 for more details).

7.2.6 CHALLENGES AND LIMITATIONS

7.2.6.1 Path Loss

While magnetic induction-based communication and positioning systems have the aforementioned advantages of low attenuation in dense media, and the ability to use the vector field for positioning, they suffer from inherent physical limitations.

One of those limitations is path loss; the magnetic field of a magnetic dipole in the near-field region decays with the inverse cube of the distance,

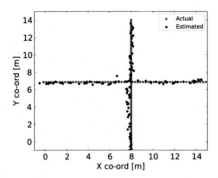

Figure 7.2.4 Plot showing 2D positioning accuracy of underground animal tracking system

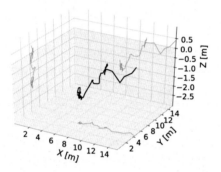

Figure 7.2.5 Detailed trajectory showing animal emerging from below ground to above ground

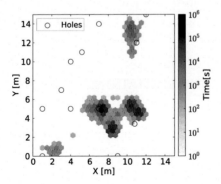

Figure 7.2.6 History of animal movements over many weeks reveals their underground burrow structure

such that $H(r) \propto r^{-3}$, corresponding to a path loss of 60 dB per decade. In contrary, the magnetic field of a Hertzian dipole radiating in the far-field region decays as $H(r) \propto r^{-1}$ (see Jin, 2010), corresponding to a path loss of only 20 dB per decade.

To compensate for a high path loss, magneto-inductive transmitters have to comprise of bulky coils able to carry large amounts of current, and receivers of highly sensitive coils with a very low noise floor, able to sense tiny fields. For communication, sophisticated signal processing techniques must be used to make the most of the limited bandwidth.

In designing a transmitter with a particular environment in mind, path loss is a major concern. More specifically, there are two major factors that must be taken into account: (a) the physical configuration/geometry of the field source, and (b) the material properties of the medium the field is established in, or propagating through.

The first factor becomes important especially in the near field, where the decay does not solely follow an inverse power law, but has multipole terms, which become more significant as one approaches the source. These decay terms are independent of material properties and frequency, and only depend on the distance.

An additional type of attenuation comes into effect when the medium is electrically conductive or magnetically permeable, or when the frequency is high, or both. This type of attenuation does not follow an inverse power law, but instead is exponential in nature. The attenuation constant, usually denoted as α, is what determines the rate of decay, and it is a function of material properties as well as operating frequency. A general expression for α is given by Jordan and Balmain (1968) as

$$\alpha = \omega\sqrt{\frac{\mu\epsilon}{2}\left(\sqrt{1+\frac{\sigma^2}{\omega^2\epsilon^2}}-1\right)} \tag{7.2.27}$$

where ω is the angular frequency, μ is the magnetic permeability, ϵ is the electric permittivity, and σ is the electrical conductivity. One may notice that indeed, α increases with frequency, electrical conductivity, and magnetic permeability. When considering conducting objects, $\epsilon = 1$ and (7.2.27) simplifies to

$$\alpha = \sqrt{\frac{\sigma\omega\mu}{2}}, \tag{7.2.28}$$

where $\delta = 1/\alpha$ is the well known skin depth.

7.2.6.2 Distortion Due to Nearby Conducting Objects

Electrically conducting objects significantly obstruct electromagnetic fields and can pose a problem in communication and positioning. In the far-field one can assume plane wave propagation, which greatly increases mathematical tractability when calculating reflection and scattering from objects. This assumption is valid when the distance of the observation point from the source is much larger than the wavelength of the radiated electromagnetic wave, which causes the electric and magnetic fields of the wave to be mutually orthogonal. In the near-field, where the distance from the source is comparable to the wavelength such that the source geometry matters, this approximation is not valid, and one does not commonly find closed form solutions that describe reflections from conducting objects, even for simple geometries.

In general, the flux density generated by the transmitter induces eddy currents within the conductors, which then radiate a secondary (scattered) flux density. If the eddy-currents can be classified as inductance limited (see Hammond, 1962), the surfaces of the conductors behave as if they were perfect electric conductors (PEC), and the magnetic field in their vicinity is as if the conductors had zero magnetic permeability (the flux density normal to the surface is close to zero). The total flux density $\mathbf{B}^t$ outside the conducting regions will then be equal to

$$\mathbf{B}^t = \mathbf{B}^i + \mathbf{B}^s, \tag{7.2.29}$$

where $\mathbf{B}^i$ is the incident flux density due to the point dipole and $\mathbf{B}^s$ is the scattered flux density due to reradiation by the conducting objects.

One way to model the scattered flux density and obtain an expression for the total field $\mathbf{B}^t$ is to use image theory (Fig. 7.2.7). In image theory, the conducting object is replaced with an image of the dipole, and the sum of the real and imaginary dipoles gives the total field (see Kypris et al., 2015).

The way to autonomously combat the problem of distortions due to conducting objects, is to (i) recognize that the field does not behave as expected, (ii) switch to a different channel model that takes into account these objects. The former can be achieved by assessing the symmetry of the dipole field. In the presence of conducting material, the energy is redistributed among different axes (while preserved in total), thus changing the symmetry of the field. Knowledge of this can be used to switch to a different channel model that takes into account these distortions (see Abrudan et al., 2015).

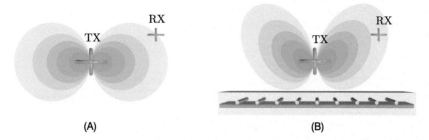

Figure 7.2.7 Representation of the magnetic field distribution of a magneto-inductive system, depicting a triaxial coil transmitter and receiver (A) in free space, and (B) over an array of metallic objects, which perturb the field symmetry (see Kypris et al., 2015)

7.2.7 CONCLUSION

As has been outlined, magneto-inductive technology has certain unique advantages that make it ideal for communication and location through many different media, ranging from concrete and soil to people and air. It does not suffer from boundary effects like acoustic and seismic technology, or is prone to multipath fading like RF techniques. MI positioning has enabled tracking of objects in environments where before it was difficult or even impossible. However, it does have a number of disadvantages, the most significant being the limited range due to rapid path loss. This fundamentally curtails its maximum achievable range. With advances in sensor technology, however, leading to miniature, solid state magnetic sensors that can approach supercooled SQUIDs, there is certainly a bright future for this MI.

REFERENCES

Abrudan, T.E., Xiao, Z., Markham, A., Trigoni, N., 2015. Distortion rejecting magneto-inductive three-dimensional localization (MagLoc). IEEE J. Sel. Areas Commun. 33 (11), 2404–2417.

Abrudan, T.E., Xiao, Z., Markham, A., Trigoni, N., 2016. Underground, incrementally deployed magneto-inductive 3-D positioning network. IEEE Trans. Geosci. Remote Sens. 54 (8), 4376–4391.

Barkand, T., Damiano, N., Shumaker, W., 2006. Through-the-Earth, two-way, mine emergency, voice communication systems. In: Industry Applications Conference, 2006. 41st IAS Annual Meeting. Conference Record of the 2006 IEEE, vol. 2, pp. 955–958.

Bedford, M., Kennedy, G., 2012. Evaluation of ZigBee (IEEE 802.15.4) time-of-flight-based distance measurement for application in emergency underground navigation. IEEE Trans. Antennas Propag. 60 (5), 2502–2510.

Costa, M., Koivunen, V., Poor, H.V., 2014. Estimating directional statistics using wavefield modeling and mixtures of von-Mises distributions. IEEE Signal Process. Lett. 21 (12), 1496–1500.

Cypriani, M., Delisle, G., Hakem, N., 2015. WiFi-based positioning in a complex underground environment. J. Netw. 10 (3).

Hammond, P., 1962. The calculation of the magnetic field of rotating machines. Part 3: eddy currents induced in a solid slab by a circular current loop. Proc. IEE C: Monographs 109 (16), 508–515.

Haverinen, J., Kemppainen, A., 2011. A geomagnetic field based positioning technique for underground mines. In: 2011 IEEE International Symposium on Robotic and Sensors Environments, ROSE, pp. 7–12.

Hawkins, W., Daku, B., Prugger, A., 2006. Positioning in underground mines. In: 32nd Annual Conference on IEEE Industrial Electronics, IECON 2006, pp. 3139–3143.

Higham, N.J., 1989. Matrix nearness problems and applications. In: Gover, M.J.C., Barnett, S. (Eds.), Applications of Matrix Theory. Oxford University Press, pp. 1–27.

Huang, X., Zhu, W., Lu, D., 2010. Underground miners localization system based on Zig-Bee and WebGis. In: 2010 18th International Conference on Geoinformatics. IEEE, pp. 1–5.

Hummel, J.B., Bax, M.R., Figl, M.L., Kang, Y., Maurer, C., Birkfellner, W.W., Bergmann, H., Shahidi, R., 2005. Design and application of an assessment protocol for electromagnetic tracking systems. Med. Phys. 32 (7), 2371–2379.

International Commission on Non-Ionizing Radiation Protection, 2010. Guidelines for limiting exposure to time-varying electric and magnetic fields (1 Hz to 100 kHz). Health Phys. 99 (6), 818–836.

Jin, J.-M., 2010. Theory and Computation of Electromagnetic Fields. Wiley, Hoboken, NJ.

Jordan, E.C., Balmain, K.G., 1968. Electromagnetic Waves and Radiating Systems, vol. 4. Prentice-Hall, Englewood Cliffs, NJ.

Kindratenko, V., 2000. A survey of electromagnetic position tracker calibration techniques. Virtual Real. 5 (3), 169–182.

Kypris, O., Abrudan, T.E., Markham, A., 2015. Reducing magneto-inductive positioning errors in a metal-rich indoor environment. In: IEEE Sensors Conference, pp. 1–4.

Markham, A., Trigoni, N., 2012a. Magneto-inductive networked rescue system (MINERS): taking sensor networks underground. In: 11th International Conference on Information Processing in Sensor Networks, IPSN 2012.

Markham, A., Trigoni, N., 2012b. Underground localization in 3-D using magneto-inductive tracking. IEEE Sens. J. 12 (6), 1809–1816.

Markham, A., Trigoni, N., Ellwood, S.A., Macdonald, D.W., 2010. Revealing the hidden lives of underground animals using magneto-inductive tracking. In: 8th ACM Conference on Embedded Networked Sensor Systems, Sensys 2010. Zürich, Switzerland.

Misakian, M., 2000. Equations for the magnetic field produced by one or more rectangular loops of wire in the same plane. J. Res. Natl. Inst. Stand. Technol. 105 (4).

Nkakanou, B., Delisle, G.Y., Hakem, N., 2011. Experimental characterization of ultrawideband channel parameter measurements in an underground mine. J. Comput. Netw. Commun. 2011.

Noonan, M.J., Markham, A., Newman, C., Trigoni, N., Buesching, C.D., Ellwood, S.A., Macdonald, D.W., 2015. A new magneto-inductive tracking technique to uncover subterranean activity: what do animals do underground? Methods Ecol. Evolut. 6 (5), 510–520.

Park, B., Myung, H., 2014. Underground localization using dual magnetic field sequence measurement and pose graph slam for directional drilling. Meas. Sci. Technol. 25 (12), 125101.

Pfeil, R., Pichler, M., Schuster, S., Hammer, F., 2015. Robust acoustic positioning for safety applications in underground mining. IEEE Trans. Instrum. Meas. 64 (11), 2876–2888.

Raab, F., Blood, E., Steiner, T., Jones, H., 1979. Magnetic position and orientation tracking system. Aerospace and Electronic Systems. IEEE Trans. AES 15 (5).

Shaffer, G., Stentz, A., Whittaker, W., Fitzpatrick, K., 1992. Position estimator for underground mine equipment. IEEE Trans. Ind. Appl. 28 (5), 1131–1140.

Silva, A., Vuran, M., 2009. Empirical evaluation of wireless underground-to-underground communication in wireless underground sensor networks. In: Proceedings of the 5th IEEE International Conference on Distributed Computing in Sensor Systems.

Squire, J., Sullivan, G., Baker, E., Flathers, G., 2009. Proof-of-concept testing of a deep seismic communication device. Soc. Mining Metallurgy Exp. 326, 97–100.

Stolarczyk, L.G., 1991. Emergency and operational low and medium frequency band radio communications system for underground mines. IEEE Trans. Ind. Appl. 27 (4), 780–790.

Stuntebeck, E., Pompili, D., Melodia, T., 2006. Underground Wireless Sensor Networks Using Commodity Terrestrial Motes. Tech. rep., Georgia Institute of Technology.

Sun, Z., Akyildiz, I., 2010. Magnetic induction communications for wireless underground sensor networks. IEEE Trans. Antennas Propag. 58 (7), 2426–2435.

Telford, W.M., Geldart, L.P., Sheriff, R.E., 1990. Applied Geophysics, 2nd edition. Cambridge University Press.

Tumanski, S., 2007. Induction coil sensors – a review. Meas. Sci. Technol. 18 (3), R31.

Wadley, T., 1946. Underground Communication by Rocks in Gold Mines on the Witwatersrand. South Africa Wet. Ny-Werheid-Navorsingsraad, Johannesburg, South Africa. Telekommunikasies Navorsing Laboratorium TRL.

Yarkan, S., Guzelgoz, S., Arslan, H., Murphy, R., 2009. Underground mine communications: a survey. IEEE Commun. Surv. Tutor. 11 (3), 125–142.

SUBCHAPTER 7.3

Integration of UAVs With Underground Sensing: Systems and Applications

Christian Wietfeld
Communication Networks Institute (CNI), TU Dortmund University, Dortmund, Germany

7.3.1 USE CASES AND REQUIREMENTS

The use of unmanned aerial vehicles (UAVs) is generally associated with tasks characterized by the three Ds: dull, dirty, and dangerous. In recent years, small unmanned aerial vehicles (UAVs) or remotely piloted aircraft systems (RPAS) equipped with all kinds of sensors have become a useful and well accepted tool for many new civil application areas, such as agriculture, inspection of critical infrastructure and emergency response (Wietfeld and Daniel, 2014; Dunbabin and Marques, 2012). This subchapter focuses on small (up to 25 kg weight) and micro UAVs (up to 5 kg weight) to be used in civilian use cases.

In first responder scenarios UAVs allow first responders to flexibly and efficiently explore disaster sites and therefore enable them to plan fast recovery and rescue actions, limiting the risks for the first responders as well as the population affected by a disaster impact. Prominent examples are the exploration of plumes (Daniel et al., 2011) as well as earthquakes or large-scale forest fires. Even the detection of nuclear substances using a swarm of UAVs has been demonstrated recently. UAVs have also become a useful day-to-day tool to inspect critical infrastructures such as high voltage grids or base stations of mobile networks.

In addition to the **aerial exploration** capabilities, UAVs serve also as a platform for **communication** access points enabling ad-hoc ground connectivity in areas that lack wide-area communication coverage. This may be because those areas are very remote or because the networking infrastructure is temporarily not available (Rohde et al., 2010).

Depending on the tasks to be performed, different UAV types (Fig. 7.3.1) are available for the integration with underground sensing systems:

- **Multicopter** (or rotary-wing) UAVs allow for a very flexible deployment (vertical and horizontal flight), but usually are limited in flight time (Daniel et al., 2011).
- **Fixed-wing** UAVs are limited to horizontal flight trajectories but typically allow for longer flight times (Frew and Brown, 2008). Because they can reach high speeds, their weather resistance (in particular against strong winds) is usually better than for multicopters or air ship-type platforms.
- **Tilt-wing** UAVs aim to combine the benefits of multicopter and fixed wing UAVs, as they support both vertical take-off and landing as well as energy-efficient gliding flight. While multicopter and fixed-wing UAVs are commercially available in a wide variety, cost-efficient tilt-wing UAVs for civil applications are still in research stage (Rohde et al., 2010).

Figure 7.3.1 Different UAV types (clockwise starting from top left): quadcopter, balloon, tilt-wing, and fixed-wing

Table 7.3.1 Overview of UAV types to be integrated with underground sensing systems

	Multicopter	Fixed Wing	Tilt-Wing	Air ship	Balloon
Range	◐	●	●	◕	◔
Endurance	◐	◕	◐	◕	●
Payload weight	◐	◐	◐	◐	◕
Steering Capabilities	●	◐	◕	◐	○
Weather resistance	◐	◕	◐-◕	○-◔	◔-◕
Small landing space	◕	◔	◕	◐	◕

○ - Very limited capabilities   ● - Excellent capabilities

- **Balloon-type**-UAVs are suitable for long-endurance deployments, but are restricted in their steering capabilities, in particular with regards to horizontal movements (Gomez et al., 2016).
- **Air ships** are generally also suited for long endurance deployments and allow for better steering performance. On the other hand, they are sensitive against strong winds (Elfes et al., 1998).

In general, all of the above mentioned UAV types are suited for the aerial exploration of potential underground sensing deployments as well as for data ferrying and data transfer services for underground sensor data. Nevertheless, the detailed system design needs to take into account the specific characteristics regarding range, endurance, payload, and steering capabilities. Table 7.3.1 aims to provide an overview of some characteris-

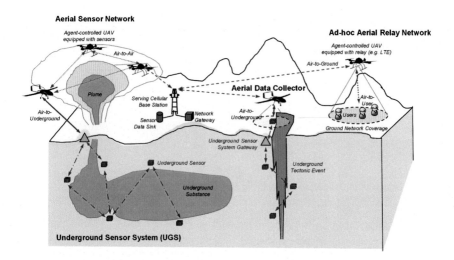

Figure 7.3.2 UAV – underground sensing system integration scenario

tics and qualitative performance levels based on the experience gained by the author with all mentioned platform types. Multicopter UAVs are very popular due to their steering capabilities, while fixed-wing and in future tilt-wing platforms score with their longer endurance and range.

The integration of UAVs with underground sensing systems can take place on various levels (see Fig. 7.3.2):

- **Deployment of underground sensing systems:** the remote sensing capabilities of UAVs can be very useful for identifying suitable deployment locations for underground sensing systems. An example would be the scanning of an area potentially affected by landslides with optical sensors to identify those critical areas, which require an in-depth, underground sensor deployment.

- **Data collection and data transfer from underground sensing systems:** UAVs are very suitable to provide wide-area connectivity for an underground sensing system in areas which are not reachable by terrestrial systems (e.g., very remote, unpopulated areas) and/or not suitable for satellite systems (e.g., in forests or chasms). While UAVs might not always be suited to provide continuous connectivity, they may be used to regularly collect data from underground sensing systems, also known as **data ferrying** (Zeng et al., 2016). In this context, low energy wireless communication technology with limited range (e.g., Bluetooth low energy with ranges of 10 m) may be used for enhanced battery lifetime. In future scenarios it may be even considered that the communication with the underground sensing system will be enabled by wireless energy transfer (transponder-type communication) from the UAV to the underground sensing system.

The above mentioned use cases and integration scenarios lead to specific system requirements for UAVs as well as for the underground sensor system:

- Underground Sensing Systems (UGS) require a **wireless communication gateway**, which enables some sort of above-ground communication, which serves as an interface to the UAV. To meet energy constraints and potentially obstructed wireless communication channels, the receive power and corresponding communication range can be limited. The communication capabilities need to take into account the UAV-type in use: especially a multicopter-type UAV is able to communicate via a very short range (several meters) for a longer time period, while a fixed wing UAV would benefit from a larger communication range when flying by the gateway to collect data.

- UAVs need to be able to fly autonomously suitable trajectories in line with the communication zone provided by the UGS-gateway and the amount of data to be transmitted. The pairing between UAV and UGS can be enabled by so-called **communication–aware UAV steerings** developed to maintain continuous connectivity within a swarm of UAVs.

Fig. 7.3.3 summarizes how the UAV service platform as described in Wietfeld and Daniel (2014) can be extended to integrate underground system features: the new application service "Data Ferrying for UGS" will be realized by adding an air-UGS-communication link and cognitive mobility features as part of the Service Control Air (running on the UAV).

The following paragraph discusses in more detail the design of the pairing process as essential component of an integrated UAV-UGS-design.

7.3.2 COMMUNICATION-AWARE PAIRING BETWEEN UAVs AND UNDERGROUND SENSING SYSTEMS

The design of communication-aware steering of UAVs is an important research area of the recent years. The design depends on the following three main areas:

- The type of the wireless communication system,
- The characteristics of the wireless air-to underground system communication channel, and

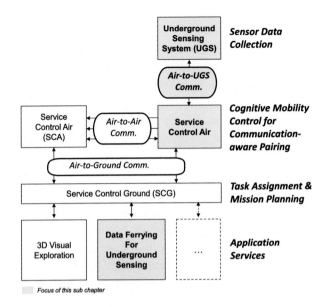

Figure 7.3.3 UAV service platform with specific underground-sensing integration features

- The steering algorithm and its parameterization.

  The following wireless communication links need to be considered (see Figs. 7.3.2 and 7.3.3):

- The **air-to-ground** (A2G) link describes the communication link between the ground station and the air-borne system. Via the A2G link mission control data as well as payload data gathered by sensors are transferred. Depending on the use case, the communication technologies in use range from local area (IEEE 802.11–Wi-Fi, IEEE 802.15.4 –ZigBee) to wide area (Cellular, in particular 3G and 4G) systems (Goddemeier et al., 2012).

- The **air-to-air** (A2A) link enables communication between UAVs for the purpose of self-organized control (collision avoidance, swarm behavior) as well as for relaying A2G link data. Wireless mesh networking technologies, such as IEEE 802.11s and extensions (Sbeiti et al., 2015), as well as ZigBee, have been considered as promising technology for A2A links (Hayat et al., 2015). In the future 4G and 5G device-to-device solutions may present an alternative option.

- The **air-to-user** (A2U) link has been introduced in [Handbook] to describe the use case, in which the UAV serves as aerial access point or base station for ground based user equipment. A2U data will be relayed via A2A and A2G links to the ground based systems, such as a network gateway. The A2U link can be realized by WLAN as well as flying LTE base stations (Gomez et al., 2016).

In the specific context of the integration of UAVs with UGSs the air-to-UGS (A2UGS) can be considered as a special case of an A2U link, in which the user equipment is realized by the previously introduced communication gateways of the UGS. The tight energy budgets caused by limitations of battery lifetimes and energy harvesting potentials require special consideration for underground sensing systems to be deployed in remote areas. Therefore low energy wireless technologies, such as ZigBee or Bluetooth Low Energy are of particular interest for the Air-to-UGS link. The short communication range of those technologies needs to be and can be compensated by appropriate UAV flight trajectories.

The **characteristics of the A2UGS wireless channel** can be partially derived from the state-of-the-art regarding the modeling of A2G channels. It has been shown in Bacco et al. (2014) that the height of the ground station antenna impacts the channel characteristics considerably. Typically, a two-way-channel model, which considers both a direct path as well as a reflection from the ground, can be applied (see Fig. 7.3.4, left). In case the antenna is positioned on the ground or only a few centimeters above ground the communication capabilities are severely impacted, as shown in measurements documented in Wang et al. (2012). This is due to strong interference between the two propagation paths as well as the impact of the ground absorbing part of the wireless signal. In case the underground sensing system shall be completely integrated in the underground without any visible gateway or antenna component above ground, the A2UGS channel is influenced by the underground material to be penetrated by the wireless signal (see Fig. 7.3.4, right). The weaker the wireless signal which can be received above ground, the more accurate the flight trajectories need to be in order to capture the signal. The detailed channel modeling can build on work addressing the subsurface channel for smart meter deployments in basements (Hägerling et al., 2014) in combination with work on the wireless underground sensor networks (Yoon et al., 2011).

With regards to **UAV steerings**, one can distinguish between macroscopic and microscopic steering (Wietfeld and Daniel, 2014). The **macroscopic steering** determines the flight path within the complete application scenario: in case of the data ferrying approach considered for the integration with underground sensing systems, the macroscopic steering ensures

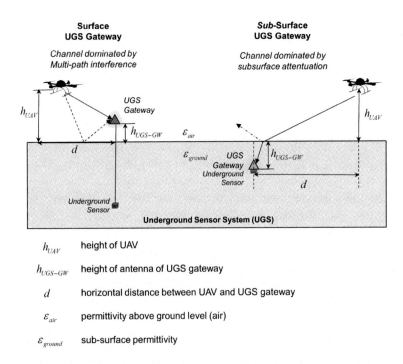

h_{UAV}	height of UAV
h_{UGS-GW}	height of antenna of UGS gateway
d	horizontal distance between UAV and UGS gateway
ε_{air}	permittivity above ground level (air)
ε_{ground}	sub-surface permittivity

Figure 7.3.4 Air-to-UGS channel modeling depending on UGS gateway deployment

that all relevant UGS communication gateways are visited with the aim to collect the complete sensor data in time. The route for such a waypoint steering would be typically determined on the basis of efficiency parameters leveraging algorithms such as Dijkstra's shortest path routing. The **microscopic steering** controls the interaction between the UAV and its direct environment.

The key challenge in the design of UAV steerings lies in the trade-off of conflicting goals: while the application tasks typically aim for a maximum range and unrestricted movements, the communication constraints call for a limitation of the distance between the different components of the communication link. A system architecture and various technical approaches, such as *cluster breathing*, *smart cubes* and *communication-aware potential fields* have been discussed in Wietfeld and Daniel (2014) with a focus on aerial exploration and connectivity provisioning tasks. For the integration of UAVs with underground sensing systems, the communication-aware potential fields approach will be further explored.

In Goddemeier et al. (2012) UAVs have been associated with potential fields, thereby creating respective attracting as well as repelling forces. The

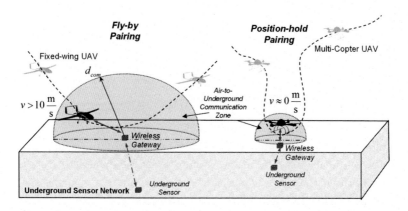

Figure 7.3.5 Pairing options depending on UAV type

communication-aware potential fields (CAPF) approach is suited to provide both micro- as well as macroscopic steering behavior: the area to be provisioned with a wireless communication service can, for example, be associated with a potential field generating an attracting force. In addition, base stations required to ensure connectivity with other networks induce attracting forces. In the context of the integration of UAVs with UGSs the **pairing between the UAV and the ground-based UGS gateway** is the specific challenge of the CAPF-steering algorithm: the wireless gateways are associated with attracting forces and therefore guide the UAVs in the direction of the gateway. In order to determine the desired steering, the location of the wireless gateways, the expected communication range d_{com}, the minimum cruising speed v, the achievable data rate r_{com}, and the expected amount of data produced by the underground sensing system D_{UGS} need to be known to the UAVs in order to overlay the different forces accordingly: in case of a limited communication range, the attracting force will be stronger to achieve a low distance during the pairing process.

Two pairing options can be identified (see Fig. 7.3.5):
- A fixed-wing is able to perform a **fly-by pairing** with a minimum cruising speed, which limits the time span available for communication with the wireless gateway of the underground sensing system.
- A multicopter-type UAV is able to gather data from the wireless gateway in **position-hold mode**, with speeds near zero enabling long communication time spans.

In case of a sufficiently large d_{com}, the fixed wing may also perform a circular movement within the communication zone, but for the purpose of this discussion, the two most general cases mentioned above are considered.

To achieve an appropriate time span for communication T_{com} during the pairing process, the UAV-specific minimum cruising speed is key. The minimum communication time is given by

$$T_{com} = \frac{d_{com}}{v_{min}} > \frac{D_{UGS}}{r_{com}}. \qquad (7.3.1)$$

While the time span available for the data transfer can be easily scaled with a multicopter platform due to its capability for holding a defined position with a speed near zero, a fixed-wing UAV may need to fly-by several times in order to gather all available data. The number of necessary fly-bys n can be calculated with

$$T_{com} = n\frac{d_{com}}{v_{min}} > \frac{D_{UGS}}{r_{com}} \Leftrightarrow n > \frac{D_{UGS}}{r_{com}}\frac{v_{min}}{d_{com}}. \qquad (7.3.2)$$

The above introduced formulas shall provide a general indication of the design considerations. At the same time it should be noted that although line-of-sight connection is often a valid assumption in UAV-based scenarios, the communication distance as well as the available data rates depend on numerous system and environmental parameters impacting the performance of the communication link, such as adaptive modulation and coding, antenna characteristics, transmit power, multipath fading and shadowing.

7.3.3 EXAMPLE: DAM MONITORING AND INFORMATION SYSTEM

Natural hazards and natural disasters with a great potential for damage typically occur suddenly and are relatively unlikely. With the increasing occurrence probability of some natural hazards caused by climate change, the need for technical measurement systems to early detect the potential for damage is rising. The German TaMIS project (Stasch et al., 2017) focuses explicitly on dams, as these man-made structures are subject to direct natural exposure. Beyond that, as a large-scale facility, they have a significant potential for damage. Potential damages associated with dams usually stem from the failure of the structure due to overtopping, seepage water flow, and deformation. As part of a risk landscape that is constantly changing due to climatic changes and human impacts (urbanization), dam (systems) require continuous monitoring with various observations. These measures need to be quantitatively combined in order to extract maximum information. The type of data collection normally differs on a case by case basis and

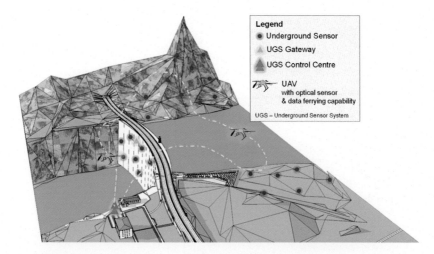

Figure 7.3.6 Dam monitoring application with UAV-assisted data ferrying for underground sensing system

can range from automated to semiautomated or manual measurement. The main challenge in this context is to capture multiparametric data sets with ground-based sensors in a multiscale environment, to combine them with intelligent sensor networks and connect them with geoprocessing, alerting, and visualization capabilities. The system architecture and information flow of the proposed sensor and measurement subsystem is depicted in Fig. 7.3.6.

In order to reliably monitor risk-related phenomena and to enhance operational decision support, time- and space-continuous sensor measurements are provided by including various independent measurement systems and sensors. To extend the state-of-the-art, TaMIS will use innovative (minimal-invasive-) sensor technology capable of detecting retention levels and enabling early detection of unexplained seepage.

Although the individual measurements for one figure of merit are significant, comprehensive conclusions or the identification of cascading influences cannot be detected using state-of-the-art solutions. Therefore, TaMIS provides a comprehensive risk management system that aims to keep the potential damage to a minimum. Monitoring relies on the observation of different system parameters that are directly or indirectly relevant for the detection of abnormal events and are instantaneously provided by the sensor and measurement infrastructure. The continuous gathering, combination, and analysis of the measured data are therefore the basis for all risk related decisions. The risk potential associated with the measured data will be verified by integrated time series statistical methods and specialized modeling

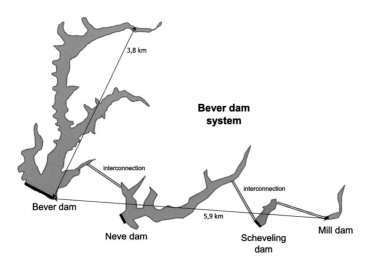

Figure 7.3.7 Overview of Bever dam system to be used for the validation of a dam monitoring system within the TaMIS project

software, in which a predicted risk situation in the operational mode can be derived by the same methods.

The concepts and technologies developed in TaMIS will be validated in a real-life environment, the so-called Bever dam system (see Fig. 7.3.7), which is a set of interconnected dams in the German Wupper area.

In addition to the original scope of the project, which initially did not yet consider the use of UAVs, the concepts outlined in this sub chapter will be validated in a real-life demonstrator. While the operator of the Bever dam system uses UAVs already for optical inspection, in the future the data ferrying service discussed in this chapter will become one backup option, in case the standard cellular systems used for inline-monitoring would not be available.

This application example shows that the integration of underground sensor systems with UAVs has – although yet in an early research state – valid potential to gain practical relevance in the future.

REFERENCES

Bacco, M., Ferro, E., Gotta, A., 2014. UAVs in WSNs for agricultural applications: an analysis of the two-ray radio propagation model. In: Sensors 2014. IEEE, pp. 130–133.
Daniel, K., Rohde, S., Goddemeier, N., Wietfeld, C., 2011. Cognitive agent mobility for aerial sensor networks. IEEE Sens. J. 11 (11), 2671–2682.

Dunbabin, M., Marques, L., 2012. Robots for environmental monitoring: significant advancements and applications. IEEE Robot. Autom. Mag. 19 (1), 24–39.

Elfes, A., Siqueira Bueno, S., Bergerman, M., Ramos Jr., J.G., 1998. A semi-autonomous robotic airship for environmental monitoring missions. In: Proc. IEEE Int. Conf. Robotics and Automation, vol. 4, pp. 3449–3455.

Frew, E., Brown, T., 2008. Airborne communication networks for small unmanned aircraft systems. In: Proceedings of the IEEE, vol. 96, no.12, pp. 2008–2027.

Goddemeier, N., Daniel, K., Wietfeld, C., 2012. Role-based connectivity management with realistic air-to-ground channels for cooperative UAVs. In: Special Issue on Networking for Unmanned Autonomous Vehicles. IEEE J. Sel. Areas Commun. 30, 951–963.

Gomez, K., et al., 2016. Aerial base stations with opportunistic links for next generation emergency communications. In: Special Issue on Critical Communications and Public Safety Networks. IEEE Commun. Mag.. Accepted for publication.

Hägerling, C., Ide, C., Wietfeld, C., 2014. Coverage and capacity analysis of wireless M2M technologies for smart distribution grid services. In: 2014 IEEE International Conference on Smart Grid Communications (SmartGridComm), pp. 130–133.

Hayat, S., Yanmaz, E., Bettstetter, C., 2015. Experimental analysis of multipoint-to-point UAV communications with IEEE 802.11n and 802.11ac. In: 2015 IEEE 26th Annual International Symposium on Personal, Indoor, and Mobile Radio Communications, PIMRC, Aug. 30–Sept. 2, pp. 1991–1996.

Rohde, S., Goddemeier, N., Wietfeld, C., Steinicke, F., Hinrichs, K., Ostermann, T., Holsten, J., Moormann, D., 2010. AVIGLE: a system of systems concept for an avionic digital service platform based on micro unmanned aerial vehicles. In: Proc. of IEEE International Conference on Systems, Man, and Cybernetics, SMC, Istanbul, Turkey.

Sbeiti, M., Goddemeier, N., Behnke, D., Wietfeld, C., 2015. PASER: secure and efficient routing approach for airborne mesh networks. In: IEEE Transactions on Wireless Communications, no. 99. p. 1.

Stasch, C., Pross, B., Gräler, B., Malewski, C., Förster, C., Jirka, S., 2017. Coupling sensor observation services and web processing services for online geoprocessing in water dam monitoring. Int. J. Digital Earth, 1–17. http://dx.doi.org/10.1080/17538947.2017.1319977. Taylor & Francis.

Wang, D., Song, L., Kong, X., Zhang, Z., 2012. Near-ground path loss measurements and modeling for wireless sensor networks at 2.4 GHz. Int. J. Distrib. Sens. Netw. 2012, 969712. http://dx.doi.org/10.1155/2012/969712. 10 pages.

Wietfeld, C., Daniel, K., 2014. Cognitive networking for UAV swarms. In: Valavanis, K.P., Vachtsevanos, G.J. (Eds.), The Handbook of Unmanned Aerial Vehicles. Springer, pp. 749–780.

Yoon, S.-U., Cheng, L., Ghazanfari, E., Pamukcu, S., Suleiman, M.T., 2011. A radio propagation model for wireless underground sensor networks. In: Proceedings of IEEE Globecom. Houston, TX.

Zeng, Y., Zhang, R., Lim Teng, J., 2016. Wireless communications with unmanned aerial vehicles: opportunities and challenges. IEEE Commun. Mag. 54 (19), 36–42.

CHAPTER 8

Underground Sensing Strategies for the Health Assessment of Buried Pipelines

**Sean M. O'Connor*, Jerome P. Lynch*, Mohammad Pour-Ghaz[†],
Srinivasa S. Nadukuru*, Radoslaw L. Michalowski*, Russell A. Green[‡],
Aaron Bradshaw[§], W. Jason Weiss[¶]**
*University of Michigan, Ann Arbor, MI, USA
[†]North Carolina State University, Raleigh, NC, USA
[‡]Virginia Tech, Blacksburg, VA, USA
[§]University of Rhode Island, Kingston, RI, USA
[¶]Oregon State University, Corvallis, OR USA

8.1 INTRODUCTION

Lifelines infrastructure includes natural gas and oil pipelines, water and sewage lines, gas and oil storage facilities, tunnels, power and communication lines, among others (Ariman and Muleski, 1981), which are vital to modern society and urbanization. As the scale of urbanization increases and societal reliance on modern infrastructure grows, the negative impacts stemming from a lifeline failure also grow. Damage to critical infrastructure lifelines, such as buried pipelines, can have potentially extreme consequences including loss of water pressure, energy supply and communication as well as secondary effects such as widespread disease due to contaminated drinking water and hindered response efforts due to lack of lifeline resources (e.g., water supply for firefighting). Among of the most serious hazards to buried pipelines are seismic and landslide events, resulting in wave propagation and permanent ground displacement (PGD). PGD hazards are typically considered to be much more severe (R. Eguchi, 1983; O'Rourke, 2005) than wave propagation. PGD can be localized to a small section of the pipeline, such as the case for surface faulting, or widespread, such as the case for large scale lateral spreading during lique-faction. Widespread PGD can result in many damage locations throughout the lateral spreading area while localized PGD can result in few damage locations but with potentially much more severe damage. Pipeline damage estimations based on wave motion and PGD metrics have been devel-

Underground Sensing.
DOI: http://dx.doi.org/10.1016/B978-0-12-803139-1.00008-4

oped. Eguchi (1983) correlated pipe-break rate with modified Mercalli intensity (MMI). O'Rourke and Ayala (1993) presented wave propagation damage rate versus peak ground velocity for various pipe types and materials. Several researchers developed empirical wave propagation damage relations for various pipe types and situations (Eidinger et al., 1995; Honegger, 1995; O'Rourke and Jeon, 1991). Empirical damage relations for PGD have also been developed (Heubach, 1995; Eidinger et al., 1995; Porter et al., 1991). The current methodologies for post-PGD pipeline assessment are predominately visual inspection from the ground level. Specifically, inconsistencies at the ground level may indicate pipeline displacement below. Above ground sensing methods can also be used such as infrared thermography (IT) and ground penetrating radar (GPR). IR thermography and GPR methods are convenient for providing images that may indicate pipe leaks or discontinuities occurring at the subsurface level (Birken and Oristaglio, 2014). However, these imaging methods can be costly, slow to deploy and operate, and require skilled technicians to operate. Imaging methods may also not provide the level of resolution required for damage localization. In-pipe methods have also been deployed, such as sending small remote devices ("smart pigs") through the inside of a pipeline to detect pipe damage. While empirical PGD-to-damage models and subsurface imaging technologies exist, better decisions could be made on subsurface pipelines after PGD if *in situ* sensing and monitoring is adopted.

Due to the importance of buried lifelines, it is critical that damage be located and diagnosed quickly so that risks to humans and property can be minimized and repairs made that minimize service outages. Monitoring systems are an obvious approach, allowing for quick assessment of damage severity and location for fast repair. However, two challenges exist for monitoring buried pipelines. First, pipelines often run for tens to hundreds of miles requiring a judicious approach to selecting where sensors are installed. Second, their subsurface location hinders the acquisition of data from installed sensors. To date, most *in situ* pipeline monitoring systems have been tethered (including traditional wired and fiber-optic sensors) with wiring installed alongside the pipeline to communicate data to a data acquisition system (Glisic, 2014). Such methods can be expensive due to the invasive installation requirements.

The primary goal of this chapter is to illustrate experimental sensing methods that can serve as the basis of future pipeline monitoring systems. Traditional and novel sensing devices are explored in addition to the use of wireless telemetry as a strategy for data acquisition from buried sensors.

Specifically, the chapter focuses its attention on customization of a sensing strategy for buried segmented concrete pipelines subjected to PGD. Segmented pipelines, and concrete segmented pipelines in particular, are common buried lifeline systems used to transport waste and storm water. They exhibit dramatic damage when exposed to large PGD events. Depending on the orientation of the pipeline with respect to the fault plane and the direction of faulting, the pipeline may experience axial forces resulting in tension or compression, as well as shear and bending. The principal failure modes in continuous pipelines are tensile rupture and local buckling (O'Rourke, 2003). Segmented pipelines failure is primarily observed as joint distress (O'Rourke, 2003). Axial forces in segmented pipelines can lead to joint pullout or bell and spigot crushing (i.e., telescoping). The sensing technologies presented will be specific to monitoring the motion of pipeline segments and to directly detect damage at the pipeline joints during PGD. Towards this end, the chapter describes the full-scale testing of buried segmented concrete pipelines at the Network for Earthquake Engineering Simulation (NEES) Lifeline Experimental and Testing Facilities at Cornell University. The tests provide an ideal venue to not only validate the performance of buried sensors for pipeline health monitoring, but the tests also advance the understanding of the evolution of damage in segmented concrete pipelines under PGD. Tasks include construction of a full-scale segmented concrete pipeline, design and installation of a sensing system for damage detection and localization, testing and validation of buried wireless communications, observation of soil–pipeline interaction during PGD, and analysis of damage evolution.

8.2 OVERVIEW OF BURIED PIPELINE SENSING TECHNOLOGY

Sensing technologies for buried pipelines fall into several categories including internal sensing, external sensing, and remote sensing (Bradshaw et al., 2009). Internal sensing utilizes sensors inside the buried pipeline to detect damage. An example of internal sensing is the use of intelligent pig devices. Pig devices were originally intended to clean the interior of the pipes but today are instrumented with sensor technologies to inspect for signs of damage including eddy current and ultrasonic transducers for crack detection and wall thickness estimations (concerned with corrosion section loss) (Liu, 2003). A clear benefit of pigs is that they can travel over long distances; this allows the expensive precision instrumentation to be quickly amortized over very large inspection distances. External sensing includes the installation of sensors on the surface of the pipeline. For example, acoustic sensors have been used to detect leaks in metallic pipelines; sudden pressure drop due to a leak causes a rarefaction wave that propagates away from the crack at the speed of sound (Mannan, 2012). These methods can detect the location of cracks but typically not their severity. Ultrasonic transducers can also be installed to introduce Lamb waves in the thin walls of metallic pipelines for direct damage detection through electromechanical impedance and pitch-catch methods (Giurgiutiu, 2008). Also included in external sensing are fiber optic sensors for measurement of global pipeline strain. There are three main principles for distributed sensing using fiber optic sensors (FOS): Rayleigh scattering (Posey et al., 2000), Raman scattering (Kikuchi et al., 1988) and Brillouin scattering (Kurashima et al., 1990). Finally, remote sensing methods are those using electromagnetic principles for aboveground noncontact measurements including IT (Inagaki and Okamoto, 1997) and GPR (Hayakawa and Kawanaka, 1998). IT makes thermal observations of the soil around a pipeline structure and searches for leaks in the pipeline while GPR uses electromagnetic wave reflection to map the soil–pipe boundary and thereby look for discontinuities. IT and GPR methods typically suffer from image resolution and user interpretation although these methods are rapidly improving.

More recent research has focused on wireless sensor systems for real-time damage detection in a variety of civil infrastructure applications (Lynch and Loh, 2006). Wireless systems are potentially beneficial for buried pipeline monitoring due to their low-cost, making it possible to deploy a large array of sensors on a fixed budget. Wireless systems also offer the opportunity to install them locally with local excavation; this is in contrast to wired solutions that often entail excavation of long sections of a pipeline for the placement of wiring. There are obvious concerns for wireless technology in buried pipeline monitoring too, including remote power and through-soil communication distances. Wireless systems have been shown to be capable of through-soil communication but soil is a high attenuation media that limits the range of buried sensors (Akyildiz and Stuntebeck, 2006; Stuntebeck et al., 2006). In addition to range, buried sensors require a power source. While batteries are feasible with long sleep cycles, eventually batteries need to be replaced. Long-term operational strategies can be adopted to maximize the life expectancy of buried sensors including deep sleep configuration until a triggering event (e.g., ground displacement, soil vibrations) occurs. Wireless sensors also possess computational power on-board, allowing real-time interrogation of sensor data and automation of damage detection. This can be especially valuable for buried sensors because data processing can be used to convert raw data into processed information that is only communicated when absolutely necessary. Also, on-board data processing has been shown to be an important part of maximizing the life expectancy of battery operated wireless sensors (Lynch et al., 2004).

8.3 SYSTEM ARCHITECTURE AND DESIGN

The remainder of this chapter describes three similar full-scale buried pipeline PGD testing programs performed at the Lifeline Experimental and Testing Facility at Cornell University between 2009 and 2011 (Kim et al., 2010; Pour-Ghaz et al., 2011; Glisic and Yao, 2012). These tests were performed to better understand the damage evolution in a segmented concrete pipeline under PGD and also to develop structural health monitoring methods for automating real-time evaluation of buried pipelines during PGD to immediately support disaster response. The three tests performed had slight variation in sensor instrumentation and pipeline construction. The following sections describe the test facility, pipeline, instrumentation, burial process, PGD actuation and data acquisition including buried wireless data collection.

8.3.1 Test Facility

Full-scale PGD testing was conducted at the Lifeline Experimental and Testing Facility at Cornell University, one of 14 George E. Brown, Jr. Network for Earthquake Engineering Simulation (NEES) research laboratories funded by the United States National Science Foundation (NSF). The facility, shown in Fig. 8.1, houses a soil storage system, a test basin, a reaction wall, large-stroke actuators, and a multi-channel National Instruments (NI) measurement system. The test basin itself consists of a stationary half and movable half so that pipeline specimens could be subject to simulated PGD. These halves were separated by a 50° angle from the longitudinal dimension of the pipeline. The test basin measures 3.4 m (134 in) wide, 13.4 m (528 in) long and 2.0 m (79 in) high. Ground faulting is simulated by controlled displacement of the north half of the test basin via four hydraulic actuators fixed to the reaction wall as seen in Figs. 8.1 and 8.2. The actuators are connected at the base of the test basin frame, causing the movable half of the test basin to slide a prescribed amount when the actuators are

(A)   (B)

Figure 8.1 NEES Lifeline Experimental and Testing Facility (Cornell University): (A) staging area for pipe instrumentation with the test basin seen to the right of the staging area; (B) displacement actuators at the base of the test basin for PGD simulation

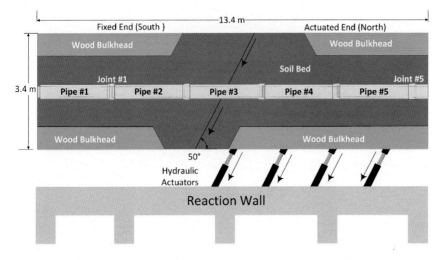

Figure 8.2 Test basin configuration and pipeline layout at the NEES Lifeline Experimental and Testing Facility

invoked. Fig. 8.2 shows an illustration of the test basin configured with a pipeline specimen subjected to PGD via hydraulic actuation. The wood bulkheads shown were introduced to reduce the amount of soil needed to bury the pipeline.

8.3.2 Pipe Segments

The three tests performed have slight variations in their pipeline designs. The 2009 segmented pipeline was constructed from five full length (and one partial length) commercial reinforced concrete pipes (Class 3/4). Each segment was 2.4 m (96 in) long, had a 30.5 cm (12 in) inside diameter, wall thickness of 6.3 cm (2.38 in) and weighed approximately 4000 N (900 lb). The pipe walls were reinforced with steel bars having reinforcement ratio 0.07 (in^2/linear ft). The compressive strength of the concrete was tested to be 27.6 MPa (4.0 ksi).

Tests conducted in 2010 and 2011 incorporated fiber reinforced concrete pipe segments. These segments were developed at Purdue University consistent with the standard Class 3/4 reinforced concrete pipe segments used in 2009. The average compressive strength of these segments was measured at 130.0 MPa (18.8 ksi). These segments were designed with a high volume fraction to achieve high fracture toughness (Pour–Ghaz et al., 2011). For the 2010 experiment, one fiber reinforced concrete segment was used (Pipe #4, see Fig. 8.2). The 2011 experiment utilized fiber reinforced concrete segments for the entire pipeline. The effect of the various pipeline configurations among the three tests will be observable in the sensor response data. Throughout this chapter, reference to the 2009, 2010 and 2011 tests are synonymous with reference to Ho-RC (homogeneous standard reinforced concrete), He-RC/FRC (heterogeneous concrete) and Ho-FRC (homogeneous fiber reinforced concrete), respectively (Pour–Ghaz et al., 2016).

8.3.3 Instrumentation

The overall measurement strategy was intended to assess damage caused by PGD induced ground rupture, determine damage evolution under PGD and explore monitoring technologies for buried pipelines. A mix of laboratory and remote health monitoring sensors were utilized including strain gages, potentiometers, acoustic emission sensors, magnetic proximity sensors, conductive surface sensors, fiber optic sensors and load cells to characterize pipeline behavior under PGD (partially seen in Fig. 8.3). Description of the use of potentiometers, strain gages, acoustic emission sensors and conductive surface tape as potential sensors for use in damage detection of buried pipelines is provided here. The fiber optic sensors used during testing are not covered in this chapter.

Relative displacement and rotation at the pipe joints were measured by installing a planar surface on the pipe interior at the spigot end and projecting five potentiometers installed on the interior of the adjoining bell end (Fig. 8.3A) onto that surface. While only three potentiometers are required to define the motion (i.e., relative longitudinal displacement and

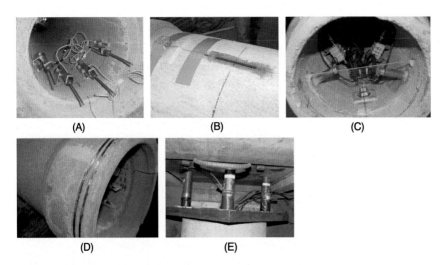

(A)   (B)   (C)

(D)   (E)

Figure 8.3 Sensor instrumentation: (A) potentiometers; (B) strain gages; (C) acoustic emission sensors; (D) conductive surface tape sensors; (E) load cells

rotation) of the pipe joints, redundant potentiometers serve to improve the estimation of joint motion. The potentiometer used was the Novotechnik TR100 linear potentiometer having a 10 cm (3.9 in) stroke range which modulates the displacement reading on a voltage output over a 5 V range. To create the planar bearing surface for the spring loaded potentiometers, a Lucite plate was installed at the spigot interior. The plate was installed at the pipe end so that when the pipe segments are attached the potentiometer shafts bear directly on the plate with the shaft moved to roughly the midpoint of its range. To prevent the potentiometer needles from contacting the adjacent pipe interior wall during rotation, the potentiometers were installed on mounting blocks approximately 7.6 cm (3 in) tall to bring the measuring probes away from the interior pipe wall and increase the range of measurable rotation. Both the mounting blocks and Lucite plates were installed using a high strength two-part epoxy. In total, 16 potentiometers were installed in 2009 and 20 potentiometers were installed in 2010 and 2011 (five at each joint from Joint #1 through Joint #4).

Metal foil strain gages were installed along each pipe segment to measure axial and bending strain behavior during transverse PGD, as well as hoop strain in the pipe bell due to compressive telescoping of the joint. Due to the nonuniform and nonhomogeneous nature of the concrete pipe, long-gage strain gages were used on the pipe body. The long-gage strain gages (Vishay N2A-06-40CBY 350 Ω metal foil gages with 10.2 cm (4 in) gage

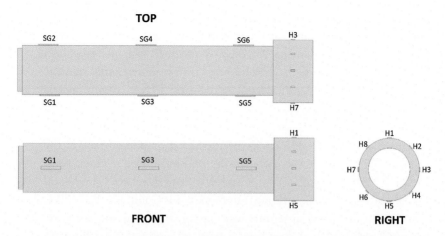

Figure 8.4 Strain gage sensor layout configuration for Pipe #1 to Pipe #5 (see Fig. 8.2). Thirty long-gage transducers (six per pipe) were used on the pipe body to while sixteen smaller gages (eight on Joint #3 and #4) were used to measure hoop strain

lengths) were arranged in pairs at three sections of each pipe body as seen in Fig. 8.4 (e.g., pair SG1 & SG2, pair SG3 & SG4, and pair SG5 & SG6). As seen in Fig. 8.4, when looking down on the installed pipe (i.e., top view) the strain gages are seen on the pipe side aligned along the longitudinal axis of the pipe segment. This strain gage configuration was designed to infer axial and bending deformations in each pipe segment during PGD. Axial strains were computed as the average strain of each pair; bending strains were computed as the difference strain of each pair (resulting in three axial and three bending strain measures for each pipe). At Joints #2 and #3, an additional eight strain gages (Vishay L2A-06-250LW 350 Ω with 0.63 cm (0.25 in) gage length) were mounted to the outside perimeter of the pipe bell to measure hoop strain in the joint under PGD, especially due to compressive telescoping. These gages can be observed again in Fig. 8.4 labeled H1–H8. Joint #2 and #3 are located on either side of the strike-slip fault. The hoop strain gages were installed at eight equidistant locations around the pipe bell. To prepare the pipe for strain gage installation, each surface was first ground smooth and filled with a very thin layer of epoxy. The epoxy layer was finely sanded and cleaned with an acid and base solution. Strain gages were applied to the smooth epoxy surface using strain gage bonding adhesive, coated with a thin layer of polyurethane, and finally covered with a wax layer to protect the gage from the soil. Fig. 8.3B shows a long-gage strain gage installed over an epoxy filler.

The potentiometers and strain gages provide direct measurement of structural responses of the buried pipeline during PGD but do not provide a direct measure of damage. To directly measure damage, especially early in the damage process, the pipelines are instrumented with acoustic emission sensors (Fig. 8.3C). The acoustic emission sensors were concentrated in locations at the joints near the fault due to the high probability of damage at these locations during PGD. In the Ho-RC pipeline, five acoustic emissions sensors were epoxy bonded to the inner surface of the pipe segment bells at Joint #2 (two sensors at the 3 and 9 o'clock positions), Joint #3 (two sensors at the 3 and 9 o'clock positions), and Joint #4 (one sensor at 12 o'clock position). In the Ho-FRC and He-RC/FRC pipeline, 12 acoustic emission sensors were installed: two sensors were installed at the 3 and 9 o'clock positions at Joint #1, Joint #4 and Joint #5 and three sensors were installed at the 3, 9 and 12 o'clock position at Joint #2 and Joint #3.

An additional direct sensing strategy for direct damage detection were conductive surface sensors installed on the circumference of the pipe segments (Fig. 8.3D). The copper surface sensors consist of low cost copper tape (0.63 cm (0.25 in) wide and 0.09 mm (3.5×10^{-3} in) thick) that is wrapped around the circumference of the pipe. When significant cracking of the pipe occurs at any point along the copper tape, the tape tears resulting in the tape no longer being conductive. This provides a binary measurement (i.e., conductive versus nonconductive) to determine when cracking in the pipe wall is sufficient to tear the tape. For the Ho-RC pipeline, a total of 24 conductive surface sensors were installed along the pipeline length: 2 around the bell end of Pipe #1, 6 along the length of Pipe #2 and Pipe #4 (2 at the spigot end, 2 in the middle of the pipe, and 2 at the bell end), and 2 around the spigot end of Pipe #5. For the Ho-FRC and He-RC/FRC pipelines, the conductive surface sensors were standardized with each pipeline having 45 sensors installed. Each pipe segment had 9 surface conductive sensors in the same installation configuration as shown in Fig. 8.5.

Additional sensor instrumentation not described here was also utilized during the Cornell project including fiber optic sensors (Glisic and Yao, 2012), magnetic opening sensors (Pour-Ghaz et al., 2011) and conductive grout (Pour-Ghaz et al., 2011). Fiber-optic sensors can be installed on the pipeline or even in the surrounding soil making for simple installation. Magnetic opening sensors are a low-cost method for detecting large relative

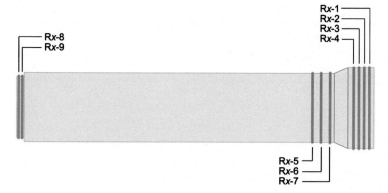

Figure 8.5 Schematic illustration of the location of the conductive surface for a typical pipe segment of the Ho-FRC and He-RC/FRC pipelines (R$x - y$ corresponds to the yth sensor position on Pipe #x)

(A)   (B)

Figure 8.6 Pipeline installation on 20 cm (8 in) thick soil bed: (A) view from the basin end; (B) view from above across the fault line

displacements at joint locations. Conductive grout is another low cost option for detecting severe displacement at joint locations (detected by grout cracking).

8.3.4 Pipeline Assembly

The assembly of the pipeline among all three tests was consistent. First, a 20.3 cm (8 in) soil bed was placed in the test basin and compacted to provide a stable foundation for the pipeline construction. Instrumented pipes were lifted into the test basin and assembled in order, starting at the south end and finishing with the cut piece on the north end (Fig. 8.6). Adjacent

Table 8.1 Soil specifications used for the three tests

Layer	Average total Density (kg/m^3)	Average water Content (%)	Average dry Density (kg/m^3)
L7 (Top)	1730	3.55	1665
L6	1745	3.60	1675
L5	1720	3.2	1660
L4	1755	3.7	1695
L3	1750	3.15	1690
L2	1750	3.05	1695
L1 (Bot)	NA	NA	NA

pipes were fit together at the bell–spigot connection with rubber gaskets placed on the spigot end of each pipe to ensure a water-tight and snug connection. Interior sensor cables were fed through each pipe prior to pipe fitting and arranged along the pipe invert, while exterior sensor cables were arranged along the side of the pipe most appropriate for the anticipated pipe displacement. After the fifth pipe was installed and the remaining length was measured, a final piece was cut and placed. The ends of the pipeline were held fixed by steel plates. At each end, load cells (Fig. 8.3E) were installed to measure the global axial forces developed in the pipeline during PGD.

After installation of the pipeline, the pipe joints were grouted. A wet grout was used at each pipe joint containing 30% fine aggregate, a 0.5 water-to-cement ratio, and a 0.5% high-range water reducer. The grout was poured into the joint void and contained by a plastic mold until the grout had set. All grout was allowed to set before backfilling. The compressive strength of the grout material was measured at 54.0 MPa (7.8 ksi).

The test basin was backfilled in 20.3 cm (8 in) lifts and compacted in accordance with field construction practice. Following compaction of each lift, the soil was checked for density and moisture using a nuclear density gage. A relatively constant soil density (1,700 kg/m^3) and water content (3.5%) were observed throughout the backfilling process (see Table 8.1). Backfilling was complete after the pipe was buried approximately 140 cm (55 in) beneath the surface. Once the final lift was compacted, a grid was laid out on the surface near the fault region to aid in visual observation of soil disturbance during PGD.

8.3.5 Data Acquisition

Data acquisition (DAQ) during testing was done using both wireless and wired data acquisition systems. All of the sensors installed were fed to National Instruments (NI) data acquisition systems (PXI chasses with SCXI data acquisition cards installed) that resided outside of the test basin. All of the pipeline sensors were fitted with wire extensions and fed through the pipeline interior (or exterior) to the end of the pipeline and finally outside of the test basin. Two NI data acquisition chasses were setup at the north and south ends of the test basin so that the sensor wiring running inside the pipeline would not have to cross the fault line. In total, the NI system collected 20 potentiometers, 46 strain gages, 8 load cells, and 4 actuators. Data was collected at 10 Hz using the NI system. The conductive surface sensors and acoustic emission sensors were interfaced to separate data acquisition systems. The resistances of the conductive surface sensors were measured using a traditional digital multimeter (Keithley Model 2000) and a multiplexer after each displacement step of the test basin. The acoustic emission sensors were interfaced to multichannel AMSY-6 (Vallen Systeme) acoustic emission data acquisition system, which collected acoustic energy based on threshold triggers.

8.3.6 Permanent Ground Displacement Simulation

The movable section of the test basin was displaced along the 50° angle of the basin creating a strike-slip fault in a displacement controlled manner by the four hydraulic large-stroke actuators fixed to the reaction wall described previously. Each displacement stage moved the test basin 2.5 cm (1 in) at a rate of 0.5 cm/s (0.2 in/s). The displacement program was scheduled for a total 12 actuation steps (30 cm PGD) along the fault line. Following each displacement stage a short period of rest was given to observe sensor data and allow the pipeline to relax. After testing, the soil was carefully excavated and a visual forensic analysis of the pipeline was conducted.

8.4 BURIED WIRELESS SENSING OF PIPELINE BEHAVIOR DURING PGD

A wireless monitoring system was implemented to develop automated real-time structural health monitoring techniques for buried pipelines during and immediately following a major PGD. The testing in the NEES Lifeline Experimental and Testing Facility provided a very unique opportunity to not only demonstrate the role of wireless sensors for monitoring segmented pipelines exposed to PGD, but more importantly, to explore the performance of wireless telemetry in underground applications. Towards this end, two types of wireless sensor nodes were utilized to collect data in the underground environment: *Narada* (Swartz et al., 2005) and *WiMMS* (Wang et al., 2005). Both platforms are similar in their sensing interface (4 channel, 16 bit analog-to-digital conversion) with the main differentiator of interest being their wireless radios. The *Narada* communicates at 2.4 GHz using the Texas Instruments CC2420 wireless transceiver whose maximum radio output power is 0 dBm. The *WiMMS* unit uses the Digi 9XCite wireless transceiver operating on 900 MHz with a maximum output power of 0 dBm. The intention behind using two wireless sensors is that each wireless sensor uses a different wireless interface communicating on different carrier frequencies; this allows the performance of the buried wireless sensor network to be evaluated as a function of the wireless communication parameters.

The voltage outputs of the potentiometers at every joint were split three ways with one signal interfaced to the laboratory NI system and the other two interfaced to wireless sensors. Three wireless sensors were used to record the potentiometers at each joint. One *WiMMS* node was used to measure four of the five potentiometers at each joint. In addition, two *Narada* wireless sensor nodes measured the output of the five potentiometers. The *WiMMS* and *Narada* nodes were enclosed in water-tight enclosures with a battery power source and installed on the outer surface of the pipelines. The enclosures were placed in a manner to ensure they were 75 cm (29.5 in) below the surface of the backfilled test basin as shown in Fig. 8.7. Two base stations were set-up outside of the test basin to collect data in real-time from the *WiMMS* and *Narada* wireless monitoring systems. The wireless transceivers interfaced to the base stations were placed on the soil surface in the vicinity of Joint #3. The wireless monitoring systems collected data at 10 Hz by each base station with the base stations running a MATLAB script that provided real-time joint displacement and rotations based on the potentiometer measurements.

8.4.1 Performance of Wireless Telemetry Underground

The main challenge to wireless sensing in underground applications is degradation of the propagated radio frequency (RF) wave due to the soil medium. Waves suffer from path loss, reflection/refraction, multi-path fading, reduced propagation velocity and noise in an underground setting (Akyildiz and Stuntebeck, 2006). Signal propagation will be largely affected by the wave frequency and soil properties, most especially soil moisture which directly affects the dielectric properties of the soil. Lower frequencies

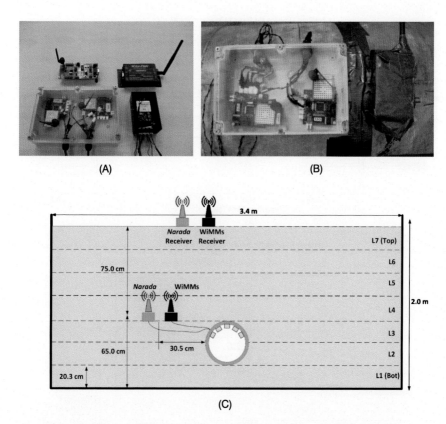

Figure 8.7 Wireless monitoring of the buried pipeline: (A) wireless equipment with *Narada* and *WiMMS* wireless nodes and receivers shown; (B) two Narada nodes in a water tight enclosure installed on the pipeline external surface prior to burial; (C) schematic of wireless sensor positioning in the test basin

will propagate with less attenuation than higher frequencies; it is therefore expected that the *WiMMS* wireless sensor nodes will perform better than the *Narada* sensor for buried sensing.

Fig. 8.8 shows the potentiometer response from the *Narada* (Fig. 8.8A) and *WiMMS* (Fig. 8.8B) wireless monitoring systems for the He-RC/FRC pipeline experiment at Joint #2. Shown is the displacement measured at each potentiometer during the complete set of PGD displacements. These plots serve as a general indicator of displacement measurement quality and the reliability of the buried wireless monitoring systems. The overall success of both wireless platforms was encouraging. As expected, the *WiMMS* platform outperformed the *Narada* platform in terms of successful data

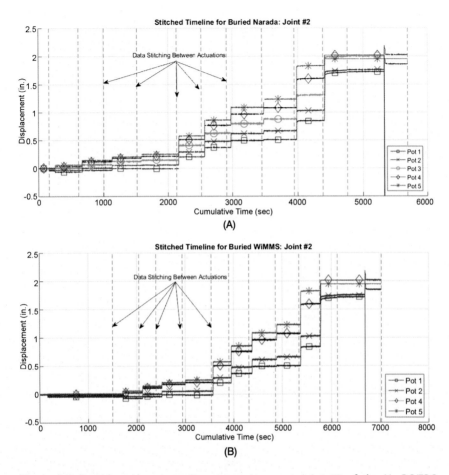

Figure 8.8 Buried wireless potentiometer response at Joint #2 of the He-RC/FRC pipeline: (A) *Narada* and (B) *WiMMS* measurements

Table 8.2 Wireless Packet Loss (%)

	Pot 1	Pot 2	Pot 3	Pot 4	Pot 5
Narada	5.3	4.2	10.0	6.5	3.4
WiMMS	0	0	NA	0	0

transmission with packet loss used as a evaluative performance metric. Table 8.2 tabulates the packet loss rate for the wireless sensors installed at Joint #2 during testing of the He-RC/FRC pipeline. The two *Narada* nodes at Joint #2 showed packet losses ranging from 3.4% to 10.0% for the five transmission channels (3.4% to 6.5% on one *Narada*, 10.0% on the other) while the *WiMMS* platform demonstrated no packet loss in all four transmission channels. The expectation that the *WiMMS* node would respond with better communication has been verified, validating the literature and theoretical expectation that a 900 MHz RF signal would suffer less attenuation than a 2.4 GHz RF signal. It should be noted that in these tests the aim was to assess the wireless transmission reliability. Hence, the wireless nodes are configured to send their data without the use of acknowledgments from the base stations. However, for real systems deployed in the field, the lost packets would still be recoverable through the use of send-acknowledge protocols.

In order to preserve battery life in the buried wireless system, data transmission was halted between actuations. These pauses in data collection were not synchronized between the *Narada* and *WiMMS* systems leading to the inconsistent time scale shown in Figs. 8.8A and 8.8B. Dashed vertical lines were added to each figure to show the start and stop time of each data collection stage for each actuation. The important piece of information to compare between the two figures is the potentiometer response collected at each actuation. These responses are nearly identical for the two systems after every actuation stage, showing that the data collected using *Narada* was reliable despite the less than perfect transmission behavior. At a high level, it is apparent from these figures that there is a growing compressive translation in Joint #2 (since each potentiometer is rising) as well a growing joint rotation (since the difference between potentiometers is rising accordingly). More detailed results on joint rotation and translation are shown later.

Although the *WiMMS* sensor outperformed the *Narada* sensor, the packet losses detailed in Table 8.2 appear mostly insignificant in Fig. 8.8A, especially for this type of data analysis. These losses would not reduce the ability of the *Narada* system to perform a condition assessment of the monitored pipeline. Further, utilizing a 2.4 GHz radio may present benefits of power reduction and longer battery life (likely a crucial design parameter for buried systems) due to shorter transmission times.

8.5 ASSESSMENT OF PIPELINE RESPONSES AND DAMAGE

Following excavation of the pipeline after PGD testing, joint damage was immediately evident in all three pipelines. The most severe damage was obvious at Joint #2 and #3 in all three pipelines; this is expected since these joints are the closest to the fault line. The type of damage observed was combined compressive telescoping and rotation. Damage to the body

Figure 8.9 Deformed Ho-FRC pipeline after excavation: (A) bird's eye views from the top of the test basin; (B) west side show pronounced rotations at Joint #2 and #3

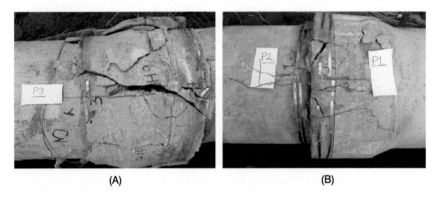

Figure 8.10 Typical damage to compressive PGD: (A) compressive failure including spalling of the bell at Joint #3 of the Ho-RC pipeline; (B) severe crack damage in bell at Joint #1 of the Ho-RC pipeline

of the pipe segments was not observed. Fig. 8.9 shows a global view of pipeline deformation following PGD in the Ho–FRC pipeline test in 2011. This type of deformation was consistent across the three years of tests. Pronounced rotation of Pipe #3 is evident, with minimal rotation of the remaining pipes. The hinge–like rotation behavior seen at Joint #2 and Joint #3 implies a significant loss of rotational stiffness at these locations. Joint #2 and Joint #3 experienced the most severe telescoping and rotational damage, with large segments of the pipe joint spalled off of the main structure, as shown in Fig. 8.10A. Severe telescoping failures were also observed at Joint #1 and Joint #4 as shown by Fig. 8.10B. The compression of the pipeline caused by PGD resulted in the spigot end of the joints be-

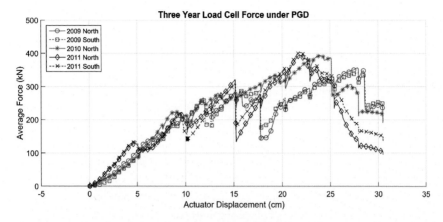

Figure 8.11 Load cell measurements recording pipeline force at the north and south ends of the test basin during PGD of the Ho-FRC pipeline

ing pushed into the bell end, forcing the bell outward and causing tensile cracks in the bell. Visually observed damage is expected to be quantitatively observed by sensor data as well; the following sections describe the findings of the sensor data.

8.5.1 Pipeline Load

The north and south ends of the test basin housed load cells to measure the load on the test basin from the pipeline under transverse PGD. The load cell measurements taken at both end of the pipeline indicate when load was building in the pipeline and when significant energy was released, possibly due to a joint failure. Fig. 8.11 shows the load from the combined load cells at the north and south ends for all three years of testing. The load cell readings from the south end during the He-RC/FRC pipeline (2010 test) is not shown due to incomplete data at that location. All three pipelines have slightly different behavior. The He-RC/FRC (2010 test) and Ho-FRC pipelines (2011 tests) both reach a maximum pipeline load of approximately 400 kN (89.9 kips) before a permanent decline in the axial load carrying capacity of the pipeline structure; the Ho-RC pipeline (2009 test) tops out closer to 350 kN (78.7 kips). Each test also shows a different point in time when a major load loss occurs. In the Ho-RC, He-RC/FRC and Ho-FRC pipelines, major load drops are observed at the seventh actuation (~17.5 cm), tenth actuation (~25 cm), and sixth actuation (~15 cm) load steps, respectively. There are some common characteristics from all pipelines as well, especially when observing the overall load carrying behavior of the pipeline from the start of PGD to the end. In each pipeline, a significant drop in force was observed between the fourth and fifth actuation (~10 to 12.5 cm) followed by a rebound in force sustained until between the sixth and seventh actuation (~15 to 17.5 cm), and again followed by a sustained rebound in load carrying capacity until a final loss of capacity between the ninth and tenth actuation (~20 to 23 cm). Strain gage data shown later are used to estimate pipe loads which could be analyzed similarly to the load cell data shown in Fig. 8.11; clearly, strain gages are more practical in a real-world setting where load cells are not feasible. Knowledge of pipeline axial loads indicate only when load carrying capacity is lost but does not inform hazard responders if the pipeline has exceeded lower performance limits associated with the evolution of damage. The other sensors are necessary to infer the health of the pipeline.

8.5.2 Joint Rotation and Translation

The *Narada* wireless system was utilized during all three pipeline tests to acquire and process relative pipe movement using the potentiometer sensors at the bell-spigot connections. Data from these sensors were used to estimate the relative rotation and translation of the pipeline at the joints. Fig. 8.12A shows the type of data visualization produced in real-time to provide information on the rotation and displacement of each joint during PGD testing in the test basin. For clarity in Fig. 8.12A, only 5 of the 12 actuator displacement stages are shown. The figure shows the evolved overall shape of the pipeline by projecting joint rotations with the numerical value of joint rotation and translation overlaid. For example, it can be seen that Joint #1 has a relative rotation of $-0.02°$ and relative compressive translation of -1.5 cm (-0.59 in) after 22.86 cm (9.0 in) of simulated PGD (actuation step #9). The highest amount of pipe rotation and translation was concentrated to the joints on either side of the fault line. Specifically, sensor measurements computed relative rotation at Joint #2 and Joint #3 of $-5.2°$ and $4.3°$, respectively, and relative translations of -2.2 cm (-0.87 in) and -1.8 cm (-0.71 in), respectively. Joint #1 and Joint #5 each showed very minimal measure of rotation. However, Joint #1 did show a significant amount of compressive telescoping of -1.50 cm (-0.59 in). Relative rotations measured by hand at the end of the tests during excavation are within 15% of the measured rotations by the wireless monitoring system. Specifically, hand measures of absolute rotation at Joint #2 and Joint #3 were $6.6°$ and $5.7°$, respectively. Some of the inaccuracy may be attributed to not

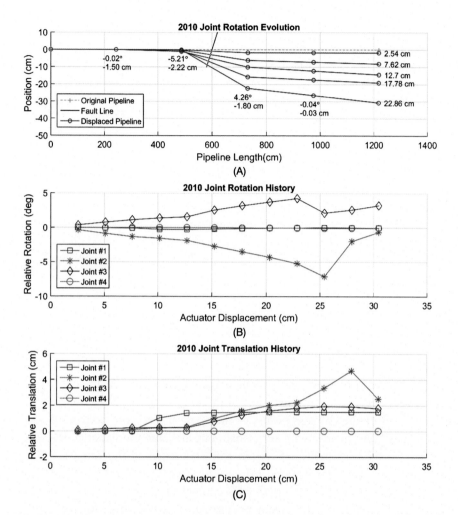

Figure 8.12 Real-time monitoring of pipe joint rotations and translations of the He-RC/FRC pipeline: (A) visualization of pipeline global displacements; (B) rotation of pipeline joints; (C) longitudinal translation of pipeline joints

recording a baseline pipeline orientation following pipeline construction (i.e., it was assumed each pipe started with a relative rotation of 0°).

Figs. 8.12B and 8.12C show the complete rotation and translation history for each joint. Fig. 8.12B shows minimal rotation in Joint #1 and #4 and large opposing rotations in Joint #2 and #3. Two distinct rates of rotation are apparent in the figure as well (changing at approximately 12 cm (4.7 in) displacement). This is an indication that some type of change in the pipeline rotational stiffness happened near this amount of PGD. From Fig. 8.12C, compressive telescoping insights can also be drawn. It is apparent that the telescoping which occurred at Joint #1 started to develop in the third actuation step (approximately 7.6 cm (3.0 in)). Also apparent is that the compressive translation increases dramatically near the same 12 cm (4.7 in) displacement mark mentioned previously, reinforcing the assumption that a major structural change occurred at this point of the PGD simulation.

Joint rotation and translation data from the three pipeline experiments are shown in Fig. 8.13 and Fig. 8.14 in order to gain insight on the damage evolution of segmented concrete pipelines. It is expected that this information will reinforce the load cell data shown previously in Fig. 8.11. For the most part, joint behavior can be categorized into two groups: (1) Joint #1 and #4 (Fig. 8.13) and (2) Joint #2 and 3 (Fig. 8.14) which are the immediate joints on both sides of the fault line.

Joint #2 and #3 are observed to gradually rotate and translate consistently throughout PGD, eventually resting a final 5 to 6° rotation. The most dramatic changes in rotation occur near the end of the PGD and happen as the result of severe joint failure at these joints (i.e., a large loss of rotational stiffness). Joint #1 and #4 show much different behavior. These joints are observed to experience a single damaging moment. At Joint #1, damage occurs between the fourth and fifth actuations (10 to 12.5 cm); at Joint #4, all of the damage occurs at the sixth and seventh actuations (15–17.5 cm). Here rotations are very small, but the large translational displacements imply telescoping failures in the later stages of the simulated PGD. Joint #1 and #4 translations are seen to reach over 2 cm (0.79 in) such as that seen in Joint #4 of the Ho-FRC pipeline. Joint #4 translation events of the Ho-FRC pipeline seem to correlate well with large force reductions in the load cell measures shown in Fig. 8.11. Initial losses in the cells near the fourth actuation step (10 cm) also seem to correlate well with Joint #1 damage events. Joint #3 shows an increased rotation and translation rate near 22.5 cm (8.9 in), which matches well with the final load drop witnessed in the Ho-FRC pipeline (see Fig. 8.11).

From the rotation and translation data shown, a damage evolution hypothesis can be made. Initially, Joint #2 and #3 begin to rotate and bear on each other as a result of the compressive displacement caused by the PGD. Telescoping and rotational damage begins between 0 and 10 cm (3.9 in) of PGD. As the compressive load in the pipeline increases, the telescoping failures occur further away from the fault location, first happening in the

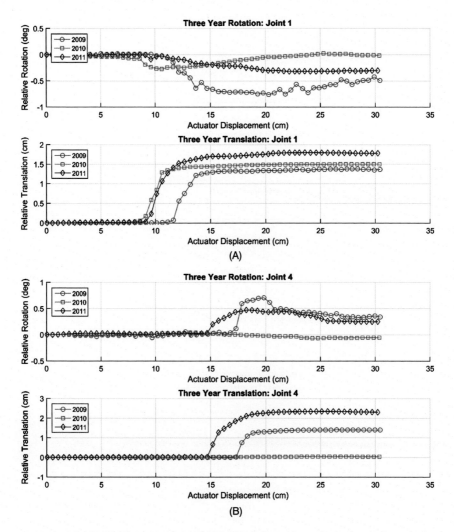

Figure 8.13 Pipeline joint rotations and translations for all three pipelines (Ho-RC, He-RC/FRC and Ho-FRC tested in 2009, 2010 and 2011, respectively): (A) Joint #1 and (B) Joint #4

fixed half at Joint #1 near 10 to 12 cm (3.9 to 4.7 in) of PGD, and later occurring at the movable half in Joint #4 near 15 to 17.5 cm (5.9 to 6.9 in) of PGD. Even after the outside joints (Joint #1 and Joint #4) have suffered telescoping failures, they continue to bear on each other resulting in further load increase. This has a final catastrophic result at the inside joints (Joint #3 and Joint #4) in addition to increased rotation.

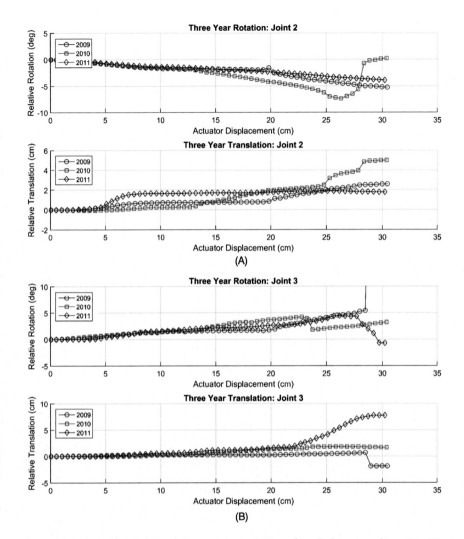

Figure 8.14 Pipeline joint rotations and translations for all three pipelines (Ho-RC, He-RC/FRC and Ho-FRC tested in 2009, 2010 and 2011, respectively): (A) Joint #2 and (B) Joint #3

8.5.3 Pipe Strain Responses

In each of the three pipelines tested, six long-gage strain gages were instrumented to each of the five pipe segments making up the pipeline as described previously. Using the measured strains, the axial load and bending strain can be estimated for each segment. In order to estimate pipe load, strain measures across each pipe segment were averaged and Hooke's

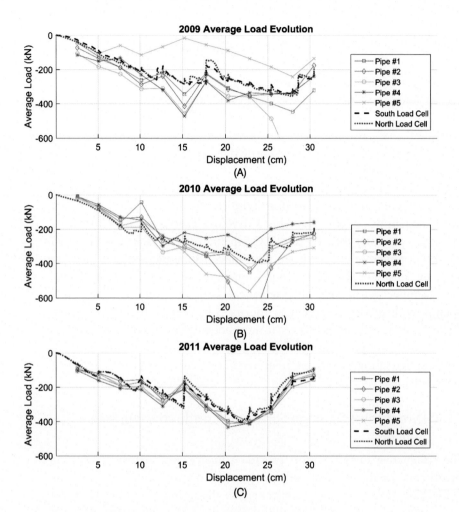

Figure 8.15 Average axial load in all pipeline segments based on strain measurements: (A) Ho-RC pipeline (2009 test); (B) He-RC/FRC pipeline (2010 test); (C) Ho-FRC pipeline (2011 test). Average pipe load from strain gage measurements compared with load cell data

law was applied where the standard reinforced concrete was given an elastic modulus of $E = 57000(\sqrt{f_c'})$, where f_c' is in psi (AASHTO LRFD 5.4.2.4). The bending strain consists of the differential strain measured at the 3 and 9 o'clock positions at each transverse location along the pipeline; bending strain is assumed to be proportional to the internal sectional moment.

Fig. 8.15 shows estimated pipe axial load for the full PGD (30.5 cm (12.0 in)) for each of the five pipes instrumented in all three years. The axial load cell force measurements are superimposed on each plot. In each of the three pipelines, the overall trend as well as the abrupt changes in axial load are captured well by the strain gage estimations for axial load. The benefit of this figure is that it demonstrates how strain gages can be used to infer axial load behavior in pipelines near a PGD. Furthermore, it draws a direct causal relationship between axial strain measurements and observed damage behavior (e.g., excessive translations and rotations). The strain gages could be implemented for field use whereas potentiometers and load cells could not be used.

Fig. 8.16 shows bending strain at each strain gage pair for all of the odd numbered actuations steps (for figure clarity). It can be observed that bending strain in the pipe segments is developed more in Fig. 8.16B (He–RC/FRC pipeline) than Pa (Ho–RC pipeline). The result in Fig. 8.16C (Ho–FRC pipeline) is even more pronounced showing bending moment behavior consistent with a continuous pipeline under PGD. More specifically, an inflection point is observed at the fault location in the Ho–FRC pipeline with positive moment in the fixed half and a negative moment in the movable half. It appears that the fiber-reinforced concrete pipeline (Fig. 8.16C) is capable of developing larger bending deformation before joint damage compared to the reinforced concrete specimens (Fig. 8.16A and 8.16B).

8.5.4 Direct Joint Damage Sensing – Conductive Surface Sensors

The conductive surface sensors provide a binary measurement indicating the absence or the presence of damage. The damage detected using the conductive surface sensors is illustrated on the average load displacement of the actuators in Fig. 8.17 for all three pipelines. The numbering scheme used in Fig. 8.17 for conductive surface sensors is based on the pipe segment number and the location of the sensor on the pipe segment: the sensors are indicated by letter R (for resistance-based) followed by two numbers separated by a dash where the first number indicates the pipe segment number and the second number indicates the sensor number on the pipe segment. For example, R4-6 indicates a conductive surface sensor installed on Pipe #4 and the sixth sensor counted from the bell section (see Fig. 8.5).

In Fig. 8.17A, the damage to the Ho–RC pipeline (2009) occurs after the second displacement step (5.08 cm displacement). Initially the damage is concentrated at the joints in the immediate vicinity of the fault line since four out of five sensors are located at the bell section of Pipes #2 and #3

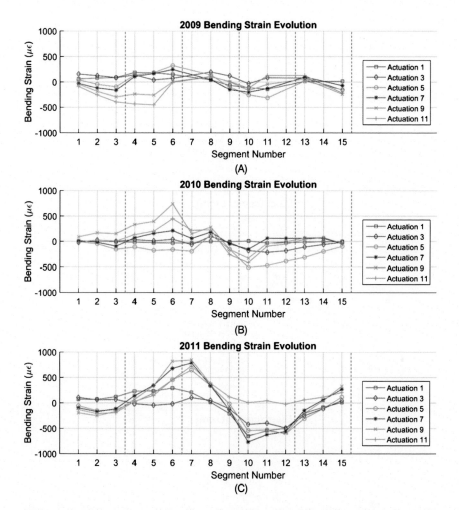

Figure 8.16 Bending strain differential at all strain measurements locations: (A) Ho-RC pipeline (2009 test); (B) He-RC/FRC pipeline (2010 test); (C) Ho-FRC pipeline (2011 test). Pipe segments delineated by dotted vertical lines and every other actuation step shown

(R2-1 and R2-2 at the bell section of Pipe #2 and R3-1 and R3-2 at the bell section of Pipe #3). Minor damage is also observed at this stage at the spigot section of Pipe #5 (R5-2). With further displacement up to 12.70 cm, more damage occurs away from the fault line (R4-6). The damage that occurs with displacement up to 17.78 cm is further away from the fault line (R4-1 and R4-2). This damage is accompanied by a drop in the actuator load. Note that this drop in the actuator load is consistent

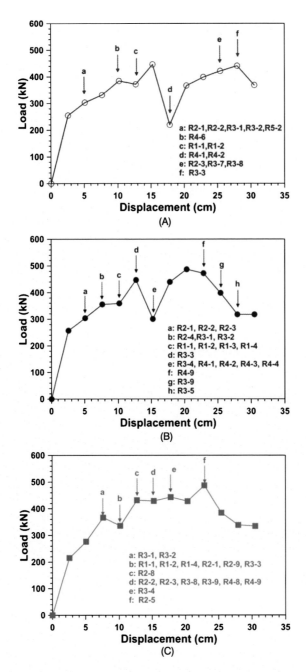

Figure 8.17 Illustration of conductive surface sensor signal on load-displacement curves: (A) Ho-RC pipeline (2009), (B) Ho-FRC pipeline (2011), and (C) He-RC/FRC pipeline (2010)

with the drop in axial load in Fig. 8.11 (2009 North and South). With actuator displacements up to 25.40 and 27.94 cm, damage again occurs at the pipe segments close to the fault line with the difference that the damage extends away from the bell section of the pipe segments and ruptures the body of the pipe segments (R2-3 and R3-3 shown by arrows "e" and "f" in Fig. 8.17A).

Fig. 8.17B illustrate the detected damage using conductive surface sensors on the Ho-FRC pipeline (2011). Note that a denser array of sensors is used in Ho-FRC pipeline compared to the Ho-RC pipeline. However, the damage propagation in Ho-FRC pipeline (Fig. 8.17B, 2011 pipeline) follows the same trend as the damage in Ho-RC pipeline (Fig. 8.17A, 2009 pipeline). Initially damage occurs in the immediate vicinity of the fault line (sensor R2-1, R2-2, R2-3, and R2-4 at the bell section of Pipe #2 and sensors R3-1 and R3-2 at the bell section of Pipe #3). Note that the detected damage in Pipes #2 and #3, Joints #2 and #3, respectively, are very consistent with the measured relative rotations in Fig. 8.14A (Joint #2) and Fig. 8.14B (Joint #3). The damage and cracking to these joints therefore are caused by the relative rotation of the joints. With further displacement in Fig. 8.17B, damage occurs away from the fault line and is observed at the bell sections of Pipe #1 (R1-1 through R1-4) and #4 (R4-1 through R4-4).

In both the Ho-RC (2009) and Ho-FRC (2011) pipelines, damage started after 5.08 cm of actuator displacement. The first damage in He-RC/FRC pipeline (2010), however, was observed after 7.62 cm of displacement (arrow "a" in Fig. 8.17C). As shown in Figs. 8.17C, damage starts at the bell section of Pipe #3 after 7.62 cm of displacement (sensor R3-1 and R3-2). With further displacement up to 10.16 cm, significant damage occurs at the bell section of Pipe #1 (R1-1 through R1-3) and further damage occurs at the bell section of Pipe #3 (R3-3) and the bell section of Pipe #2 (R2-1). This damage is accompanied by a drop in actuator load (arrow "b" in Fig. 8.6C) consistent with the results in Fig. 8.11 (2010 North). With increased displacement up to 15.24 cm, more damage occurs at the joints in the immediate vicinity of the fault line. In this heterogeneous system, pipe rupture (i.e., propagation of damage away from the bell section and rupture of the pipe segment) occurs after 22.86 cm of displacement on Pipe #2 (R2-5).

Overall, the results of the conductive surface sensors are consistent with other sensor and sensing methods and this low-cost sensor can provide some information about the occurrence and propagation of the damage.

8.5.5 Direct Joint Damage Sensing – Acoustic Emission

Fig. 8.18 shows the results of acoustic emission for the three pipelines tested. The results are expressed as average cumulative acoustic energy (per sensor) versus actuator displacement. In all pipelines, Joints #2 and #3 are at the immediate vicinity of the fault line. Clearly, in all pipelines, highest acoustic energy is captured at Joints #2 and #3. In all graphs in Fig. 8.18, acoustic emission is recorded, at Joints #2 and #3, at the first displacement step, while no cracking is detected by conductive surface sensors at the first displacement step (Fig. 8.17). The captured acoustic energy in the first displacement step at Joints #2 and #3 (of all pipelines) is likely due to tightening of the joint and/or microcracking of the grout or pipe segment.

Fig. 8.18A suggests that the majority of the damage in Ho-RC pipeline (2009 pipeline) is accumulated at the joints in the vicinity of the fault line (Joints #2 and #3) and approximately the same level of damage occurred at these two joints. Note that the results of acoustic emission are qualitative. The damage to these joints also occurred approximately at the same displacement steps. Fig. 8.18B, however, indicates that in Ho-FRC pipeline (2011 pipeline) damage is more concentrated at Joint #3. Fig. 8.18B also indicates that Joints #1 and #2 experience approximately the same level of damage. Fig. 8.18C indicates that in He-RC/FRC pipelines (2010 pipeline) a significant damage occurred at Joint #2 and Joints #1 and #3 experienced approximately the same level of damage. These observations are consistent with visual observations.

8.6 CONCLUSIONS

Due to the importance of buried lifelines to modern society and urbanization, it is critical that damage be located and diagnosed quickly in the event of a ground disruption. Permanent monitoring systems are an obvious approach to quick damage localization and assessment. This chapter illustrates experimental sensing methods that can serve as the basis for future monitoring systems. Of specific focus is a sensing strategy for segmented buried concrete pipelines, commonly used to transport waste and storm water, subjected to PGD. Damage to buried segmented concrete pipelines is typically concentrated to the joints during PGD. As such, the sensing technologies presented are focused on observing relative motion and direct damage detection of the joint areas. Wireless sensors were used as part of the DAQ system to validate wireless telemetry in underground sensing. The technologies presented were assessed on the full-scale testing of

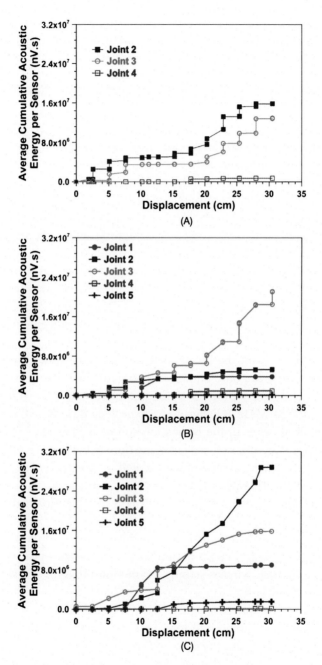

Figure 8.18 Average cumulative acoustic energy per sensor at each joint of the pipelines (A) Ho-RC pipeline (2009), (B) Ho-FRC pipeline (2011), and (C) He-RC/FRC pipeline (2010)

a buried segmented concrete pipeline to PGD at the Network for Earthquake Engineering Simulation (NEES) Lifeline Experimental and Testing Facilities at Cornell University. These tests served to validate the performance of the buried sensors for pipeline monitoring and to understand the evolution of damage in segmented buried concrete pipelines during PGD.

Three PGD tests were performed over three years on three pipeline variations: (1) a pipeline of standard reinforced concrete pipes (Ho-RC); (2) a pipeline of standard reinforced concrete pipes and a fiber reinforced concrete pipe (He-RC/FRC); and (3) a pipeline of fiber-reinforced concrete pipes (Ho-FRC). Instrumentation used to detect loads, displacements, motion and damage included strain gages, potentiometers, acoustic emission sensors, magnetic proximity sensors, conductive surface sensors, fiber optic sensors, and load cells. Wireless sensors were installed at each joint to measure potentiometers for relative joint rotation and translation. Conventional wired data acquisition systems were used to measure all other sensors. The movable section of the test basin at the NEES facility displaced along a 50° angle in a displacement controlled manner. Twelve displacement steps of 2.5 cm (1.0 in) were made for a total displacement of 30 cm.

The performance of the buried wireless sensors was shown for two wireless platforms, differing only in transmission frequency. The first is the *Narada* platform, using a 2.4 GHz RF signal and the second is the *WiMMS* platform, using a 900 MHz RF signal. As expected, the 900 MHz signal showed less attenuation in the buried environment than the 2.4 GHz signal. Packet loss rates for the *Narada* were shown between 3.4% and 10.0% while the *WiMMS* platform demonstrated no packet loss in transmission. Although the *WiMMS* outperformed the *Narada*, the packet loss is mostly insignificant for the type of data analysis being performed. The lower power consumption of the 2.4 GHz *Narada* is an important consideration for buried sensors requiring long battery life. Also, in a real-life implementation, lost packets could be retrieved by making repeated calls for transmission. Overall, wireless sensing for the test program was shown to be largely successful.

Following PGD, the buried pipeline was excavated to assess damage. Damage was largely concentrated to joint locations, with the most severe damage occurring at the joints closest to the fault line (Joints #2 and #3) in all three years. The type of damage observed was combined compressive telescoping and rotation, leading to tensile cracks and spalling of the pipe bell as well as crushing of the spigot. A hinge-like rotation of Joint

#2 and #3 shows a complete loss of rotational stiffness in the pipeline in the fault region. Joint #2 and #3 closest to the fault line showed major combined compressive telescoping and rotational damage while joints #1 and #4 further from the fault line showed mostly telescoping damage.

Sensor measurements confirm visual observations of pipeline damage and provide insight into the damage evolution of segmented concrete pipelines subject to PGD. Relative rotation and translation measurements showed an accuracy of within 15% of the rotations measured by hand at the end of the tests during excavation. Some of this error may be contributed to an assumption that the pipeline was installed with 0° rotation between adjacent pipes. Rotation and translation measures showed consistent results in terms of the damage evolution for the three years of testing leading to a damage evolution hypothesis. Initially, Joint #2 and #3 begin to rotate and bear on each other as a result of the compressive displacement caused by the PGD. Telescoping and rotational damage begins at Joints #2 and #3 between 0 and 10 cm (3.9 in) of PGD. As the compressive load in the pipeline increases, the telescoping failures occur further away from the fault location, first happening in the fixed half at Joint #1 near 10 to 12 cm (3.9 to 4.7 in) of PGD, and later occurring at the movable half in Joint #4 near 15 to 17.5 cm (5.9 to 6.9 in) of PGD. Even after the outside joints (Joint #1 and Joint #4) have suffered telescoping failures, they continue to bear on each other resulting in further load increase. This has a final catastrophic result at the inside joints (Joint #3 and Joint #4) in addition to increased rotation.

Strain gages were used to infer axial load behavior in pipelines near a PGD and to draw a direct causal relationship between axial strain measurements and observed damage behavior (e.g., excessive translations and rotations). Strain gages are a practical implementation for field use whereas potentiometers and load cells are not. In each of the three pipelines, the overall trend as well as the abrupt changes in axial load were captured well by the strain gage estimations for axial load. Strain gages were also used to estimate pipe bending behavior during PGD. It was shown that the development of bending strain in the pipe segments increased from the Ho-RC pipeline to the He-RC/FRC pipeline to the Ho-FRC pipeline, respectively, where the pronounced bending moment behavior in the Ho-FRC pipeline is consistent with a continuous pipeline under PGD. An inflection point was observed at the fault location in the Ho-FRC pipeline with positive moment in the fixed half and a negative moment in the movable half showing that the fiber reinforced concrete pipeline is capable of developing larger bending deformation before joint damage compared to the reinforced concrete specimens.

Conductive surface sensors and acoustic emission measurements successfully captured direct damage detection. Both sensing methods showed results that coincide with the visual observations and other sensor measurements. Conductive surface sensors shows damage starting as early as 5.08 cm (2.0 in) of actuator displacement, starting at the bell section of Joints #2 and #3. With further displacement up to 10.16 cm (4.0 in), significant damage occurs at Joints #1 and #4, and further damage occurs at the bell section of Joints #2 and #3. With increased displacement up to 15.24 cm, more damage occurs at the joints in the immediate vicinity of the fault line. Acoustic emission showed to capture damage earlier than conductive surface sensors. Acoustic emission is recorded at Joints #2 and #3 in the first 2.5 cm (1.0 in) of displacement, before conductive surface sensors detect cracking. Acoustic emission results showed that the majority of damage occurred in the joints closes to the fault (Joints #2 and #3), which is consistent with visual observations. For the Ho-RC pipeline, acoustic emission showed approximately the same level of damage in Joints #2 and #3. For the Ho-FRC pipeline damage is more concentrated at Joint #3, and Joints #1 and #2 experience approximately the same level of damage. For the He-RC/FRC pipeline a significant damage occurred at Joint #1 and Joints #2 and #3 experienced approximately the same level of damage. These observations are consistent with visual observations.

The work presented in this chapter illustrates buried pipeline monitoring methods successful in identifying pipeline behavior and damage evolution due to PGD. Strain gage measures align closely with load cell measures and are a pragmatic sensor approach for field implementation. Direct damage sensing approaches successfully identify where and when significant damage occurred in the test pipeline. Real-time rotation and translation measures were successfully demonstrated using a buried wireless sensor system, showing accurate estimates of pipeline deformation which can be used to infer damage levels. The validation of wireless telemetry encourages future use for field applications. Buried wireless sensors offer a low cost opportunity to densely distribute sensors in regions of high risk for PGD and provide a means for real-time damage information for first responders.

ACKNOWLEDGMENTS

This work was supported by the National Science Foundation (NSF) under the NEES program (Grant CMMI-0724022). Any opinion, findings, and conclusions or recommendations expressed in this material are those of the authors and do not necessarily reflect the views of the National Science Foundation. The large-scale testing was performed at the Large-Scale Pipeline Testing Facility at Cornell University that is a node in the George E. Brown, Jr. Network for Earthquake Engineering Simulation (NEES). The authors thank the Cornell staff for their outstanding help with the experimental program. In particular, the help of Mr. Tim Bond and Mr. Joe Chipalowsky is greatly appreciated.

REFERENCES

Akyildiz, I.F., Stuntebeck, E.P., 2006. Wireless underground sensor networks: research challenges. Ad Hoc Netw. 4, 669–686.

Ariman, T., Muleski, G., 1981. A review of the response of buried pipelines under seismic excitations. Earthq. Eng. Struct. Dyn. 9 (2), 133–152.

Birken, R., Oristaglio, M., 2014. Mapping subsurface utilities with mobile electromagnetic geophysical sensor arrays. In: Wang, M.L., Lynch, J.P., Sohn, H. (Eds.), Sensor Technologies for Civil Infrastructures: Applications in Structural Health Monitoring, vol. 2. Woodhead Publishing, London, UK.

Bradshaw, A.S., daSilva, G., McCue, M.T., Kim, J., Nadukuru, S.S., Lynch, J.P., Michalowski, R.L., Pour-Ghaz, M., Weiss, J., Green, R.A., 2009. Damage detection and health monitoring of buried concrete pipelines. In: International Symposium on Prediction and Simulation Methods for Geohazard Mitigation. Kyoto, Japan, pp. 473–478.

Eidinger, J.M., Maison, B., Lee, D., Lau, B., 1995. East bay municipal utility district water distribution damage in Loma Prieta earthquake. In: Proceedings of the 4th US Conference on Lifeline Earthquake Engineering. San Francisco, CA, pp. 240–247.

Eguchi, R.T., 1983. Seismic vulnerability models for underground pipes. In: Earthquake Behavior and Safety of Oil and Gas Storage Facilities, Buried Pipelines and Equipment. ASME, New York, NY, pp. 368–373.

Giurgiutiu, V., 2008. Structural Health Monitoring with Piezoelectric Wafer Active Sensors. Academic Press, Amsterdam, The Netherlands.

Glisic, B., Yao, Y., 2012. Fiber optic methods for health assessment of pipelines subjected to earthquake-induced ground movement. Struct. Health Monit. 11 (6), 696–711.

Glisic, B., 2014. Sensing solutions for assessing and monitoring pipeline systems. In: Wang, M.L., Lynch, J.P., Sohn, H. (Eds.), Sensor Technologies for Civil Infrastructures: Applications in Structural Health Monitoring, vol. 2. Woodhead Publishing, London, UK.

Hayakawa, H., Kawanaka, A., 1998. Radar imaging of underground pipes by automated estimation of velocity distribution versus depth. J. Appl. Geophys. 40 (1), 37–48.

Heubach, W.F., 1995. Seismic damage estimation for buried pipeline systems. In: Proceedings, 4th US Conference on Lifeline Earthquake Engineering. ASCE, Reston, VA, pp. 312–319.

Honegger, D.G., 1995. An approach to extend seismic vulnerability relationships for large diameter pipelines. In: American Society of Civil Engineers. New York, NY, pp. 320–327. Report No. CONF-9508226.

Inagaki, T., Okamoto, Y., 1997. Diagnosis of the leakage point on a structure surface using infrared thermography in near ambient conditions. NDT Int. 30 (3), 135–142.

Kurashima, T., Horiguchi, T., Tateda, M., 1990. Distributed temperature sensing using stimulated Brillouin scattering in optical silica fibers. Opt. Lett. 15 (18), 1038–1040.

Kikuchi, K., Naito, T., Okoshi, T., 1988. Measurement of Raman scattering in single-mode optical fiber by optical time-domain reflectometry. IEEE J. Quantum Electron. 24 (10), 1973–1975.

Kim, J., O'Connor, S., Nadukuru, S., Lynch, J.P., Michalowski, R., Green, R.A., Pour-Ghaz, M., Weiss, W.J., Bradshaw, A., 2010. Behavior of full-scale concrete segmented pipelines under permanent ground displacements. In: Proceedings of SPIE 7650, Health Monitoring of Structural and Biological Systems. San Diego, CA.

Liu, H., 2003. Pipeline Engineering. CRC Press, Boca Raton, FL.

Lynch, J.P., Loh, K.J., 2006. A summary review of wireless sensors and sensor networks for structural health monitoring. Shock Vib. Dig. 38 (2), 91–128.

Lynch, J.P., Sundararajan, A., Law, K.H., Kiremidjian, A.S., Carryer, E., 2004. Embedding damage detection algorithms in a wireless sensing unit for operational power efficiency. Smart Mater. Struct. 13 (4), 800.

Mannan, S. (Ed.), 2012. Lees' Loss Prevention in the Process Industries: Hazard Identification, Assessment and Control. Butterworth-Heinemann.

O'Rourke, M.J., Ayala, G., 1993. Pipeline damage due to wave propagation. J. Geotech. Eng. 119, 1490–1498.

O'Rourke, M.J., 2003. Buried pipelines. In: Earthquake Engineering Handbook. CRC, Boca Raton, FL. Chapter 23.

O'Rourke, T.D., 2005. Soil-structure interaction under extreme loading conditions. In: 13th Spencer J. Buchanan Lecture. Texas A&M University, College Station, TX.

O'Rourke, T.D., Jeon, S., 1991. Factors affecting the earthquake damage of water distribution systems. In: Proceedings of the Fifth US Conference on Lifeline Earthquake Engineering. In: Technical Council on Lifeline Earthquake Engineering Monograph, vol. 6. ASCE, New York, NY, pp. 379–388.

Posey, R., Johnson, G.A., Vohra, S.T., 2000. Strain sensing based on coherent Rayleigh scattering in an optical fiber. Electron. Lett. 36 (20), 1688–1689.

Porter, K.A., Scawthorn, C., Honegger, D.G., O'Rourke, T.D., Blackburn, F., 1991. Performance of water supply pipelines in liquefied soil. In: Proceedings, 4th US–Japan Workshop on Earthquake Disaster Prevention for Lifeline Systems. In: NIST Special Publication, vol. 840. US Dept. of Commerce, Gaithersburg, MD, pp. 3–17.

Pour-Ghaz, M., Nadukuru, S.S., O'Connor, S., Kim, J., Michalowski, R.L., Bradshaw, A.S., Green, R.A., Lynch, J.P., Poursaee, A., Weiss, W.J., 2011. Using electrical, magnetic and acoustic sensors to detect damage in segmental concrete pipes subjected to permanent ground displacement. Cem. Concr. Compos. 33 (7), 749–762.

Pour-Ghaz, M., Wilson, J., Spragg, R., Nadukuru, S.S., Kim, J., O'Connor, S.M., Byrne, E.M., Sigurdardottir, D.H., Yao, Y., Michalowski, R.L., Lynch, J.P., Green, R.A., Bradshaw, A.S., Glisic, B., Weiss, W.J., 2016. Performance and damage assessment of plain and fiber reinforced segmental concrete pipelines subjected to transverse permanent ground displacement. In: Structure and Infrastructure Engineering. Taylor & Francis (under review).

Stuntebeck, E.P., Pompili, D., Melodia, T., 2006. Wireless underground sensor networks using commodity terrestrial motes. In: 2nd IEEE Workshop on Wireless Mesh Networks, vol. 3(7), pp. 112–114.

Swartz, R.A., Jung, D., Lynch, J.P., Wang, Y., Shi, D., Flynn, M.P., 2005. Design of a wireless sensor for scalable distributed in-network computation in a structural health monitoring system. In: Proceedings of the 5th International Workshop on Structural Health Monitoring. Stanford, CA.

Wang, Y., Lynch, J.P., Law, K.H., 2005. Design of a low-power wireless structural monitoring system for collaborative computational algorithms. In: Proceedings of SPIE: Smart Structures and Materials. San Diego, CA.

CHAPTER 9

Outlook: Advanced Hybrid Sensing for Preemptive Response

Sibel Pamukcu, Liang Cheng
Lehigh University, Bethlehem, PA, USA

9.1 INTRODUCTION

With the development of sensing techniques and big-data analysis, underground sensing for preemptive response is becoming more and more feasible and prevalent. To provide preemptive response to underground events, which may develop over a long period of time, sensing systems should be lightweight in terms of deployment and maintenance costs, low-power or power-sustainable, and environmentally rugged.

Synergistic cyber-physical systems (CPS), which integrate hybrid sensing with new concepts and paradigms such as crowdsensing, may proactively detect and localize underground hazards. The new cyber-physical system should not only perform as a feasible and reliable tool for post-event damage management, but also as a community-involved warning system that proactively detects and locates potential damage by monitoring the onset of events or conditions leading up to a likely hazard in order to facilitate agile mitigation decisions for public safety and perform faster damage management.

A hybrid sensing system is one that can integrate signals from multiple sensor types and systems to localize an event. For example, long-term low-resolution underground sensory blanket of a wireless sensor network (WSN) or a fiber-optic sensor network (FSN) can be used with short-term high-resolution agile and cognitive GPR for accurate event or damage localization underground (Haykin, 2006; Ghazanfari et al., 2012; Glisic and Yao, 2012; Glisic, 2014; Pour-Ghaz et al., 2011; Yoon et al., 2012). Furthermore, continuous engagement of local community with crowdsensing functions via smart devices can help shape connected communities. In a plausible scenario, (i) a sensory blanket (i.e., WSN or FSN) delivers time-space continuous data to map the evolution of an under-

Underground Sensing.
DOI: http://dx.doi.org/10.1016/B978-0-12-803139-1.00009-6

ground hazard as it triggers signals in the general area of a pre-event, (ii) a targeted sensor system (i.e., GPR) detects location of any damage, applicable to buried infrastructure and the surrounding soil/rock mass, and (iii) the crowd sensing then provides large-scale metadata for pre- and post-event sensing and serves as an interface for the CPS to interact with the community. Once a warning and a general area of the underground hazard event is detected by the WSN and/or crowd sensing, the agile GPR can be deployed to locate the event/damage spot with speed and precision.

A synergistic CPS, created by integration of hybrid sensing and social science solutions, aimed at preemptive event detection, localization, and management, can be generalized to many smart city applications to monitor underground hazards or the deteriorating subsystems to yield profound economical and societal impacts. In this chapter we first discuss fiber-optic underground sensor networks that are suitable for preemptive response applications. We then discuss ideas of crowdsensing, wireless signal networks, and future research on advanced hybrid sensing for preemptive response to underground events/objects. It should be noted that detailed applications of fiber-optic sensing and related developments are provided in Chapters 2.1 and 6, and newer paradigms of sensing are discussed in Chapters 2.2 and 7 of this book.

9.2 FIBER-OPTIC (FO) UNDERGROUND SENSOR NETWORKS

The fiber optic sensors have the potential to create distributed sensing capability for selectively detecting a physical, chemical or biological variable by spatial and temporal acquisition over large distances in the underground environment. Traditionally, fiber-optic sensors have several advantages including lightweight, small size, passivity, low-power requirement, environmental ruggedness (i.e., resistance to electromagnetic interference, environmental impact and corrosion), good concealment, large bandwidth, high sensitivity and flexibility, and large-scale monitoring capability (Krohn, 1988; Agrawal, 1995). These attributes have allowed fiber-optic sensors to replace traditional devices for measurement and monitoring of many indices, including rotation, acceleration, electric and magnetic fields, temperature, pressure, acoustics, vibration, strain, humidity, viscosity, pH, gas and chemical content, both aboveground and underground. These sensors use the optical fiber either as the sensing element (intrinsic sensors), or as a means of relaying signals from a remote sensing area to a signal processor (extrinsic sensor), or both. They can be installed in areas normally inaccessible by conventional sensors, and they can be interfaced with data communication systems posing no risk of electric shock in live measurements.

Sensors based on fiber optic cable functions make use of the following important features of the cable to sense the environment:

(1) optical loss – intrinsic and extrinsic energy loss properties;

(2) change in refractive index – index profile in radial direction and the reduction of index fluctuation along the axial direction;

(3) change in shape – cross-sectional shape and size, the surface finish and the fluctuation of the size along axial direction.

Most fiber-optic sensors use energy loss principles (i.e., changes in optical power in linearly positioned wave-guides) for chemical detection. These can be limited for distributed applications, particularly in underground environments, as the energy will deplete over the length of the fiber with every encounter of a detectable event or index that engages the light energy. Other sensors use the changes in refractive index and/or cross sectional size of the fiber cable that change the light backscattering property (i.e., light intensity, phase, polarization state, or light frequency) in optical fibers, which meet space requirements of distributed sensing for civil infrastructure and environment (Horiguchi et al., 1995; Kee et al., 2000; Bao et al., 2001; Ohno et al., 2001; Shi et al., 2009; Qazi et al., 2012). Fiber-optic sensor networks that generate a signal due to the changes to the refractive index or the shape of the fiber are often optical wave-guides that sense the target index along the linear positioning of a continuous optical fiber. There are three main optical wave-guides used as distributed sensors:

(i) Short-Gauge Sensors – Fiber Bragg grating (FBG) spectrometry. Fiber cladding is marked periodically along the fiber with laser to create local, narrow-band pass filters that are sensitive to the surrounding environment or events at the location of a filter (Chen, 2013).

(ii) Long-Gauge Sensors – Long-period grating (LPG) spectrometry. Codirectional coupling of light from the propagating fiber mode to cladding modes produces a series of attenuation bands in the fiber transmission spectrum. The resonance wavelengths of attenuation at the LPG locations will be sensitive to local conditions such as ambient refractive index, temperature or hydrostatic pressure surrounding the LPG (Chen, 2013).

(iii) OTDR based Sensors – Brillouin, stimulated Brillouin, Raman and Rayleigh scattering; and evanescent wave. These sensors use OTDR (optical time domain reflectometry) principles to detect the location of the frequency shift of the reflected light, and the magnitude of the frequency shift to detect the magnitude of the measured index (i.e., temperature, strain, vibration) along the linear positioning of the sensing fiber. There are several different types of OTDR based optical sensors such as phase-sensitive optical time domain reflectometer, polarization-optical time domain reflectometer (Linze et al., 2012), optical frequency domain reflectometer, as well some combinations of interferometric and backscattering-based techniques (Liu et al., 2016). Evanescent wave (EW) technique is also based on OTDR, where the fiber cladding is modified chemically to interact with the environment while the pulse travels partially through the cladding and is reflected back from the location of sensing along the fiber (Perez et al., 2013). The EW sensors demand large optical power, due to the cumulative energy loss at the points of contact with the chemicals.

9.2.1 Fiber-Optic Chemical Sensors for Underground Measurements

In general, chemical sensing using fiber-optic techniques involves the interaction of an incident beam over the analyte directly or an interface agent (i.e., thin film or deposition) on the fiber yielding *absorbed*, *reflected*, *fluorescent* or *surface plasmon resonance* signal that can be translated into presence and concentration of a target chemical. Optical sensors employ an optical transduction technique that "translates" the chemical variable into an optical signal change such as a light intensity, wavelength, polarization or phase change (McDonagh et al., 2008; Grattan and Meggitt, 1999). Optical chemical sensors can be categorized, as direct or reagent-mediated sensing systems.

(i) Absorbance-Based Techniques

Absorption in a gas or liquid, may be characterized by Beer–Lambert law, or simply the Beer law, where the intensity of the absorbance light (i.e., absorbance or optical density) is related to the molar absorptivity, ε (L mol^{-1} cm), and the concentration, C (mol L^{-1}) of the absorbing species and the absorption path length (cm). There exists a linear relation between the absorbance and the concentration of the chemical element and the incident beam. In most of the absorbance based optical chemical sensors the interface agent changes its absorbance according to the concentration of the specific chemical it comes to contact with (Antico et al., 1999; Guo et al., 2006). Some of absorption-based sensors use interface materials that change color in presence of a target chemical (Balaji et al., 2006; Prabhakaran et al., 2007). The instrumentation of absorption-based sensors often requires a monochromatic light source and a photodetector that measures the intensity of the change in absorbance.

(ii) Reflectance-Based Techniques

Chemical reactions could lead to changes in the complex refractive index of some materials; hence they can be used as transducers. When these materials, commonly in the form of a thin film, are illuminated with an appropriated light the signal will be partially or totally reflected. However, this reflectance will change when the layer is in contact with a specific chemical that it reacts with. The reflected signal can be used to deduce, directly or indirectly, the concentration of the chemical. The reflectance-based techniques are used in optical fiber schemes mostly as extrinsic type sensors (Yusof and Ahmad, 2003; Guillemain et al., 2009). The material sensitive to a specific chemical is directly deposited over the fiber tip or in a film that will be illuminated by an optical fiber. The reflected signal is collected by the same fiber, which propagates along the fiber to a photo detector, where it is analyzed and converted to a calibrated concentration measurement.

(iii) Fluorescence-Based Techniques

Some materials have the property of being fluorescent when they are illuminated with a light source of appropriated wavelength. The fluorescence is the optical radiation generated when electrons of an atom or molecule return from the excited to the ground state after absorption of a photon from an excitation light source. The intensity of the fluorescent signal (IF) is proportional to the intensity of light absorbed by the sample, therefore it is possible to establish a direct relation between the intensity of the fluorescent signal and the concentration of an absorbing material. Although the intensity of the fluorescence is directly proportional to the concentration of the absorbing species, the decay time of the fluorescence signal is more frequently used for sensing purposes because this parameter is less sensitive to source fluctuations, interference from ambient light or drift due to aging of detector. Fluorescence-based fiber–optic sensing is used frequently to detect the presence of contaminants in water, mostly as an extrinsic method (Mayra et al., 2008; Achatz et al., 2011; Aksuner, 2011).

(iv) Surface Plasmon Resonance-Based Techniques

Surface plasmon resonance (SPR) is a quantum optical-electrical phenomenon produced by the interaction of light with a metal surface. The surface plasmon is a charge density oscillation that exists at a metal–dielectric interface. The plasmon propagates in a direction parallel to the metal–dielectric interface in the boundary of the metal and the external medium. These oscillations are very sensitive to any change in the optical refractive index of the material at the boundary (Maier, 2007). In the optical domain, the surface plasmon excitation is observed as an intensity transmission loss of light at a specific wavelength. For heavy metal detection a sensitive thin film layer on the optical fiber interacts with the target metal in contact, which leads to the change of the refractive index of the layer and the shift of the peak wavelength of the transmitted light. SPR is the most sensitive refractometric method, so it is possible to detect very small traces of heavy metals (Forzani et al., 2007; Lin and Chung, 2009; Fen et al., 2012, 2013; Fen and Yunus, 2013).

The interface reagent used in most fiber-optic chemical sensors is a molecularly imprinted polymer coat (i.e., optical film) on the optical fiber (Wolfbeis, 2008; Phillips et al., 2003). This layer or coat works as a transducer when in direct contact with the target chemical, and triggers an optical signal either due to optical loss or change in physical properties of the fiber. For example, polyelectrolyte gels of cross-linked three dimensional networks of monomers possess high swelling capability due to solvent sorption. The amount of swelling is known to be a string function of pressure, temperature, ion concentrations and pH changes (Siegel, 1993; Siegel et al., 1988; Matsuo and Tanaka, 1988). Their swelling and kinetics depend on parameters such as the degree of cross-linking (Skouri et al., 1995), external salt molarity (Yin et al., 1992), and the degree of gel ionization rule (Katchalsky and Michaeli, 1995). The monomer, N-isopropylacrylamide (NIPAM), forms hydrogels which can swell when exposed to water. Incorporating NIPAM into a polymer composition would also lead to the formation of a thermosensitive polymer coating since poly(NIPAM) exhibits a strong phase transition above a critical temperature (Dong and Pamukcu, 2012 and 2014). Similarly, polymer latexes with desired functional moieties can be used to coat optical fibers. At high concentrations, latex could function as an alkali swellable-latex whereby the latex particle size, and coating swellability, would increase dramatically upon neutralization in aqueous solutions of high pH (e.g., >10) which would trigger a sensor response.

Among the different types of optical fiber devices used are evanescent wave (EW) fiber sensors, hetero-core fibers, U-bend fibers, fiber Bragg (FBG) and long-period gratings (LPG), and fibers with active doped cladding (Kocincova et al., 2007). Some of the substances that have been detected and/or quantified using optical FOSs are volatile organic compounds (alcohols, formaldehydes, methane, ketones, ammonia, CO_x, O_2, and H_2), and some metallic ions, including, Al, Ca, Cu, Fe, Hg, Pb, and Zn (Jerónimo et al., 2007; Wang et al., 2011; Qazi et al., 2012). The use of hetero-core LPG sensors have been explored and developed to some extend for potential sensing of chemicals in the environment, including the subsurface (Sekia et al., 2007; Monzon-Hernandez and Martinez-Escobor, 2009; Akita et al., 2010; Wang et al., 2011; Qazi et al., 2012; Wang, 2012).

Although highly successful fiber-optic sensor systems for chemical and biological detection in water and sediments have been developed and applied, most remain as point sensors (Chen, 2013; Perez et al., 2013). The few developments of fiber-optic sensors for distributed and space-time continuous sensing is currently at bench or floor scale (Wang, 2014; Qazi et al., 2012). The BOTDR or LPG based chemical sensing use molecularly imprinted polymer or a chemically reactive layer bonded to the fiber (Haupt and Moshbach, 2000, Phillips et al., 2003; Wolfbeis, 2008; Pervizpour and Pamukcu, 2011). Some OTDR based sensors have also been shown capable of spatially distributed, temporally continuous, and functionally selective sensing of environmental parameters in soil and water (Potyrailo and Hieftje, 1998; Moran et al., 2000; Yuan et al., 2001; Buerck et al., 2001; Texier et al., 2005; Pamukcu et al., 2006; Turel and Pamukcu, 2006; Anastasio et al., 2007; Cui et al., 2011; Angulo-Vinuesa et al., 2012; Rajeev et al., 2013). Pending more research and development, both of these systems are yet to be deployed in the field for underground environmental hazard monitoring. Two examples of LPG fiber distributed sensing for environmental indices are discussed below.

A distributed fiber-optic soil pH sensor based on a hetero-core structure was developed and tested by Garcia et al. (2014). The fiber was coated with an acrylic polymer (AP) doped with Prussian blue complex. In this design, the pH changes of the surrounding medium produced a change in the refractive index of the AP layer. The pH changes were then observed incrementally in the hetero-core transmission signal. The results of a sensitivity analysis for soil pH changes normalized with in-air reference showed

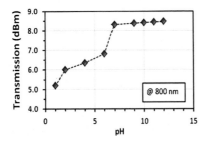

Figure 9.1 Transmission of APE/AP coated hetero-core LPG fiber optic sensor with soil pH changes (Garcia et al., 2014)

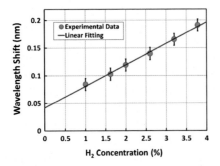

Figure 9.2 The plot of LPG wavelength shift for H_2 of different concentrations (Wang, 2012)

good sensitivity to pH values lower than 7. The most characteristic signals at 800 nm was plotted independently as shown in Fig. 9.1.

Similarly, a technique for fully distributed fiber–optic hydrogen sensing was demonstrated based on a traveling LPG in a single–mode fiber coated with a platinum (Pt) catalyst layer (Wang et al., 2011). The traveling LPG was generated by an acoustic pulse propagating along the fiber. The Pt-coated fiber section was heated by the thermal energy released from Pt-assisted combustion of H_2 and O_2. The resulted temperature change gave rise to a measurable wavelength shift in the transmission optical spectrum of the traveling LPG when it passed through the pretreated fiber segment. Fig. 9.2 shows the H_2 detection at different wavelengths shifts of LPG. Both of the sensor designs appear to be viable to detect chemicals with adjusted coating materials of selective functionality.

A prototype of a distributed water sensor was developed based on BOTDR (Brillouin optical time domain refrectometry) for temporally and spatially continuous sensing in soil water content in underground environ-

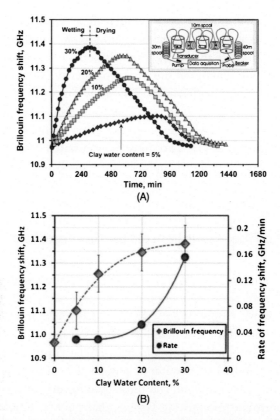

Figure 9.3 Evolution of Brillouin frequency shift of distributed fiber coupled with water reactive polymer sleeves embedded in wet clay samples. (A) Variation of Brillouin frequency shift with time of embedment in wet clay (Garcia et al., 2014). (B) Variation of Brillouin frequency shift and its time rate with clay water content

ments (Texier et al., 2005; Anastasio et al., 2007). In this application, the fiber was bonded with molecularly imprinted polymer (AEP60 hydrophilic polymer) sleeves (1 cm length; 0.3 cm diameter) that worked as transducers with mass gain when in direct contact with soil water. The fiber–polymer sleeve assemblies were embedded in wet clays of different water contents as shown in the inset sketch of Fig. 9.3. They were left in clay until fully wet, and then taken out to air dry, while the Brillouin frequency shift was recorded at several time intervals, as shown in Fig. 9.3A. The water swelling of the polymers sleeves exerted tangential stresses on the fiber at the bonded interface. The slight "pinch" or "tug" on the fiber was sufficient to cause tensile stress and axial straining of the fiber hence a change in its refractive properties locally where the sleeve was bonded. This, in turn, generated a shift in fiber characteristic Brillouin scatter frequency of light at that location. Fig. 9.3B shows the magnitude and the rate of change of the Brillouin frequency shift with increasing water content. The nonlinearity is attributed to the swelling kinetics of the thick polymer sleeve elements, as well as other design-related inefficiencies. The study proposed to improve the polymer architecture and adjust the thickness of the sleeve to obtain an optimum sensor response time when exposed to a given chemical or solvent (Pervizpour and Pamukcu, 2011). The study demonstrated that principles of BOTDR can be coupled with principles of molecularly imprinted polymer films used in other optical sensors to develop a spatially and temporally continuous distributed liquid content or chemical sensor.

9.2.2 Distributed Fiber-Optic Sensors for Underground Sensing

Time and space continuous health monitoring of civil infrastructures, particularly those underground is highly important for the safety of the public from unforeseen hazards. Some of the critical underground infrastructure are the utilities including those that transport water, sewage, oil, gas, chemicals, electric power, communications and mass media content, as well as tunnels that transport and/or store water or sewage and tunnels that transport hydropower, traffic, rail, freight, and those used for dry storage. There are multiple underground monitoring systems that provide time-ordered point measurements at strategically placed devices along these structures. Point monitoring will work well if the damage location is predictable with some degree of probability. Alternatively, a large number of monitoring devices need to be placed to provide an acceptable degree of resolution in damage detection. Due to the large spatial distribution of the underground transport facilities, the application of conventional methods that can only provide point measurements remain limited. Distributed methods that use pressure, chemical and thermal indicators and/or magnetic, acoustic or wireless sensor networks are available for select underground infrastructure such as pipelines. The shortcoming of these methods is that they cannot function as pre-warning systems but help localize damage or leakage only after the event occurs. Distributed optical waveguide based sensors have been shown versatile for not only damage detection for underground infrastructure, but also as prewarning systems for underground hazards and perimeter security. These sensor systems can provide vital information along the entire linear positioning of the fiber which can go up to several hundred kilometers, and with abilities of prewarning, high sensitivity and accurate localization (Liu et al., 2016; Glisic and Yao, 2012; Glisic, 2014).

Most of the optical–waveguide based sensing methods applicable underground use OTDR. Optical time domain reflectometry (OTDR) consists of sending a powerful light pulse and subsequently observing pulse losses of the reflected light due to scatter originating from local physical inhomogeneities along the fiber. The inhomogeneities on the fiber can be built-in as the fiber Bragg gratings (FBG) which are laser markings on the fiber intended to create local narrow band pass filters sensitive to the measured index (i.e., environmental parameters, strain, temperature) (LeBlanc et al., 1996; Schulz et al., 1998). The inhomogeneity can either be built-in, as the FBGs and LPGs, or they can occur due to local interactions of the fiber with the surrounding environment. When a pulse of light propagates through an optical fiber, part of the backscattered light is known as Brillouin scattering. This scatter results from the interaction between light photons and acoustic waves (phonons). Brillouin scattered light has a peak frequency shifted from the incident pulse light frequency. Brillouin frequency is characteristic to the wave guide directly proportional to its acoustic velocity and refractive index. The characteristic Brillouin frequency shifts at locations where variations in temperature or strain alter the acoustic velocity or the refractive index in the fiber. The traveling pulse of light is then scattered back with an alteration in its the characteristic Brillouin frequency from every point along the fiber where an alteration change in its acoustic velocity and/or refractive index is encountered. The magnitude of the Brillouin frequency change or shift can then be interpreted as the magnitude of the target index (i.e., temperature, strain) with reference to a null value of equilibrium. Using the velocity of light pulse in the fiber, the time domain information is converted into location, or distance, l, from the detector. The spatial resolution, w, is determined by the pulse width and the velocity of light in the fiber.

Similar measurements can be made using Raman and Rayleigh backscattering of pulse light in wave guides (Froggatt and Moore, 1998). Brillouin frequency–based technique is different than intensity–based technique such as Raman. Brillouin technique has been shown to be inherently higher in accuracy with long term stability, since intensity-based techniques suffer from high sensitivity to drifts. Several of the distributed sensing techniques based on Brillouin or Raman backscattering are briefly defined below.

(i) Stimulated Brillouin Scattering (SBS)

Brillouin scattering can be optically stimulated leading to the greatest intensity of the scattering mechanism and consequently an improved signal-to-noise ratio. For intense beams (e.g., laser light) traveling in a medium such as an optical fiber, the variations in the electric field of the beam itself may produce acoustic vibrations in the medium via electrostriction. The beam may undergo Brillouin scattering from these vibrations, usually in opposite direction to the incoming beam, a phenomenon known as stimulated Brillouin scattering (SBS). In SBS, researched and developed for the last two decades, the strain profile along the sensing fiber can be extracted by adjusting the frequency difference between a probe (continuous light) and a pump wave. In the SBS technique, as in a null detector, the pump and probe are initially de-tuned to a frequency that is slightly greater than the Brillouin frequency of the fiber. Therefore, in unstressed fiber, the base line remains flat resulting in a self–referenced sensor. Local straining of the fiber results in a measurable change in its local acoustic properties, hence a Brillouin frequency change, calibrated to detect the magnitude of the strain. The location of the generated signal is determined by time domain reflectometry.

(ii) Brillouin Optical Time Domain Analysis (BOTDA)

Horiguchi and his colleagues reported the first distributed sensor, called Brillouin optical time-domain analysis (BOTDA), using a pulsed pump wave and a continuous wave (CW) Stokes probe in a counter propagation scheme (Horiguchi and Tateda, 1989). It was a double ended access to the light source and detection system for long sensing lengths. The pumping pulse light is launched at one end of the fiber and propagates in the fiber, while the CW light is launched at the opposite end of the fiber and propagates in the opposite direction. The pump pulse generates backward Brillouin gain in a single-mode fiber, amplifying a counter propagating signal when its optical frequency falls in the SBS gain spectrum. The frequency shift is then directly related to the acoustic properties of the fiber in which the waves interact. Single-laser modulated pulse base reflection BOTDA is simpler than the conventional BOTDA system as it requires access to one end of the fiber. In this configuration, the probe is generated by the modulated pump base reflection, where the pump and the probe waves are simultaneously connected to an electro-optic modulator (EOM) (Cui et al., 2009).

(iii) Brillouin Optical Frequency Domain Analysis (BOFDA)

The pump beam in BOFDA systems is not pulsed; rather its intensity is sinusoidally modulated. The modulated pump light and the backscattered light signals are fed to a vector network analyzer, which calculates a complex transfer function by recording the amplitude and phase of the modulation induced by SBS on the Stokes beam intensity, over a range of modulation frequencies (Garus et al., 1996; Bernini et al., 2004, 2012). This function is then converted by the inverse fast Fourier transform, which gives the distribution of strain and temperature along the fiber. The main advantage of the method is that Stokes and pump waves preactivate a stationary (narrowband) acoustic wave. As a result, Brillouin gain curve keeps its natural linewidth leading to high accuracy and spatial resolution (Bernini et al., 2012).

(iv) Brillouin Time Domain Reflectometry (BOTDR)

BOTDR was introduced by Kurashima et al. (1990, 1993). They analyzed the spontaneous Brillouin backscattered light instead of the Brillouin amplification signal using coherent detection. Niklès et al. (1996) presented a distributed Brillouin sensor consisting of a single laser and a single fiber based on a pump-and-probe technique. Using only one end of the sensing fiber is the advantage of the BOTDR system. The drawback is that the spontaneous Brillouin scattering signal is much weaker than that of BOTDA and the spatial resolution is limited to the phonon lifetime.

(v) Raman Optical Time Domain Reflectometry (ROTDR)

In ROTDR one-dimensional optical radar provides an echo scan of the entire length of an optical fiber at Raman Stokes and anti-Stokes lines. A short laser pulse is sent along the fiber and backscattered Raman light is detected with high temporal resolution. This Raman light contains the information about the sensed index (i.e., temperature) along the fiber.

Distributed optical fiber sensing is already a mature technology used to measure strain and temperature in underground utilities and tunnels (Iten et al., 2015). There are various standards developed for terminology, application, safety, maintenance and analysis of distributed fiber optic sensing of underground utilities and tunnels (ASTM F3092, F3079, F2462, F2350, F2349, F2303, F2233).

Recently, fiber-optic sensor networks (FSNs) have also been shown to be versatile for sensing underground vibrations and dynamic strains in high resolution over long distances (Hotate and Ong, 2003; Juarez and Taylor, 2007; Peng et al., 2013; Martins et al., 2015). Simultaneous measurement of distributed temperature and discrete dynamic strain fields are useful indices for several applications, including leakage detection in oil and gas wells and underground pipelines (i.e., oil, water, gas), and tracking environmental and security hazards (Rajeev et al., 2013; Zhou et al., 2017). Liu et al. (2016) reviewed various technologies of distributed fiber-optic vibration sensing including interferometric and backscattering-based sensing technologies, such as phase-sensitive optical time domain reflectometer, polarization–optical time domain reflectometer, optical frequency domain reflectometer, as well as some combinations of the interferometric and backscattering-based techniques.

Research and development is continuing by many to overcome practical issues such as processing of weak detection signals, enhancing system reliability and positioning accuracy in complex environments, and implementing high-speed and real-time signal processing for broad application of the technology in practice (Liu et al., 2016). With ongoing advancements in optical all methods and attention given to updating of the related standards for field applications, fiber-optic sensor networks are poised to become one of the most versatile innovations among all measurement tools for field monitoring, particularly for underground. A few relevant examples of fiber-topic sensor applications and related development projects for subsurface are discussed below.

9.2.2.1 Infrastructure Health Measurements

The past two decades has seen significant development in application of distributed fiber-optic sensor networks for bridge health monitoring, detection of pipeline leakages and cracks on civil infrastructure (Bao and Chen, 2012; Song et al., 2014). Although the majority of the fiber-optic sensor network applications reported have been on health monitoring of above ground infrastructure (i.e., bridges), successful research and development has also been accomplished for pipeline and tunnel monitoring and health assessment underground. Most of these works has been discussed in detail and with case studies in the previous chapters of this book (i.e., by Soga et al., in Chap. 6.1, and O'Connor et al. in Chap. 8). Only a few other relevant cases are provided below.

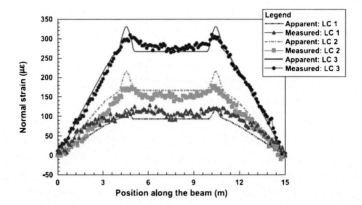

Figure 9.4 Comparison of the BOTDR measured strains with the theoretical results for three loading cases of a beam with simulated cracks. The distributed measurements indicated slight bumps in the vicinity of the simulated cracks, more pronounced for the largest load amplitude (Feng et al., 2013)

(i) Deformation of Secant Pile Wall

Brillouin optical time domain reflectometry (BOTDR) was used to obtain the full deformation profile of a secant pile wall during construction of an adjacent basement in London (Mohamed et al., 2007). By installing optical fiber down opposite sides of the pile, the distributed strain profiles obtained could be used to give both the axial and lateral movements along the pile. Measurements obtained from the BOTDR were found in good agreement with inclinometer data from the adjacent piles.

(ii) Detection of Structural Cracks

A large scale experimental program was undertaken to develop a method for accurate detection of simulated cracks and their locations on a 15-m-long steel beam using BOTDR fiber-optic distributed sensors. The discontinuities in the strain distribution based on the theoretical analysis of the beam provided the means to accurately pinpoint the location of the simulated cracks. The distortion effect of the BOTDR system due to averaging of the strains over the available spatial resolution masked the influence of the strain discontinuities. Optomechanical relationships were used to simulate the effect of spatial resolution on the theoretical results. The apparent strains obtained in this way were compared with the BOTDR measured values (Feng et al., 2013) as shown in Fig. 9.4. Other works on crack detection involved the use of Brillouin optical correlation domain analysis (BOCDA) in fiber-reinforced concrete (Imai et al., 2010). The sensor system was shown to detect minute cracks before visual recognition during a beam-bending test which was carried out to detect crack-induced strain distribution during loading.

(iii) Measurement of Strain Profiles on Structural Components

Strain profiles along the steel girders of a continuous slab-on-girder bridge subjected to diagnostic load testing were obtained using a BOTDR system (Matta et al., 2008). A 1.16 km long sensing circuit was installed onto the web of four girders. These strain profiles were converted into deflection profiles and validated against discrete deflection measurements performed with a high-precision total station system. Structural health assessment of the girder was performed by comparing the BOTDR strain profiles with the results of three-dimensional finite-element analysis of the bridge super-structure.

(iv) Measurement Of Leakage on Oil/Gas Pipelines

The distributed fiber-optic vibration sensors are frequently investigated for pipeline leakage detection. Vibration and temperature sensing fiber-optic networks have been proposed and used in practice for detection and localization of pipeline leakages (Huang et al., 2007; Feng et al., 2009; He et al., 2011; Frings and Walk, 2011). The application and use of commercially available pipeline surveillance systems of distributed fiber-optic methods that provide pipeline integrity management including real-time leakage detection and threat identification are standardized (i.e., ASTM F2349, 2350).

(v) Borehole Profiling

Distributed fiber-optic sensors have been tested for borehole seismic profiling (Frignet et al., 2014; Hartog et al., 2014; Dixon et al., 2014). Particularly useful in oil and gas industry, borehole characterization of the underground formation, monitoring the production flow and hydraulic fracture profiling can be accomplished with distributed fiber optic sensors that can gather information on a single fiber (Hartog et al., 2014).

9.2.2.2 Dynamic Strains and impact Wave Measurements in Soil

A number of studies on dynamic measurements of vibration underground have been carried out successfully using SBS, BOTDA, slope assisted BOTDA (SA-BOTDA) or phase sensitive OTDR (Φ-OTDR) (Zhang and Bao, 2008; Bernini et al., 2009; Lu et al., 2010; Cui et al., 2011;

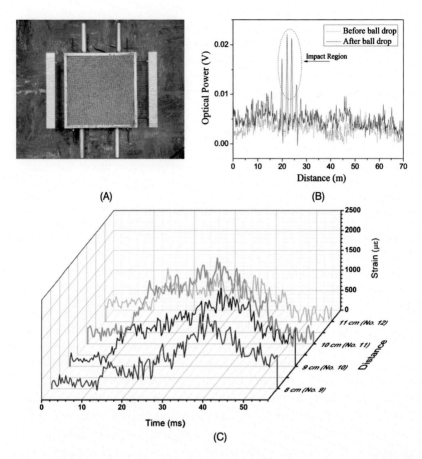

Figure 9.5 Impact test set up and the dynamic test results obtained using a BOTDA distributed fiber-optic sensing system (Cui et al., 2015). (A) Continuous fiber fence lay-out in sand packed test box. (B) Variation of optical power along the length of the continuous fiber showing large changes at the region of impact. (C) Distributed measurement of dynamic strains over the impacted region of the fiber strands Nos. 9, 10, 11, 12

Liang and Lin, 2012; Peled et al., 2013; Masudi et al., 2013; Lin et al., 2015; Zhou et al., 2015; Ren, 2016). Among these studies dynamic ranges between 2.5 Hz to 160 MHz were demonstrated with spatial resolutions ranging from 1 cm to 1.5 m over 50 m to 10 km of sensing fiber.

For instance, a pulse–based BOTDA technique which used a short pulse and the slope of the stimulated Brillouin spectrum was proposed for distributed vibration sensing by Cui et al. (2011). Subsequently, a fiber–optic fence operated with this technique was demonstrated to measure the internal dynamic strain response of a sand bed caused by surface impact (Cui et al., 2015). A packed sand bed simulated the surroundings of embedded infrastructure and the impact was generated by a ball-drop (7.2-cm diameter, weighing 192.3 g) from a fixed height of 2 m. An aluminium frame was used to affix and embed 30 strands of continuous fiber at 1 cm grid intervals, 2 cm below the surface of the sand bed (305 × 305 × 102 mm), as shown in Fig. 9.5A. Parallel placement of the fiber strands separated by 1-cm spacing provided the capability of 1 cm spatial resolution of measurement in the orthogonal direction to fiber axis. Fig. 9.5B shows the time domain sweep of optical power change, represented in linear distance over the fiber before and after the ball drop. The signal is based on the 11.20-GHz fixed modulation frequency and is recorded for the total length of the fiber (70 m). In Fig. 9.5B, the black colored trace (i.e., the dot trace) is the Brillouin intensity signal record before the ball drop. The red colored trace (i.e., the solid trace) is the Brillouin intensity signal record after the ball drop. In general, the two traces are similar except for the central four peaks representing the ball impact region over the fiber fence. These peaks reflect the strains caused by the ball impact on strands numbered as Nos. 9, 10, 11 and 12 (shown in Fig. 9.5A also). Fig. 9.5C presents the recording of time and space variation of the calibrated dynamic strains from the impacted region of the fibre strands numbered 9 through 12. Dynamic changes in several milliseconds and in 1 cm location resolution were obtained. The implications of the measurement system is promising for high resolution distributed monitoring of vibrations and impact waves on and in the vicinity of buried infrastructure where other monitoring techniques may not provide the continuous spatial and temporal coverage required.

A similar development was also investigated to determine the features of a stress wave propagating through high plasticity oily clay upon a blunt impact on its surface. A BOTDA based fiber-optic fence and an array of FBG sensors capable of measuring strains with spatial resolution of few centimeters and temporal resolution on the order of milliseconds for the BOTDA and micro seconds for the FBG array were employed (Pamukcu et al., 2013). In the first phase of this work, involving a BOTDA based fiber-optic mesh sensor fence, a standard impact test (NIJ – National Institute of Justice Standard–0101.06, Ballistic Resistance of Body Armor) was employed to create impact on a packed clay sample. The NIJ test calls for the free fall of a steel sphere, with a diameter of 63.5 mm and a mass of 1043 g, dropped from 2 m of height onto the smooth surface of a packed clay bed. The clay used in this procedure is Roma Plastillina No. 1 (RP #1) oil based modeling (ballistic) clay, which consists of oil, waxes, and clay

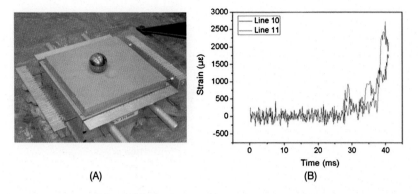

(A)   (B)

Figure 9.6 Ball drop impact test on clay instrumented with BOTDA based fiber-optic sensing fence (Pamukcu et al., 2013). (A) Fiber fence layout in clay packed test box and ball drop position. (B) Dynamic strains measured over the ball drop position coincident to fiber strand lines 10 and 11

minerals. The clay was packed into a $305 \times 305 \times 102$ mm ($12 \times 12 \times 4$ in) aluminum tray. A distributed fiber-optic fence with 1 cm separation between its strands was placed 2 cm below the impact surface, as shown in Fig. 9.6A. A reflective laser trigger system was placed close to the top of the clay box to trigger the data acquisition for the BOTDA setup. The temporal record of strains over the fiber-optic strands numbered 10 and 11 are given in Fig. 9.6B. The record shows that the event is triggered at 0 ms, but the effect of the impact at 2 cm below the clay surface takes place at about 27 ms and lasts only about 13 ms, when the fiber strains peak in the range of 1700–2700 $\mu\varepsilon$. A simple mathematical relationship between the vertical and horizontal deformations in clay upon a standard NIJ ball-drop was used to estimate axial fiber strain from the vertical deformation in the clay. The computed fiber strains based on the assumed geometry of ball-fiber-clay interaction was 0.0030 (3000 $\mu\varepsilon$) on average.

In the second phase of the investigation a three point FBG array was embedded 5 cm below the clay surface. Impact response was acquired simultaneously at each one of the FBG points placed in an equilateral triangular formation of 60 mm base and 30 mm hypotenuse as shown in Fig. 9.7A (Pamukcu et al., 2014). The time resolution of the acquired signal in response to a standard ball drop (NIJ standard) ranged from 20 to 40 μs, with a distinct difference of time of arrival of stress wave between the nearest (1543 and 1544 nm FBG) and farthest FBG (1546 nm FBG) node from the point of impact, as shown in Fig. 9.7B. These results were encouraging for accurate determination of the transition time and mag-

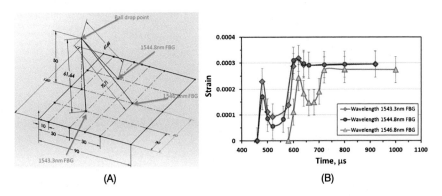

(A)   (B)

Figure 9.7 Ball drop impact test on clay instrumented with FBG fiber-optic sensing array (Pamukcu et al., 2014). (A) Sketch of FBG layout in clay packed test box and ball drop position. (B) Dynamic strains measured simultaneously at the three FBG points coincident with the ball drop

nitude of the traveling stress waves through the model clay as a result of projectile impact on its surface.

Thirdly, an FBG array instrumented clay head form, representing inside of a helmet, was tested for small projectile impact response, as shown in Fig. 9.8A (Pamukcu et al., 2013). An air-gun was used to fire a 1 cm diameter steel ball onto the clay surface while the embedded FBG recorded the time evolution of the internal strains. The array was able to record simultaneous data at a temporal resolution of 20 μs, which was sufficient to capture all the early strain peaks, as shown in Fig. 9.8B. The intensity of the peak strains were correlated with the arrival time of the peak. An indirect, nonlinear relationship was observed between the two parameters, which were expected due to geometric and material damping between the impact point and the measurement location. Using constrained surface fitting, the measured strains could be extrapolated over the three data points to develop a rough visualization of the strain field at the FBG embedment depth, and its temporal variation. This approach was based on the assumption that the strain vector field $\varepsilon(x, y, z)$ could be described by a relatively smooth function of the three spatial coordinates $\{x, y, z\}$. In this case the entire field could be modeled as an interpolating surface whose parameters were assessed based on the measured values at the three FBG points and appropriate boundary conditions. Fig. 9.8C shows the strain fields obtained at two distinct time stamps following the projectile impact. Although still under development, the results show potential for distributed fiber-optic

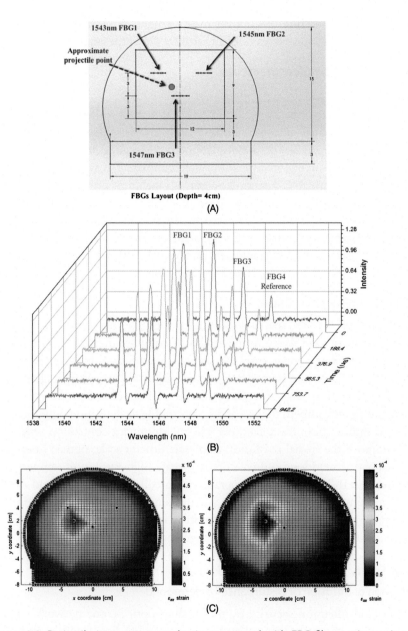

Figure 9.8 Projectile impact test on clay instrumented with FBG fiber-optic sensing array (Pamukcu et al., 2014). (A) Sketch of FBG layout in clay packed test form and projectile impact position. (B) The simultaneous measurement of signal intensity at different time stamps at the three FBG points in the clay form during projectile impact from an air-gun. (C) Visuals of strain fields at two different time instants developed using constrained curve fitting and the FBG fiber-optic array data of the projectile impact tests

techniques capable of high resolution measurement of internal dynamic strains and traveling stress waves resulting from a projectile impact.

9.2.2.3 Under Water Measurements

(i) Subsea

BOTDR based fiber-optic temperature sensor was used to detect pipeline leakage using thermal changes along a subsea oil pipeline network (Madabhushi et al., 2015). Predictions from a mathematical model and results from an experimental investigation were compared. It was concluded that the optical fiber cable detection system is capable of providing an accurate and rapid assessment of the location of a leak along a subsea pipeline using the temperature gradients detected by the system. The resolution of the thermal traces showed that even very small length (10 mm) of the fiber exposed to relatively low temperature differentials (10 °C) caused signal peaks that were easily distinguishable from the baseline measurements.

(ii) Water Column and Streambed

Infrared imagery and high spatial resolution distributed fiber-optic temperature sensors were used to study the effects of complex thermal dynamics on endangered dwarf wedgemussel species at a groundwater discharge bank along the upper Delaware River (Briggs et al., 2013). Using the spatial and temporal temperature data gathered by the fiber-optic sensors in the water column and the streambed, the researchers were able to determine whether discrete or diffuse groundwater inflow was the dominant control on refugia. To improve the spatial resolution to the centimeter scale, optical fibers were wrapped around longitudinal cores to create high-resolution temperature sensors (HRTS). Typical temperature data collected over a period of time in midsummer is shown in Fig. 9.9. This same concept was used again to record high-resolution temperature profiles in snowpacks, fluid flux in the streambed, and shallow aquifer exchange and stratification in ponds.

9.2.2.4 Safety and Security Applications

(i) Detection of Intrusion or Disturbance Underground

Perimeter and intrusion security by use of fiber-optic sensor networks at critical public sites such airports and seaports, or life-line facilities such as water distribution and electric power plants, or nuclear facilities and military bases have been proposed and introduced into practice within the last decade (Liang and Lin, 2012; Tu et al., 2012; Li et al., 2012; Juarez and Taylor, 2007; Peng et al., 2014; Wu et al., 2012, 2014; Li et

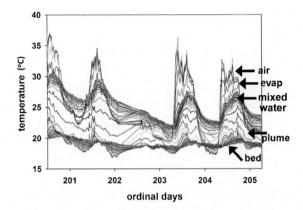

Figure 9.9 The 0.014-m spatial resolution fiber-optic temperature sensor data collected at HRTS3A over 5 days along a vertical profile installed through the air, water, and streambed domains. Each line on the plot depicts temperature at a specific elevation on the high resolution temperature sensor (Briggs et al., 2013)

al., 2013). These sensors, mostly concealed as "electronic fences" serve as prewarning systems for disturbance, intrusion and or environmental hazard. For example, China boasts a national border monitoring system which has been in operation along a 220 km long national borderline north of China since 2012 (Wu et al., 2014). Similarly, researchers have developed a BOTDR based method of detecting covert tunneling activity at depths of more than 60 ft below ground surface. The method was developed as part of a project to create an underground fence aimed at preventing the creation of tunnels between secured borders. The method uses wavelet decomposition of BOTDR signals to detect distortions along the fiber due to small tunnel activity at up to 10–15 m from the fiber through the surrounding soil (Linker and Klar, 2013).

(ii) Detection of Methane in Underground Coal Mine

Sequential multipoint fiber-optic gas cells connected by a fiber-optic cable were used to determine methane concentrations in an underground coal mine environment (Shemshad et al., 2009). The measurements were used in an analytical formulation to determine the average gas concentration of the methane gas in the area where the fiber optic cells were located.

(iii) Detection of Stray Current in Coal Mine

Current leakage, termed as stray current, from coal mine DC traction supply can cause electric explosions, electric shock, or corrosion of the buried metal pipes and cables. A distributed fiber optic current sensing system based on Faraday magneto-optic effect was tested on a simulated rail system to determine its effectiveness in detecting stray currents in coal mining operations. Optical fiber cable was spirally wound around an electricity conductor. The conductor was loaded at various levels of current creating magnetic field, which would cause the polarization plane of a polarized light in the fiber rotate proportionally. The measured changes in the polarization angle could then be correlated directly with the applied current (Cai et al., 2011).

(iv) Detection of Strain and Temperature in Nuclear Waste Repositories

Fiber-optic measurements based on BOTDR and ROTDR were proposed and successfully qualified to use as temperature and strain monitoring sensors in the French geological repository for high- and intermediate-level long-lived nuclear wastes with gamma radiation and hydrogen release (Delepine-Lesoille et al., 2012). The efficiency of simultaneous Raman and Brillouin scattering measurements to provide both the distributed strain and temperature measurements were demonstrated.

Fiber-optic sensor networks, either used in a distributed or single point configuration, provide great advantages for large-scale monitoring in underground environment due to their versatility and relatively cost-effective manner by which they may be applied. There are already multiple standards and specifications developed for part of the practice. Furthermore, rapidly ongoing advancements and innovations on waveguide technologies offer multitude of solutions that can be applied for underground sensing for preemptive warning and response in an effective manner.

9.3 FUTURE RESEARCH ON ADVANCED HYBRID SENSING FOR PREEMPTIVE RESPONSE

9.3.1 Crowdsensing for Preemptive Response to Underground Events

Crowdsensing, participatory sensing, or community sensing is a sensing process that involves a crowd or community for data harvesting. It has been becoming a valuable and complementary approach to event monitoring, situation-evolution prediction, and post-event analyses with the proliferation of mobile devices (e.g., smart phones, wearables, and tablets with various types of sensors such as cameras, microphones, GPS receivers, accelerometers, and temperature sensors) and the advancement of big-data research. For example, researchers have been successful in using a real-time picture sharing network with millions of users and pictures to localize urban events (Giridhar et al., 2017).

Depending on the involvement level of individual participant in the data harvesting process, crowdsensing or community sensing may be carried out in the form of participatory sensing (Dutta et al., 2009), where individuals actively contribute sensory data such as taking a picture of the embankment, or opportunistic sensing, where individuals involvement is minimal in contributing sensory data such as location data reporting without explicit individual actions (Ganti et al., 2011). With the facilitation of mobile applications, participatory sensing may become opportunistic, e.g., continuous road pavement condition sampling using accelerometers on smart phones and data reporting to data server(s).

Crowdsensing have been used to detect potholes as well as sense bumps in the driveways. For example, a distributed sensing system, *Nericell* (Mohan et al., 2008), using mobile applications to collect accelerometer and GPS data locally by mobile devices and process data locally by the mobile devices and/or remotely by the data server depending on the resource limitations such as energy capacity, bandwidth availability, and computation capability, and data-related concerns such as participant privacy, information security, and data integrity, has been successfully used to detect road potholes and their locations. We envision that this type of crowdsensing may be used for preemptive response to sinkholes and other geohazards under development when integrated with big-data analysis. Sinkholes, big or small, result in economic and sometimes human life losses. For instance, on February 23, 2007, a sinkhole due to sewage pipe ruptures collapsed in Guatemala City, forming a very large, deep circular hole with vertical walls and killing five people. The occurrence of sinkholes resulting from bedrock roof collapses is rare, in comparison with the occurrence of sinkholes induced by downward migration of unconsolidated deposits. Their occurrences may be predicted if crowdsensing and big-data analysis of subsurface environment would be available. For instance, researchers have analyzed the spatial and temproal characteristics of the social media feeds responsing to a 5.8 magnitude earthquake occurred on the US East Coast on August 23, 2011 from an online news and social networking service through which milions of daily users post and interact with messages, and concluded that the crowdsensing offers comparable results in a timely manner that complement other sources of data to enhance situational awareness (Crooks et al., 2013).

Even though crowdsensing leveraging interest in intelligent sensing for citizen science (O'Grady et al., 2016) may have advantages over infrastructure-based sensing such as ease of deployment, involvement of community, potentially lower cost, it presents a couple of technical and social challenges including (i) sensing region characterization, (ii) sensing density quantification, (iii) sensing accuracy, and (iv) incentive scheme (Cardone et al., 2013). When using crowdsensing for preemptive response to underground events, these issues should be addressed well before its deployment.

Crowdsensing may be integrated with other sensing techniques such as remote sensing and environmental monitoring to achieve success in preemptive response to underground events. For example, data provided by satellite imaging services and rainfall data reported by weather services may be used in a hybrid sensing system with crowdsensing to predict landslide events. A recent survey of remote sensing platforms (e.g., satellite, airborne, drone, vehicular, and static platforms) and sensors (Toth and Jóźków, 2016) argues that crowdsensing data will provide high-quality geospatial information in the future that will match that of the current state-of-the-art mapping technologies. The "Crowdsensing for Observation from Satellite" project sponsored by European Space Agency (ESA) has investigated how citizen science and crowdsensing impact upon earth observation (Mazumdar et al., 2017). The authors have identified three critical areas that must be addressed for a significant impact, which can be a good reference when integrating crowdsensing with remote sensing, namely (i) data governance, standardization, trust, (ii) social presence, support and education, and (iii) technical implementation.

9.3.2 Pipeline Monitoring With Hybrid Sensing Using WSiN, GPR, and Crowdsensing

The issue of aging infrastructure that may experience more frequent failures, which then disrupt the delivery of water, electricity, natural gas, and oil to residents, has become an increasing concern among city administrators and service providers. Tragedies have occurred due to failed infrastructure such as the natural gas explosion that killed people and destroyed homes because underground infrastructure cracked. Existing approaches to pipeline damage management, which include external sensing, internal sensing, and remote sensing (Bradshaw et al., 2009), still leave great room for research to achieve the preemptive response goal.

(i) Wireless signal networks (WSiN)

Previous research on wireless signal networks (Yoon et al., 2012) has proven that the variation of link quality between wireless transceivers can be used as an effective sensing mechanism which reflects characteristics of geo-media subjected to various geo-events. The related research results include (i) an accurate and simple radio propagation model for underground environments (Yoon et al., 2011), (ii) a window-based minimum distance classifier based on Bayesian decision theory to classify geo-events using the measured signal strength as a main indicator of geo-events, (iii) and a proof-of-concept lab-scale system for real-time global subsurface monitoring based on the wireless signal network concept, the model, and the classifier (Yoon et al., 2012).

Comparing the theoretical estimations of the underground radio propagation and the measured data from lab experiments, the theoretical model fits the measured data well within a 3.45 dBm deviation or with an accuracy of 96.3% on average. A window-based minimum distance classifier based on Bayesian decision theory can be designed to detect and classify geo-events using the measured signal strength as a main indicator of geo-events. The window-based classifier for wireless signal networks has two steps: event detection and event classification. With the event detection, the window-based classifier classifies geo-events on the event occurring regions that are called a classification window.

Event detection (window selection). Assume that if the strength of the received signal does not change from previously collected data at a specific location among N transceiver deployment locations, then there is no new geo-hazard or event in the soil medium and the event classification is not required. Thus, it is important to detect the region where the event occurs with a simple classification such as two-category case (ω_1, event; ω_2, no-event) in which ω_1 can be assigned to a binary value 1 and ω_2 to a binary value 0. The subsurface event ω_1 at kth position ($1 \leq k \leq N$) can be detected by the signal strength deviation from existing M sample average at the nth sensing time t_n as follows where ζ is the deviation criterion, which can be empirically decided based on the measured data that has small variation in soil:

$$\left| x_k(t_n) - \left[\sum_{j=n-1}^{n-M} x_k(t_j) \right] / M \right| geq \zeta$$

Event classification in selected window. Assume that there are N positions to sense geo-events in underground wireless signal networks. Let

$\{\omega_1, \omega_2, \ldots, \omega_c\}$ be the finite set of c states of events. $P(\omega_j)$ describes the prior probability that the event is in state ω_j. The variability of a measurement in probabilistic terms is expressed as x which is considered a random variable whose distribution depends on the state of event which is expressed as $p(x|\omega_i)$. If there is an observation x for which $P(\omega_i|x)$ is greater than $P(\omega_j|x)$, there would be a higher possibility that the true state of the event is ω_i. Thus, choosing ω_i minimizes the probability of error.

In the classification with more than one measurement, the scalar x is replaced by the feature vector, which is in an N-dimensional Euclidean space R^N. The posterior probability $P(\omega_i|)$ can be computed from $p(|\omega_i)$ by the Bayes formula:

$$P(\omega_i|\vec{x}) = \frac{p(\vec{x}|\omega_i)P(\omega_i)}{[(\vec{x})}$$

The Bayes formula shows that by observing the value of we can convert the prior probability $P(\omega_i)$ to the a posteriori probability $P(\omega_i|\vec{x})$ – the probability of the state of event being ω_i given that the feature value of $\vec{x}$ has been measured. $p(\vec{x}|\omega_i)$ is called likelihood of ω_i with respect to $\vec{x}$. The Bayes decision rule emphasizes the role of the posterior probabilities, and the evidence factor $p(\vec{x})$ is unimportant as far as making a decision is concerned. To represent event classifiers, a set of discriminant functions $g_i(\vec{x})$ is used, where $i = 1, \ldots, c$. The measured received signal at wireless transceivers is assumed to be independent and of a normal (or Gaussian) distribution. The discriminant functions are simplified as follows:

$$g_i(\vec{x}) = -\frac{\|\vec{x} - \vec{\mu}_i\|^2}{2\sigma^2} + \ln P(\omega_i)$$

where $\| \bullet \|$ denotes the Euclidean norm, where $\vec{x}$ is an N-dimensional column vector, μ is the N-dimensional mean vector, σ is the variance, and this forms the minimum distance classifier.

The window–based classification method is evaluated with a water leakage experiment in which the data has been measured in laboratory experiments. The results show that the detection and classification method based on wireless signal network can detect and classify subsurface events (Yoon et al., 2012).

(ii) Low-Frequency Radio for Field-Scale Wireless Signal Networks

Future research is needed to transform the WSiN approach from 1-meter-scale lab environments to dozen- to hundred-meter-scale field environments by using low-frequency radio generated by research platforms and benchmarking the existing model of underground sensing.

The USRP (Universal Software Radio Peripheral) platform may be a good candidate system to operate in a wide range of low frequencies and may be customized to be suitable for underground sensing. USRP is a hardware platform for making software radios in which radio communication components (e.g., mixers, filters, amplifiers, modulators/demodulators, detectors, etc.) are implemented by means of software.

Using the existing underground radio propagation model, the communication radius of the 2.4 GHz transceiver and 433 MHz transceiver are about 20 and 30 cm in wet clay type soil (electrical conductivity, 78 mS; relative permittivity, 30) sampled at Lehigh University test site. To overcome the high attenuation and extend the communication radius in soil, three parameters can be controlled such as using high transmission power, high gain antenna, and low radio frequency. In the cases of high transmission power and high gain antenna, the improvement of communication radius is not sufficient for practical underground monitoring applications. If the underground transceivers use a low frequency, the communication radius can be extended more efficiently. There are suitable low frequency bands for underground communication such as LowFER (low-frequency experimental radio) of 160–190 kHz and low frequency ISM (industrial, scientific and medical) bands of 6.78 and 13.56 MHz. ISM bands are radio bands reserved internationally for the "license exempt" use of radio frequency for industrial, scientific and medical purposes. LowFER is a license-free form of two-way radio communications practiced on frequencies below 500 kHz. One permits the use of up to 1 W of power and a 15 m antenna between 160~190 kHz with no license requirement. Thus 20~25 m communication may be achieved even in wet clay underground with 1 W Tx power and the same MICA's antenna gain as shown in Fig. 9.10.

An advanced hybrid sensing system may be developed for proactive damage localization of underground pipeline infrastructure by integrating the long-term low-resolution underground monitoring by real-time wireless signal networks (WSiNs), short-term high-resolution damage localization by agile and cognitive ground-penetrating radar (GPR), and continuous engagement of local community through citizen-scientist initiatives with crowdsensing using applications on their smart devices. The WSiN and GPR delivers time-space continuous data to (i) map the evolution of a hazard as it triggers calibrated signals from pre-event, and (ii) detect

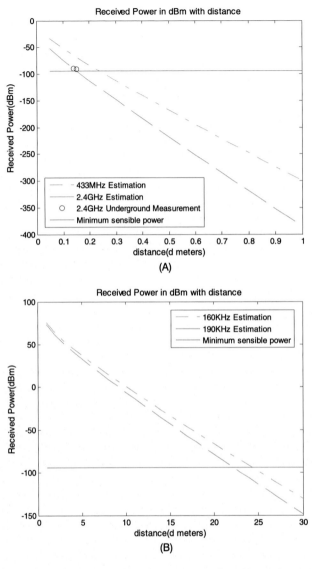

Figure 9.10 (A) Underground radio signal attenuation of 2.4 GHz and 433 MHz sensors. (B) Underground Radio Signal Attenuation of 160 kHz and 190 kHz

location of any damage, applicable to buried pipelines and the surrounding soil/rock mass. The citizen-scientist initiatives for the crowdsensing may provide large-scale meta-data for pre- and post-event sensing from the community. Once a warning and a general area of the underground pipeline damage or its subsequent leakage is detected by the WSiN and/or crowdsensing, the agile GPR can be deployed to locate the damage spot with speed and precision.

9.3.3 Land-Mine Detection Using Hybrid EM and Seismic-Acoustic Sensing

Electromagnetic (EM) sensors have been used to detect, map, and locate underground utilities in the field. For example traditional low to medium-frequency EM sensors have been used to detect buried land-mines where the EM frequency ranges from tens to hundreds of kHz. Ground penetrating radar uses wideband EM waves of frequencies in GHz (ultrahigh and superhigh frequencies) for underground sensing. The categorization of EM spectrum is based on the ITU (International Telecommunication Union) designation. Recent research (Heinz et al., 2014, 2015) shows that high to very high (VH)-frequency EM waves have been evaluated for detecting metal/anomaly. More work may further the research and development of long-range and low-power underground sensors using high to VH EM waves.

A specific application area of such research is to develop agile and accurate methods of land-mine detection for military and humanitarian projects. A hybrid sensing system integrating EM sensors with seismic-acoustic sensing may lead to new research and development results in this area. Existing research (Donskoy, 1998; Donskoy et al., 2002) has shown that seismic-acoustic techniques may differentiate acoustically compliant container articles (e.g., land mine) from the surrounding soil. The idea is to use seismic or airborne acoustic waves to stimulate the vibration of buried objects, then use laser or microwave vibrometers to remotely measure the soil surface vibration, and analyze the measurement to obtain the object's vibration signature. The underlying principle is that the soil mass on top of a compliant container makes a mass–spring system and the linear and non-linear vibration responses of compliant containers (e.g., land mines) and other buried objects (e.g., rocks) could have distinct signatures. In fact, recent research (Wetherington and Steer, 2013) also shows that vibrations of a buried object underground may be detected by GPR or other similar EM sensors.

REFERENCES

Achatz, D.E., Ali, R., Wolfbeis, O.S., 2011. Luminescent chemical sensing, biosensing, and screening using upconverting nanoparticles. Top. Curr. Chem. 300, 29–50.

Agrawal, G.P., 1995. Nonlinear Fiber Optics. Academic Press Inc., San Diego, CA.

Akita, S., Sasaki, H., Watanabe, K., Seki, A., 2010. A humidity sensor based on a hetero-core optical fiber. Sens. Actuators B 147, 385–391.

Aksuner, N., 2011. Development of a new fluorescent sensor based on a triazolothiadiazin, derivative immobilized in polyvinyl chloride membrane for sensitive detection of lead (II) ions. Sens. Actuators B, Chem. 157, 162–168.

Anastasio, S., Pamukcu, S., Pervizpour, M., 2007. BOTDR for detection of chemical and liquid content FMGM 2007: field measurements in geomechanics 2007. In: Jerry DiMaggio, P.E., Osborn, Peter (Eds.), Geotechnical Special Publication, vol. 175, ASCE, September 2007, pp. 1–12.

Angulo-Vinuesa, X., Martin-Lopez, S., Nuño, J., Corredera, P., Ania-Castañon, J.D., Thévenaz, L., González-Herráez, M., 2012. Raman-assisted Brillouin distributed temperature sensor over 100 km featuring 2 m resolution and 1.2C uncertainty. J. Lightwave Technol. 30 (8), 1060–1065.

Antico, E., Lerchi, M., Rusterholz, B., Achermann, N., Badertscher, M., Valiente, M., Pretsch, E., 1999. Monitoring Pb^{2+} with optical sensing films. Anal. Chim. Acta 388, 327–338.

ASTM F2233, Guide for Safety, Access Rights, Construction, Liability, and Risk Management for Optical Fiber Networks in Existing Sewers.

ASTM F2303, Practice for Selection of Gravity Sewers Suitable for Installation of Optical Fiber Cable and Conduits.

ASTM F2349, Practice for Operation and Maintenance of Integrated Natural Gas Pipelines and Optical Fiber Systems.

ASTM F2350, Practice for Selection of Natural Gas Pipelines Suitable for Installation of Optical Fiber Systems.

ASTM F2462, Practice for Operation and Maintenance of Sewers with Optical Fiber Systems.

ASTM F3079-14, 2014. Standard Practice for Use of Distributed Optical Fiber Sensing Systems for Monitoring the Impact of Ground Movements During Tunnel and Utility Construction on Existing Underground Utilities, ASTM International, West Conshohocken, PA. www.astm.org.

ASTM F3092-14, 2014. Standard Terminology Relating to Optical Fiber Sensing Systems, ASTM International, West Conshohocken, PA. www.astm.org, 2014.

Balaji, T., Sasidharan, M., Matsunaga, H., 2006. Naked eye detection of cadmium using inorganic–organic hybrid mesoporous material. Anal. Bioanal. Chem. 384, 488–494.

Bao, X., DeMerchant, M., Brown, A., Bremner, T., 2001. Tensile and compressive strain measurement in the lab and field with the distributed Brillouin scattering sensor. J. Lightwave Technol. 19, 1698–1704.

Bao, X., Chen, L., 2012. Recent progress in distributed fiber optic sensors. Sensors 12, 8601–8639.

Bernini, R., Minardo, A., Zeni, L., 2004. Stimulated Brillouin scattering frequency-domain analysis in a single-mode optical fiber for distributed sensing. Opt. Lett. 29 (17), 1977–1979.

Bernini, R., Minardo, A., Zeni, L., 2009. Dynamic strain measurement in optical fibers by stimulated Brillouin scattering. Opt. Lett. 34, 2613–2615.

Bernini, R., Minardo, A., Zeni, L., 2012. Distributed sensing at centimeter-scale spatial resolution by BOFDA: measurements and signal processing. IEEE Photonics J. 4 (1), 48–56. http://dx.doi.org/10.1109/JPHOT.2011.2179024.

Bradshaw, A.S., daSilva, G., McCue, M.T., et al., 2009. Damage detection and health monitoring of buried concrete pipelines. In: Oka, Murakami, Kimoto (Eds.), Prediction and Simulation Methods for Geohazard Mitigation, pp. 473–478.

Briggs, M.A., Voytek, E.B., Day-Lewis, F.D., Rosenberry, D.O., Lane, J.W., 2013. Understanding water column and streambed thermal refugia for endangered mussels in the Delaware river. 1. Environ. Sci. Technol. 47 (20), 11423–11431. 9 p.

Buerck, J., Roth, S., Kramer, K., Mathieu, H., 2001. OTDR distributed sensing of liquid hydrocarbons using polymer-clad optical fibers. In: TDR 2001, Int. Sym., and Workshop on TDR for Innovative Geotechnical Applications, vol. 2, pp. 496–509.

Cai, Lihua, Li, Wei, Fang, Haifeng, 2011. Monitoring method of stray current in coal mine on optical fiber sensing technique. Appl. Mech. Mater. 43, 457–462. http://www.scientific.net/AMM.43.457, 2011.

Cardone, Giuseppe, Foschini, Lica, Bellavista, Paolo, et al., 2013. Fostering participation in smart cities: a geo-social crowdsensing platform. IEEE Commun. Mag., 112–119.

Chen, X., 2013. Current developments in optical fiber technology. In: Harun, S.W., Arof, H. (Eds.), Optical Fibre Gratings for Chemical and Bio − Sensing. ISBN 978-953-51-1148-1. Chapter 8.

Crooks, A., Croitoru, A., Stefanidis, A., Radzikowski, J., 2013. #Earthquake: Twitter as a distributed sensor system. Trans. GIS 17, 124–147. http://dx.doi.org/10.1111/j.1467-9671.2012.01359.x.

Cui, Q., Pamukcu, S., Xiao, W., Guintrand, C., Toulouse, J., Pervizpour, M., 2009. Distributed fiber sensor based on modulated pulse base reflection and Brillouin gain spectrum analysis. Appl. Opt. 48 (30), 5823–5828.

Cui, Q., Pamukcu, S., Xiao, W., Pervizpour, M., 2011. Truly distributed fiber vibration sensor using pulse base BOTDA with wide dynamic range. IEEE Photonics Technol. Lett. 23, 1887–1889.

Cui, Q., Pamukcu, S., Pervizpour, M., 2015. Impact wave monitoring in soil using dynamic fiber sensor based on stimulated Brillouin scattering. Sensors 15, 8163–8172. http://dx.doi.org/10.3390/s1504081632015.

Delepine-Lesoille, S., Phéron, X., Bertrand, J., Pilorget, G., Hermand, G., Farhoud, R., Ouerdane, Y., Boukenter, A., Girard, S., Lablonde, L., Sporea, D., Lanticq, V., 2012. Industrial qualification process for optical fibers distributed strain and temperature sensing. J. Sens. 2012. http://www.hindawi.com/journals/js/2012/369375/abs/.

Dixon, T., Herzog, H., Twinning, S., Paulsson, B.N.P., Thornburg, J., He, R., 2014. A fiber optic borehole seismic vector sensor system for high resolution CCUS site characterization and monitoring. Energy Proc. 63, 4323–4338.

Dong, Y., Pamukcu, S., 2014. Thermal, electrical conduction in unsaturated sand controlled by surface wettability. Acta Geotech., 1–9. http://dx.doi.org/10.1007/s11440-014-0317-0. Springer.

Dong, Y., Pamukcu, S., 2012. Preparation of polymer coated sands with reversible wettability triggered by temperature. In: GeoCongress 2012. In: Geotechnical Special Publication, vol. 225. ASCE, pp. 4446–4455.

Donskoy, Dimitri M., 1998. Nonlinear vibro-acoustic technique for landmine detection. Proc. SPIE 3392, 211.

Donskoy, Dimitri, Ekimov, Alexander, Sedunov, Nikolay, Tsionskiy, Mikhail, 2002. Nonlinear seismo-acoustic land mine detection and discrimination. J. Accoust. Soc. Am. 111, 2705.

Dutta, Prabal, Aoki, Paul M., Kumar, Neil, et al., 2009. Demo abstract: common sense: participatory urban sensing using a network of handheld air quality monitors. In: Proceedings of ACM SenSys, pp. 349–350.

Fen, Y.W., Mahmood, W., Yunus, M., Yusof, N.A., 2012. Surface plasmon resonance optical sensor for detection of Pb^{2+} based on immobilized p-tert-butylca-lix[4]arene-tetrakis in chitosan thin film as an active layer. Sens. Actuators B, Chem. 171–172, 287–293.

Fen, Y.W., Yunus, W.M.M., 2013. Utilization of chitosan – based sensor thin films for the detection of lead ion by surface plasmon resonance optical sensor. IEEE Sens. J. 13, 1413–1418.

Fen, Y.W., Yunus, W.M.M., Talib, Z.A., 2013. Analysis of Pb(II) ion sensing by crosslinked chitosan thin film using surface plasmon resonance spectroscopy. Optik 124, 126–133.

Feng, H., Zhu, L., Jin, S., Zhou, Y., Zeng, Z., Zhuge, J., 2009. Modeling of pipeline leakage detection and prewarning system for locating error analysis based on Jones matrix. J. Jpn. Pet. Inst. 52, 114–119.

Feng, X., Zhou, J., Sun, C., Zhang, X., Ansari, F., 2013. Theoretical and experimental investigations into crack detection with BOTDR-distributed fiber optic sensors. J. Eng. Mech. 139 (12), 1797–1807. 11 p.

Forzani, E.S., Foley, K., Westerhoff, P., Tao, N., 2007. Detection of arsenic in groundwater using a surface plasmon resonance sensor. Sens. Actuators B, Chem. 123, 82–88.

Frignet, B., Hartog, A.H., Mackie, D., Kotov, O.I., Liokumovich, L.B., 2014. Distributed vibration sensing on optical fibre: field testing in borehole seismic applications. Proc. SPIE, 9157.

Frings, J., Walk, T., 2011. Distributed fiber optic sensing enhances pipeline safety and security. Oil Gas Eur. Mag. 3, 132–136.

Froggatt, M., Moore, J., 1998. High-spatial-resolution distributed strain measurement in optical fiber with Rayleigh scatter. Appl. Opt. 37, 1735–1740.

Ganti, Raghu K., Ye, Fan, Lei, Hui, 2011. Mobile crowdsensing: current state and future challenges. IEEE Commun. Mag., 32–39.

García, J.A., Monzón, D., Martínez, A., Pamukcu, S., García, R., Bustos, E., 2014. Optical Fibers to Detect Heavy Metals in Environment: Generalities and Case Studies 427. In: Hernandez-Soriano (Ed.), Environmental Sciences: Environmental Risk Assessment of Soil Contamination. InTech, NY. ISBN 978-953-51-1235-8. http://dx.doi.org/10.5772/57285. Chapter 15.

Garus, D., Krebber, K., Schliep, F., Gogolla, T., 1996. Distributed sensing technique based on Brillouin optical-fiber frequency-domain analysis. Opt. Lett. 21 (17), 1402–1404.

Ghazanfari, E., Pamukcu, S., Yoon, S., Suleiman, M.T., Cheng, L., 2012. Geotechnical sensing using electromagnetic attenuation between radio transceivers. Smart Mater. Struct. 21 (12).

Giridhar, Prasanna, Wang, Shiguang, Abdelzaher, Tarek, Kaplan, Lance, George, Jemin, Ganti, Raghu, 2017. On localizing urban events with instagram. In: Proceedings of IEEE Infocom.

Glisic, B., Yao, Y., 2012. Fiber optic methods for health assessment of pipelines subjected to earthquake-induced ground movement. Struct. Health Monit. 11 (6), 696–711.

Glisic, B., 2014. Sensing solutions for assessing and monitoring pipeline systems. In: Wang, M.L., Lynch, J.P., Sohn, H. (Eds.), Sensor Technologies for Civil Infrastructures: Applications in Structural Health Monitoring, vol. 2. Woodhead Publishing, London, UK.

Grattan, K.T.V., Meggitt, B.T., 1999. Optical Fiber Sensor Technology, Chemical and Environmental Sensing, vol. 4. Kluwer Academic Publishers.

Guillemain, H., Rajarajan, M., Sun, T., Grattan, K.T.V., 2009. A self-referenced reflectance sensor for the detection of lead and other heavy metal ions using optical fibres. Meas. Sci. Technol. 20, 045207.

Guo, L., Zhang, W., Xie, Z., Lin, X., Chen, G., 2006. An organically modified sol–gel membrane for detection of mercury ions by using 5, 10,15, 20-tetraphenylporphyrin as a fluorescence indicator. Sens. Actuators B, Chem. 119, 209–214.

Hartog, A., Frignet, B., Mackie, D., Clark, M., 2014. Vertical seismic optical profiling on wireline logging cable. Geophys. Prospect. 62, 693–701.

Haupt, K., Mosbach, K., 2000. Molecularly imprinted polymers and their use in biomimetic sensors. Chem. Rev. 100, 2495.

Haykin, S., 2006. Cognitive radar: a way of the future. IEEE J. Signal Process., 30–40.

He, C., Zheng, X., Luo, J., Hang, L., Xu, X., Huang, H., 2011. Research on leak detection system for pipeline network based on distributed optical fiber sensing array technology. Opt. Technol. 37, 76–79.

Heinz, Daniel C., Brennan, Michael L., Steer, Michael B., Melber, Adam W., Cua, John T., 2014. High to very high-frequency metal/anomaly detector. Proc. SPIE 9072, 907209.

Heinz, Daniel C., Brennan, Michael L., Steer, Michael B., Melber, Adam W., Cua, John T., 2015. Phase response of high to very high frequency metal/anomaly detector. Proc. SPIE 9454, 94540H.

Horiguchi, T., Tateda, M., 1989. BOTDA—nondestructive measurement of single-mode optical fiber attenuation characteristics using Brillouin interaction: theory. J. Lightwave Technol. 7, 1170–1176.

Horiguchi, T., Shimizu, K., Kurashima, T., Tateda, M., Koyamada, Y., 1995. Development of a distributed sensing technique using Brillouin scattering. J. Lightwave Technol. 13, 1296–1302.

Hotate, K., Ong, S.S., 2003. Distributed dynamic strain measurement using a correlation-based Brillouin sensing system. IEEE Photonics Technol. Lett. 15, 272–274.

Huang, S.-C., Lin, W.-W., Tsai, M.-T., Chen, M.-H., 2007. Fiber optic in-line distributed sensor for detection and localization of the pipeline leaks. Sens. Actuators A, Phys. 135, 570–579.

Imai, M., Nakano, R., Kono, T., Ichinomiya, T., Miura, S., Mure, M., 2010. Crack detection application for fiber reinforced concrete using BOCDA-based optical fiber strain sensor. J. Struct. Eng. 136 (8), 1001–1008.

Jerónimo, Rajah P., Araújo, A., Montenegro, M. Conceição B.S.M., 2007. Optical sensors and biosensors based on sol–gel films. Talanta 72 (1), 13–27.

Iten, M., Spera, Z., Jeyapalan, J.K., Duckworth, G., Inaudi, D., Bao, X., Noether, N., Klar, A., Marshall, A., Glisic, B., Facchini, M., Jason, J., Elshafie, M., Kechavarzi, C., Miles, W., Rajah, S., Johnston, B., Allen, J., Lee, H., Leffler, S., Zadok, A., Hayward, P., Waterman, K., Artieres, O., 2015. In: Pipelines 2015: Recent Advances in Underground Pipeline Engineering and Construction, pp. 1655–1666.

Juarez, J.C., Taylor, H.F., 2007. Field test of a distributed fiber-optic intrusion sensor system for long perimeters. Appl. Opt. 46, 1968–1971.

Katchalsky, A., Michaeli, I., 1995. Polyelectrolyte gels in salt solutions. J. Polym. Sci. 15 (69).

Kee, H.H., Lees, G.P., Newson, T.P., 2000. All-fiber system for simultaneous interrogation of distributed strain and temperature sensing by spontaneous Brillouin scattering. Opt. Lett. 25, 695.

Kocincova, A., Borisov, S., Krause, C., Wolfbeis, O., 2007. Fiber-optic microsensors for simultaneous sensing of oxygen and pH, and of oxygen and temperature. Anal. Chem. 79, 8486–8493.

Krohn, D.A., 1988. Fiber Optic Sensors: Fundamental and Applications. Instrument Society of America, ISA, Research Triangle Park, North Carolina.

Kurashima, T., Horiguchi, T., Tateda, M., 1990. Thermal effects of Brillouin gain spectra in single-mode fibers. IEEE Photonics Technol. Lett. 2, 718–720.

Kurashima, T., Horiguchi, T., Izumita, H., Furukawa, S., Koyamada, Y., 1993. Brillouin optical-fiber time domain reflectometry. IEICE Trans. Commun. E76-B, 382–390.

LeBlanc, M., Huang, S.Y., Ohn, M., Measures, R.M., Guemes, A., Othonos, A., 1996. Distributed strain measurement based on a fiber Bragg grating and its reflection spectrum analysis. Opt. Lett. 21 (17), 1405–1407.

Li, X., Sun, Q., Wo, J., Zhang, M., Liu, D., 2012. Hybrid TDM/WDM-based fiber-optic sensor network for perimeter intrusion detection. J. Lightwave Technol. 30, 1113–1120.

Li, Y., Liu, T., Wang, S., Liu, K., Lv, D., Jiang, J., Ding, Z., Chen, Q., Wang, B., Li, D., et al., 2013. All fiber distributed long-distance perimeter security monitoring system with video linkage function. J. Optoelectron. Laser 24, 1752–1757.

Liang, T.C., Lin, Y.L., 2012. Ground vibrations detection with fiber optic sensor. Opt. Commun. 285, 2363–2367.

Lin, T.-J., Chung, M.-F., 2009. Detection of cadmium by a fiber-optic biosensor based on localized surface plasmon resonance. Biosens. Bioelectron. 24, 1213–1218.

Linker, R., Klar, A., 2013. Detection of tunnel excavation using fiber optic reflectometry: experimental validation. In: Proc. SPIE 8709, Detection and Sensing of Mines, Explosive Objects, and Obscured Targets XVIII, 87090X (June 7, 2013).

Linze, N., Mégret, P., Wuilpart, M., 2012. Development of an intrusion sensor based on a polarization-OTDR system. IEEE J. Sens. 12, 3005–3009.

Lin, W., Liang, S., Lou, S., Sheng, X., Wang, P., Zhang, Y., 2015. A novel fiber-optic distributed disturbance sensor system with low false alarm rate. Infrared Laser Eng. 44, 1845–1848.

Liu, Xin, Jin, Baoquan, Bai, Qing, Wang, Yu, Wang, Dong, Wang, Yuncai, 2016. Distributed fiber-optic sensors for vibration detection. Sensors 2016 (16), 1164. http://dx.doi.org/10.3390/s16081164.

Lu, Y., Zhu, T., Chen, L., Bao, X., 2010. Distributed vibration sensor based on coherent detection of phase-OTDR. J. Lightwave Technol. 28, 3243–3249.

Madabhushi, S.S.C., Elshafie, M.Z.E.B., Haigh, S.K., 2015. Accuracy of distributed optical fiber temperature sensing for use in leak detection of subsea pipelines. J. Pipeline Syst. Eng. Pract. 6 (2). 2015.

Maier, S.A., 2007. Plasmonics Fundamental and Applications. Springer.

Martins, H., Martin-Lopez, S., Corredera, P., Ania-Castanon, J.D., Frazao, O., Gonzalez-Herraez, M., 2015. Distributed vibration sensing over 125 km with enhanced SNR using phi-OTDR over an URFL cavity. J. Lightwave Technol. 33, 2628–2632.

Masoudi, A., Belal, M., Newson, T.P., 2013. A distributed optical fibre dynamic strain sensor based on phase-OTDR. Meas. Sci. Technol. 24, 085204.

Matsuo, E.S., Tanaka, T., 1988. Kinetics of discontinuous volume phase transition of gels. J. Chem. Phys. 89 (3), 1695.

Matta, F., Bastianini, F., Galati, N., 2008. Distributed strain measurement in steel bridge with fiber optic sensors: validation through diagnostic load test. J. Perform. Constr. Facil. 22 (4), 264–273. July/August 2008, 10 p.

Mayra, T., Klimant, I., Wolfbeis, O.S., Werner, T., 2008. Dual lifetime referenced optical sensor membrane for the determination of copper (II) ions. Anal. Chim. Acta 462, 1–10.

Mazumdar, Suvodeep, Wrigley, Stuart, Ciravegna, Fabio, 2017. Citizen science and crowdsourcing for Earth observations: an analysis of stakeholder opinions on the present and future. Remote Sens. 9 (1), 87. http://dx.doi.org/10.3390/rs9010087.

McDonagh, C., Burke, C.S., MacCraith, B.D., 2008. Optical chemical sensors. Chem. Rev. 108, 400–422.

Mohamad, H., Bennett, P., Soga, K., et al., 2007. Monitoring tunnel deformation induced by close-proximity bored tunneling using distributed optical fiber strain measurements FMGM 2007: field measurements in geomechanics 2007. In: Jerry DiMaggio, P.E., Osborn, Peter (Eds.), Geotechnical Special Publication, vol. 175. ASCE. ISBN 9780784409404, pp. 1–13.

Monzon-Hernandez Luna-Moreno, D., Martínez-Escobar, D., 2009. Fast response fiber optic hydrogen sensor based on palladium and gold nano-layers. Sens. Actuators B, Chem. 136 (2), 562–566.

Moran, M., Thursby, C., Pierce, G., Culshaw, S., Graham, B., 2000. Distributed fiber optic sensors for humidity and hydrocarbon detection. Proc. SPIE 3986, 342–351.

Niklès, M., Thevenaz, L., Robert, P., 1996. Simple distributed fiber sensor based on Brillouin gain spectrum analysis. Opt. Lett. 21, 758–760.

O'Grady, M.J., Muldoon, C., Carr, D., et al., 2016. Mob. Netw. Appl. 21, 375.

Ohno, H., Naruse, H., Kihara, M., Shimada, A., 2001. Industrial applications of the BOTDR optical fiber strain sensor. Opt. Fiber Technol. 7, 45–64.

Mohan, Prashanth, Padmanabhan, Venkata N., Ramjee, Ramachandran, 2008. Nericell: rich monitoring of road and traffic conditions using mobile smartphones. In: Proceedings of ACM SenSys, pp. 323–336.

Pour-Ghaz, M., Nadukuru, S.S., O'Connor, S., et al., 2011. Using electrical, magnetic and acoustic sensors to detect damage in segmental concrete pipes subjected to permanent ground displacement. Cem. Concr. Compos. 33 (7), 749–762.

Prabhakaran, D., Nanjo, H., Matsunaga, H., 2007. Naked eye sensor on polyvinyl chloride platform of chromo-ionophore molecular assemblies: a smart way for the colorimetric sensing of toxic metal ions. Anal. Chim. Acta 601, 108–117.

Pamukcu, S., Texier, S., Toulouse, J., 2006. Advances in water content measurement with distributed fiber optic sensor. In: GeoCongress 2006, vol. 187, pp. 7–12.

Pamukcu, S., Naito, C., Pervizpour, M., Cui, Q., Trasborg, P.A., Medina, C.I., Mentzer, M.A., 2013. Distributed Fiber-Optic Sensing and Numerical Simulation of Shock Wave Response of Manufactured Clay – Phase II: Software Model and Breadboard. Technical Report ARL-TR-6400. Army Research Laboratory.

Pamukcu, S., Naito, C., Tiffany, J.L., 2014. Fiber-Optic Sensor Array to Determine Dynamic Projectile Interactions in Helmet Testing. Final Report, to U.S. Army Research Laboratory, Survivability/Lethality Analysis Directorate, RDRL-SLB-D, Contract W911QX-12-T-0018.

Peled, Y., Motil, A., Kressel, I., Tur, M., 2013. Monitoring the propagation of mechanical waves using an optical fiber distributed and dynamic strain sensor based on BOTDA. Opt. Express 21, 10697–10705.

Peng, F., Wang, Z., Rao, Y.-J., Jia, X.-H., 2013. 106 km fully-distributed fiber-optic fence based on P-OTDR with 2nd-order Raman amplification. In: Proceedings of the Optical Fiber Communication Conference/National Fiber Optic Engineers Conference. 17 March 2013, Anaheim, CA, USA.

Peng, F., Wu, H., Jia, X.-H., Rao, Y.-J., Wang, Z.-N., Peng, Z.-P., 2014. Ultra-long high-sensitivity Φ-OTDR for high spatial resolution intrusion detection of pipelines. Opt. Express 22, 13804–13810.

Pérez, M.A., González, O., Arias, J.R., 2013. Optical fiber sensors for chemical and biological measurements. In: Harun, S.W., Arof, H. (Eds.), Current Developments in Optical Fiber Technology. Chapter 10.

Pervizpour, M., Pamukcu, S., 2011. Characterization of a surrogate swelling polymer as a functional sensor for distributed liquid sensing. Sens. Actuators A, Phys. 168, 242–252. http://dx.doi.org/10.1016/j.sna.2011.04.026.

Phillips, C., Jakusch, M., Steiner, H., Mizaikoff, B., Fedorov, A.G., 2003. Model-based optimal design of polymer-coated chemical sensors. Anal. Chem. 75, 1106–1115.

Potyrailo, R.A., Hieftje, G.M., 1998. Optical time of flight chemical detection, spatially resolved analyte mapping with extended – length continuous chemically modified optical fibers. Anal. Chem. 70, 1453–1461.

Qazi, H.H., Mohammad, A.B., Akram, M., 2012. Recent progress in optical chemical sensors. Sensors 12, 16522–16556.

Rajeev, P., Kodikara, J., Chiu, W.K., Kuen, T., 2013. Distributed optical fibre sensors and their applications in pipeline monitoring. Key Eng. Mater. 558, 424–434.

Ren, Meiqi, 2016. Distributed Optical Fiber Vibration Sensor Based on Phase-Sensitive Optical Time Domain Reflectometry. MS Thesis. Ottawa-Carleton Institute for Physics, University of Ottawa, Ottawa, Canada.

Schulz, E., Udd, E., Seim, J., McGill, G., 1998. Advanced fiber-grating strain sensor systems for bridges, structures, and highways. Proc. SPIE 3325, 212–221.

Sekia, A., Katakuraa, H., Kaib, T., Igab, M., Watanabeb, K., 2007. A hetero-core structured fiber optic pH sensor. Anal. Chim. Acta 582 (1), 154–157.

Shemshad, J., Aminossadati, S.M., Kizil, M.S., 2009. A sequential multipoint fibre optic methane sensing system. In: Proceedings of the 9th IASTED International Conference on Wireless and Optical Communications. WOC 2009, pp. 6–10.

Shi, B., Sui, H., Liu, J., Zhang, D., 2009. The BOTDR-based distributed monitoring system for slope engineering. In: Engineering Geology for Tomorrow's Cities, Geological Society, London. Eng Geology Special Publications 22.

Siegel, R.A., 1993. Hydrophobic weak polyelectrolyte gels: studies of swelling equilibria and kinetics. Adv. Polym. Sci. 109, 233.

Siegel, R.A., Falamarzian, M., Firestone, B.A., Moxley, B.C., 1988. pH-controlled release from hydrophobic/polyelectrolyte copolymer hydrogels. J. Control. Release 8, 179.

Skouri, R., Schosseler, F., Munch, J.P., Candau, S.J., 1995. Swelling and elastic properties of polyelectrolyte gels. Macromolecules 28, 197.

Song, J., Li, W., Lu, P., Xu, Y., Chen, L., Bao, X., 2014. Long-range high spatial resolution distributed temperature and strain sensing based on optical frequency-domain reflectometry. IEEE Photonics J. 6, 1–8.

Texier, S., Pamukcu, S., Toulouse, J., 2005. Advances in subsurface water-content measurement with a distributed Brillouin scattering fibre-optic sensor. In: Proc. SPIE, vol. 5855, OFS-17, Bruges, Belgium, pp. 555–558.

Toth, Charles, Jóźków, Grzegorz, 2016. Remote sensing platforms and sensors: a survey. ISPRS J. Photogramm. Remote Sens. 115, 22–36.

Tu, D., Xie, S., Jiang, Z., Zhang, M., 2012. Ultra long distance distributed fiber-optic system for intrusion detection. Proc. SPIE, 8561.

Turel, M., Pamukcu, S., 2006. Brillouin scattering fiber optic sensor for distributed measurement of liquid content and geosynthetic strains in subsurface. In: Geo-Shanghai. In: Geotechnical Special Publication, vol. 149, Shanghai, PRC, pp. 72–79.

Wang, D.Y., Gong, J., Wang, Y., Wang, A., 201. Fully-distributed fiber-optic hydrogen sensing using acoustically-induced long-period grating. IEEE Photonics Technol. Lett. 23 (11), 733.

Wang, Y., 2012. Fiber-Optic Sensors for Fully-Distributed Physical, Chemical and Biological Measurement. Doctoral Dissertation, Electrical Engineering. Virginia Polytechnic Institute and State University, Blacksburg, VA.

Wang, Z.N., Zeng, J.J., Li, J., Fan, M.Q., Wu, H., Peng, F., Zhang, L., Zhou, Y., Rao, Y.J., 2014. Ultra-long phase-sensitive OTDR with hybrid distributed amplification. Opt. Lett. 39, 5866–5869.

Wetherington, Joshua M., Steer, Michael B., 2013. Sensitive vibration detection using ground-penetrating radar. IEEE Microw. Wirel. Compon. Lett. 23, 680.

Wolfbeis, O., 2008. Fiber-optic chemical sensors and biosensors. Anal. Chem. 80, 4269–4283.

Wu, H., Wang, Z., Peng, F., Peng, Z., Li, X., Wu, Y., Rao, Y., 2014. Field test of a fully distributed fiber optic intrusion detection system for long-distance security monitoring of national borderline. Proc. SPIE, 9157.

Wu, H.-J., Wang, J., Wu, X.-W., Wu, Y., 2012. Real intrusion detection for distributed fiber fence in practical strong fluctuated noisy backgrounds. Sens. Lett. 10, 1557–1561.

Yin, Y.L., Prud'homme, R.K., Stanley, F., 1992. Relationship between poly(acrylic acid) gel structure and synthesis. In: Harland, A.J., Prud'homme, R.K. (Eds.), Polyelectrolyte Gels. In: ACS Symp. Series, vol. 480. ACS, Washington, DC. Chapter 6.

Yoon, Suk-Un, Cheng, Liang, Ghazanfari, Ehsan, Pamukcu, Sibel, Suleiman, Muhannad T., 2011. A radio propagation model for wireless underground sensor networks. In: Proceedings of IEEE Globecom. Houston, TX.

Yoon, Suk-Un, Ghazanfari, Ehsan, Cheng, Liang, Pamukcu, Sibel, Suleiman, Muhannad T., 2012. Subsurface event detection and classification using wireless signal networks. Sensors 12 (11).

Yuan, J., El-Sherif, M.A., MacDiarmid, A.G., Jones Jr., W.E., 2001. Fiber optic chemical sensors using a modified conducting polymer cladding. In: Proceedings of SPIE Advanced Environmental and Chemical Sensing Technology, vol. 4205, pp. 170–179.

Yusof, N.A., Ahmad, M., 2003. A flow-through optical fibre reflectance sensor for the detection of lead ion based on immobilized gallocynine. Sens. Actuators B, Chem. 94, 201–209.

Zhang, Z., Bao, X., 2008. Distributed optical fiber vibration sensor based on spectrum analysis of polarization-OTDR system. Opt. Express 16, 10240–10247.

Zhou, D., Dong, Y., Wang, B., Jiang, T., Ba, D., Xu, P., Zhang, H., Lu, Z., Li, H., 2017. Slope-assisted BOTDA based on vector SBS and frequency-agile technique for wide-strain-range dynamic measurements. Opt. Express 25 (3), 1889–1902.

Zhou, L., Wang, F., Wang, X., Pan, Y., Sun, Z., Hua, J., Zhang, X., 2015. Distributed strain and vibration sensing system based on phase-sensitive OTDR. IEEE Photon. Technol. Lett. 27, 1884–1887.

INDEX

Printed in the United States
By Bookmasters